MATHEMATICAL PROGRAMMING IN THEORY AND PRACTICE

MATHEMATICAL PROGRAMMING IN THEORY AND PRACTICE

Proceedings of the NATO Advanced Study Institute
Figueira da Foz, Portugal June 12-23, 1972
Organized by A.S. Gonçalves

Editors

P. L. HAMMER
University of Waterloo
Waterloo, Ontario, Canada

and

G. ZOUTENDIJK
University of Leiden
Leiden, The Netherlands

1974

NORTH-HOLLAND PUBLISHING COMPANY – AMSTERDAM • OXFORD
AMERICAN ELSEVIER PUBLISHING COMPANY, INC. – NEW YORK

Library of Congress Catalog Card Number: 74-81329
ISBN North-Holland: 0 7204 2819 x
ISBN American Elsevier: 0 444 10741 x

PUBLISHERS:

NORTH-HOLLAND PUBLISHING COMPANY – AMSTERDAM
NORTH-HOLLAND PUBLISHING COMPANY, LTD. – OXFORD

SOLE DISTRIBUTORS FOR THE U.S.A. AND CANADA:

AMERICAN ELSEVIER PUBLISHING COMPANY, INC.
52 VANDERBILT AVENUE
NEW YORK, N.Y. 10017

PRINTED IN THE NETHERLANDS

CONTENTS

PREFACE

Linear Programming is today a well-established field of activity.
Born just after the Second World War due to military necessities, developed since thanks to its tremendous applicability, it has constantly led to new discoveries, to a widening of its sphere of uses, to numerous extensions. Major extensions of linear programming - motivated by real life problems - are aimed at lifting the rigid requirements of linearity and of continuity. Nonlinear and discrete programming are fields of rapid and profound development today.
The fast progress of these fields created the necessity of periodic meetings among those interested in them. These meetings are meant to evaluate the progress made, to discuss new results and to disseminate knowledge to newcomers in the field. The latest of these meetings was held in Figueira da Foz, in Portugal, with the support of NATO. The organizer of this meeting was Professor A. Gonçalves.
The present volume represents the proceedings of that meeting.
Due to the fact that the fields of nonlinear and discrete programming do overlap in many points and that some papers have to do with both areas we have not tried to divide the book into distinct parts. However, we have tried to order the papers in such a way that a gradual transition from one area to the other and then to applications was the result.
We express our sincere thanks to NATO for supporting the meeting, to Professor Gonçalves and the other members of the committee for organizing it, to the University of Coimbra for the cordial reception, to Professor Vajda and to the entire Program Committee for selecting the lecturers, to all the lecturers for their valuable contributions, to all the participants for their enthusiastic discussions and to North-Holland Publishing Company for its willingness to publish the Proceedings.

Amsterdam, May 1974.

The Editors.

Mathematical Programming in Theory and Practice,
P.L. Hammer and G. Zoutendijk, (Eds.)

TEST OF OPTIMALITY

S. VADJA
University of Birmingham, England

Consider the two dual problems: (*)

Minimize $c'x$
subject to $Ax \geqslant b$

Maximize $b'y$
subject to $A'y = c$
$y \geqslant 0$

For all x and y which satisfy their repective contraints, we have

$$b'y \leqslant x'A'y = c'x$$

so that, if $b'y_o = x_o'A'y_o = c'x_o$, where $y_o \geqslant 0$, $Ax_o \geqslant b$, $A'y_o = c$, then x_o minimizes $c'x$, subject to $Ax \geqslant b$.

In other words, if a vector $y_o \geqslant 0$ exists, such that $c - A'y_o = 0$ and $(b-Ax_o)'y_o = 0$, and if x_o satisfies $Ax \geqslant b$, then x_o minimizes $c'x$, subject to the last inequality.

The existence of y as described is <u>sufficient</u> for x_o to be optimal

From the Duality Theorem of linear programming we know that if a (finite) x_o exists which minimizes $c'x$ subject to $b-Ax \leqslant 0$, then there exists (as the solution of the dual problem) a vector $y_o \geqslant 0$, such that $c - A'y_o = 0$ and $c'x_o = y_o'Ax_o = b'y_o$, i.e. $(b-Ax_o)'y_o = 0$.

In other words, for x_o to be optimal, it is <u>necessary</u> for a vector y_o as described to exist.

We introduce the Lagrangean Function

$$\Phi(x,y) = c'x + b'y - y'Ax$$

and can then write

$$c - A'y = \nabla_x \Phi(x,y).$$

After these preliminaries we proceed to the case of non-linear programming and consider the minimization of $f(x)$, subject to $g(x) \leqslant 0$, where $g(x)$ is a vector function with components $g_i(x)$, $i = 1,..,m$, and x is an n-vector.

(*) Transposition is denoted by a prime

We introduce the Lagrangean

$$\Phi(x,y) = f(x) + y'g(x).$$

In the linear case we had $f(x) = c'x$, and $g(x) = b - Ax$. The two theorems which we have stated could then be written as follows :

Sufficiency Theorem: x_o minimizes $f(x)$ subject to $g(x) \leqslant 0$, if a vector $y_o \geqslant 0$ exists such that

$$\nabla_x \Phi(x_o,y_o) = 0, \quad \text{and } y_o'g(x_o) = 0.$$

Necessity Theorem: If x_o minimizes $c'x$ subject to $g(x) \leqslant 0$, then there exists a vector $y_o \geqslant 0$ such that

$$\nabla_x \Phi(x_o,y_o) = 0, \text{ and } y_o'g(x_o) = 0.$$

It is the aim of the following study to examine if these theorems are still valid in the case of non-linear (but differentiable) functions, either without or with further conditions on $f(x)$, on $g(x)$, or on both.

Some counter-examples will, in fact, show that the theorems are certainly not true in general, without qualification.

First, let us see whether the Sufficiency Theorem holds when we want to minimize $f(x) = -x^2$, subject to $g(x) = x \leqslant 0$.

We have now

$$\nabla_x \Phi(x_o,y_o) = -2x_o + y_o, \quad y_o'g(x_o) = x_o y_o$$

The conditions of the Sufficiency Theorem can be satisfied with $x_o = y_o = 0$, but this x_o, though feasible, is clearly not optimal: the theorem, as stated, is not valid.

However, it holds if $f(x)$ as well as $g(x)$ (i.e. all its component functions) are convex, as will now be shown:

If $h(x)$ is convex, then (cf. Wolfe-1967) if $x^{(1)}$ and $x^{(2)}$ are two points.

$$h(x^{(1)}) \leqslant h(x^{(2)}) - (x^{(2)} - x^{(1)})'\nabla_x h(x^{(1)}).$$

Let $h(x) = f(x) + y_o'g(x)$, $x^{(1)} = x_o$, and $x^{(2)}$ any x for which $g(x) \leqslant 0$. Then

$$f(x_o) + y_o'g(x_o) \leqslant f(x) + y_o'g(x) - (x-x_o)'\nabla_x \left[f(x_o) + y_o'g(x_o)\right] .$$

If $y_o'g(x_o) = 0$, $y_o \geqslant 0$, $g(x) \leqslant 0$, $\nabla_x \Phi(x_o,y_o) = 0$, then it follows that

$$f(x_o) \leqslant f(x).$$

There are other conditions which make the Sufficiency Theorem true, but we shall not deal with them here. Instead, we turn to the Necessity Theorem and consider the following (by now classical) example (cf. Kuhn and Tucker-1950).

Minimize x_1,
subject to $g_1(x) = x_2 - x_1^3 \leqslant 0$, $g_2(x) = -x_2 \leqslant 0$.

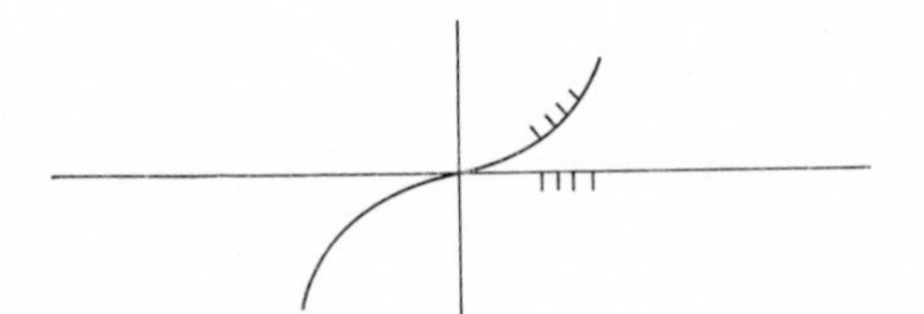

Figure 1.

The point $x_o = (0,0)$ is optimal. We have

$$\Phi(x,y) = x_1 + y_1(x_2 - x_1^3) - y_2 x_2$$

so that

$$\frac{\partial \Phi(x_o, y_o)}{\partial x_1} = 1 - 3x_{o1}^2 y_{o1} = 1$$

and

$$\frac{\partial \Phi(x_o, y_o)}{\partial x_2} = -y_{o2}$$

$\nabla_x \Phi(x_o, y_o) = 0$ would imply $1 = 0$, and there is no y_o which could put this right.

Before introducing qualifications which make the Necessity Theorem valid (qualifications which the example just given does, of course, not satisfy), we exhibit the geometrical meaning of the condition

$$\nabla_x \Phi(x_o, y_o) = 0, \text{ i.e. } \quad \nabla f(x_o) = -y_o' \nabla g(x_o)$$

together with $y_o \geqslant 0$.

If the optimal point x_o is in the interior of the feasible region, then by differential calculus $\nabla f(x_o) = 0$, and $y_o \geqslant 0$ exists, as required, viz. $y_o = 0$.

If x_o is on the boundary, assume

$g_i(x_o) = 0$ for $i = 1, \ldots, k$. (In the figure below $k = 2$.)

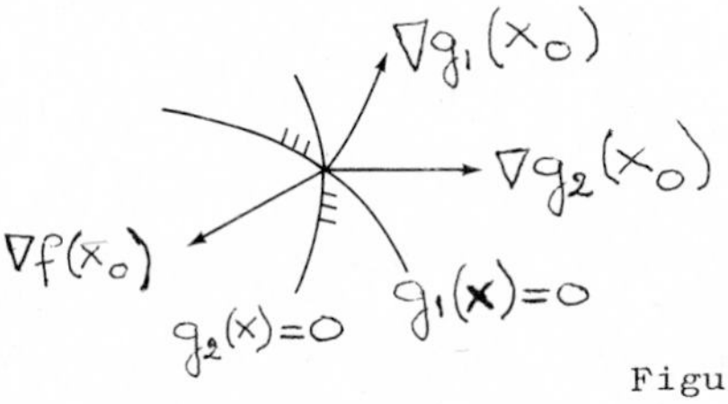

$\nabla g_1(x_o)$

$\nabla g_2(x_o)$

$\nabla f(x_o)$

$g_2(x) = 0$ $\quad g_1(x) = 0$

Figure 2.

The gradient of $f(x)$ at x_o points into the interior, if there is one, because we minimize at x_o. The negative gradients of $g_i(x)$ at x_o point also into the interior, because there $g_i(x) < 0$. Hence the equation demands that the gradient $f(x_o)$ be in the cone spanned by the negative gradients $-\nabla g_i(x_o)$.

This condition was not satisfied in the example just given, where $\nabla f(x_o) = (1,0)'$, $\nabla g_1(x_o) = (0,1)'$, $\nabla g_2(x_o) = (0,-1)'$.

We Introduce now two concepts which are convenient for the formulation of constraint qualifications.

A vector u is <u>locally constrained</u> at $\bar{x}$, if

$$u' \nabla g_i(\bar{x}) \leqslant 0 \quad \text{when } g_i(\bar{x}) = 0.$$

Clearly, all such vectors form a closed convex cone, the 'linearizing' cone.

A vector v is <u>attainable</u> at $\bar{x}$, if there exists a vector function $\psi(t)$ such that $\psi(0) = \bar{x}$, $g_i(\psi(t)) \leqslant 0$ for all i when $0 \leqslant t \leqslant t_1 (> 0)$, and $d\psi(0)/dt = v$. More vaguely, a vector is attainable, if it is possible to penetrate some distance into the interior of the feasible region, starting off in its direction.

The attainable vectors at a point form a cone, but not necessarily a convex one, unless the feasible region is convex. For instance, if the constraints are $-x_1 \leqslant 0$, $-x_2 \leqslant 0$, $x_1x_2 \leqslant 0$, then the feasible region consists of the origin and the two positive half-axes, and only these are attainable.

A vector which is attainable at $\bar{x}$, where $g_i(\bar{x}) = 0$ for $i = 1,..,k$ is locally constrained at $\bar{x}$.

Proof: Let v be attainable at $\bar{x}$ and let the corresponding $\psi(t)$ be given. Then

$$g_i(\psi(0)) = g_i(\bar{x}) = 0 \quad (i = 1,..,k)$$
$$g_i(\psi(t)) \leqslant 0 \text{ for } 0 \leqslant t \leqslant t_1 \quad (\text{all } i)$$

so that $dg_i(\psi(0))/dt \leqslant 0$ for $i = 1,..,k$.

But $$dg_i(\psi(t))/dt = \sum_j \frac{\partial g_i(\psi(t))}{\partial \psi_j(t)} \cdot \frac{d\psi_j(t)}{dt}.$$

At t = 0, this is $$\sum \frac{\partial g_i(\bar{x})}{\partial x_j} v_j = \nabla g_i(\bar{x})'v.$$

Since we know that for $i = 1,..,k$ this is non-positive, v is by definition locally constrained.

The converse is not necessarily true. This is obvious, because the locally constrained vectors form always a convex cone, while the attainable vectors need not. For instance, in the counter-example illustrated in figure 1, the attainable vectors at the origin are $(w,0)$, $w \geqslant 0$, but the vector $(-1,0)$ is also locally constrained at that point.

However, if all locally constrained vectors at a (locally) optimal point x_o are also attainable there (this is the Kuhn- Tucker constraint qualification), then the Necessity Theorem holds. We shall call this statement the Kuhn-Tucker Theorem (Kuhn and Tucker - 1950).

Proof: Let $g_i(x_o) = 0$ for $i = 1,..,k$, and let u be locally constrained at x_o, so that $u'\nabla g_i(x_o) \leqslant 0$ $(i = 1,..,k)$.

If the Kuhn-Tucker constraint qualification holds, then u is also attainable at x_o, and a suitable $\psi(t)$ exists. Then

$$\frac{df(x)}{dt} = \sum_j \frac{\partial f(\psi(t))}{\partial \psi_j(t)} \cdot \frac{d\psi_j(t)}{dt}.$$

At $t = 0$ this is $\quad \frac{df(x_o)}{dt} = u'\nabla f(x_o)$

which is $\geqslant 0$, because x_o is a (local) minimum of $f(x)$.

We have thus found that all vectors u which satisfy

$$u'\nabla g_i(x_o) \leqslant 0 \qquad (i = 1,..,k)$$

satisfy also

$$-u'\nabla f(x_o) \leqslant 0.$$

It follows from the Theorem of Farkas (see Vajda-1967) that there must exist a vector $y_o \geqslant_k 0$ such that

$$\nabla f(x_o) = -\sum_{i=1}^{k} y_{oi} \nabla g_i(x_o).$$

If we introduce $y_{oi} = 0$ for $i = k+1,..,m$, then we can write

$$f(x_o) = -\sum_{i=1}^{m} y_{oi} \nabla g_i(x_o), \quad \text{i.e.} \nabla_x \Phi(x_o,y_o) = 0.$$

We have, moreover, $y_o'g(x_o) = 0$, because in each term of this scalar product at least one of the factors is zero. Thus the Kuhn-Tucker Theorem is proved.

It might be useful to remark that the Necessity Theorem stipulates the existence of a vector y_o, as described, whatever the function $f(x)$ is, which we minimize.

It is possible that for a special function $f(x)$ such y_o exists, even if the Necessity Theorem is not generally true for the given $g(x)$. For instance, if we wanted to minimize x_2 (rather than x_1) subject to $x_2 - x_1^3 \leqslant 0$, $-x_2 \leqslant 0$, then $x_o = (0,0)$ is again optimal and

$$\Phi(x,y) = x_2 + y_1(x_2 - x_1^3) - y_2x_2, \text{ and } \nabla_x \Phi(x_o,y_o) = 0$$

means

$$-3x_1^2y_{o1} = 0, \qquad 1 + y_{o1} - y_{o2} = 0$$

which can be solved, when $x_1 = 0$, by $y_1 = 0$, $y_2 = 1$.

We add two remarks about the constraint qualifications of Kuhn and Tucker.

1. If the constraints are linear, then the qualification always holds. Let $g_i(x) = b_i - A_i'x \leqslant 0$ $(i=1,..,m)$, and $b_i - A_i'x_o = 0$ for $i = 1,..,k$. If u is locally constrained at x_o, then $-u'A_i \leqslant 0$, and taking $\Psi(t) = x_o + tu$, we have $b_i - A_i'(x_o+tu) \leqslant 0$ (since $t \geqslant 0$, $A_i'u \geqslant 0$), so that u is attainable and the Kuhn-Tucker constraint qualification is satisfied.

2. Whether the qualification holds, or not, depends on the manner in which the feasible region is defined, not on the region itself. Thus, if in the counter-example of figure 1 we add the constraint $-x_1 \leqslant 0$, then the feasible region is unaltered, but (-1,0) is not locally constrained anymore, the Kuhn-Tucker constraint qualification is satisfied, and the Necessity Theorem is valid.

- - - - - - - - - - - - - - - -

II

We have seen that if the Kuhn-Tucker constraint qualification is satisfied, then the Necessity Theorem holds. We are now led to ask whether it is necessary for the latter result that the qualification should be satisfied, or if in fact a weaker qualification could serve the same purpose.

The cone of attainable vectors is not necessarily convex. We call a vector "weakly attainable" at $\bar{x}$, if it lies in the closure of the convex hull of attainable vectors at $\bar{x}$.

For instance, if the constraints are

$$-x_1 \leqslant 0, \quad -x_2 \leqslant 0, \quad x_1x_2 \leqslant 0$$

then the feasible region consists of the origin and the positive half-axes, the latter contain the attainable vectors, and the weakly attainable vectors are all those pointing into the positive quadrant.

Obviously, every attainable vector is also weakly attainable, and the converse is true if the feasible region is closed and convex.

Every weakly attainable vector is locally constrained. This follows from the fact that each attainable vector is in the closed and convex cone of locally constrained vectors. Consequently the closed convex cone of weakly attainable vectors is contained in the former, but it might be strictly contained.

However, if the two cases are indentical, i.e. if all locally constrained vectors at a (locally) optimal point x_o are also weakly attainable there (this is the <u>Arrow-Hurwicz-Uzawa constraint qualification</u>, see Arrow-Hurwicz-Uzawa-1961), then the Necessity Theorem holds.

Proof: In the proof of the Kuhn-Tucker Theorem we have first shown that if a vector u, locally constrained at x_o, is also attainable there, then it is in the cone of vectors z for which $z' \nabla f(x_o) \leqslant 0$. This cone is closed and convex, and it contains therefore also all those vectors in the closure of the convex hull of attainable vectors at x_o. We can then again apply the Theorem of Farkas to prove our statement.

Abadie-1967 has introduced yet another constrain qualification, which is also weaker than that of Kuhn and Tucker. To describe it, we need the concept of a <u>vector tangent</u>.

The vector u is a vector tangent at $\bar{x}$, if there exists an infinite sequence of feasible points $x^{(p)}$ converging to $\bar{x}$, and a sequence of non-negative scalars t_p, such that

$$\lim_{p=\infty} t_p(x^{(p)}-\bar{x}) = u.$$

The vector tangents at $\bar{x}$ form a closed, non-empty cone, which is not necessarily convex, unless the feasible region is convex. This is shown in Abadie-1967, but we are here concerned with the following properties:

1. An attainable vector is a vector tangent at the same point.

Proof: Let u be attainable at $\bar{x}$, and let the corresponding $\Psi(t)$ be given.

$$\text{Let } x^{(p)} = \Psi(s_p), \; t_p = 1/s_p, \text{ so that}$$

$$\lim_{p=\infty} t_p(x^{(p)} - \bar{x}) = \lim_{p=\infty} \frac{\Psi(s_p) - \Psi(0)}{s_p} = \frac{d\Psi(0)}{dt} = u.$$

The converse is not necessarily true; a vector tangent need not be (even weakly) attainable. Also, a weakly attainable vector need not be a vector tangent.

2. If g(x) is differentiable, then every vector tangent is locally constrained at the same point.

Proof: Let $g_i(\bar{x}) = 0$ for $i = 1,..,k$. Then for such i, we have by Taylor's Theorem

$$g_i(x^{(p)}) = (x^{(p)}-\bar{x})' \nabla g_i(\bar{x}) + \varepsilon_{ip} |x^{(p)} - \bar{x}|$$

where $x^{(p)}$ is the sequence in the definition of the vector tangent u, and $\lim_{p=\infty} \varepsilon_{ip}=0$.

Thus, for large p,

$t_p g_i(x^{(p)})$ converges to $t_p(x^{(p)}-\bar{x})' \nabla g_i(\bar{x})$, which converges to $u' \nabla g_i(\bar{x})$.

Now suppose that u were not locally constrained at $\bar{x}$, so that $u' \nabla g_i(\bar{x}) > 0$ for some i. Then it would follow that $g_i(x^{(p)}) > 0$ for large enough p, and $x^{(p)}$ would not be feasible, contradicting the assumptions.

The converse is not necessarily true. For instance, in our example illustrated in figure 1, the vector (-1,0) is locally constrained, but it is not a vector tangent.

However, if all locally constrained vectors at a (locally) optimal point x_o are also vector tangents there (this is the Abadie constraint qualification), then the Necessity Theorem holds, provided f(x) and g(x) are differentiable.

Proof: Let $g_i(x_o) = 0$ for i = 1,2,..,k and $u' \nabla g_i(x_o) \leqslant 0$ for these i. Thus u is locally constrained at x_o, and by assumption it is also a vector tangent. We have sequences $x^{(p)}$ and t_p as above, and when $x^{(p)}$ is sufficiently near to x_o, then $f(x^{(p)}) - f(x_o) \geqslant 0$, because of the local optimality of x_o.

But $f(x^{(p)}) - f(x_o) = (x^{(p)} - x_o)' \nabla f(x_o) + \varepsilon_p |x^{(p)}-x_o|$ with $\lim_{p=\infty} \varepsilon_p = 0$,

hence

$$(x^{(p)}-x_o)' \nabla f(x_o) \geqslant -\varepsilon_p |x^{(p)}-x_o| .$$

As p increases, and $x^{(p)}$ remains near x_o, $u' \nabla f(x_o) \geqslant 0$.

We use now again Farkas' Theorem to prove the existence of $y_{oi} \geqslant 0$ (i=1,..,k) and introduce $y_{oi} = 0$ for i = k+1,..,m. Thus we obtain once more

$$\nabla_x \Phi(x_o,y_o) = 0, \quad y_o' g(x_o) = 0.$$

If the constraints are linear, say $g_i(x) = b_i - A_i' x \leqslant 0$ (i=1,..,m), and $b_i - A_i' x_o = 0$ for i = 1,..,k, then u is locally constrained at

x_o if $u'\nabla g_i(x_o) = -u'A_i \leqslant 0$. Taking $x^{(p)} = x_o + tu/p$, $t_p = p/t$, $p > 0$, we have

$$\lim_{p=\infty}(x^{(p)}-x_o) = 0, \quad \lim_{p=\infty} t_p(x^{(p)}-x_o) = u,$$

$$g_i(x_o+tu/p) = g_i(x_o) + tu'A_i/p$$

which is, for small t, $\leqslant 0$, so that u is a vector tangent, and the Abadie constraint qualification is satisfied.

As we have said, the vector tangents form a closed non-empty cone, but not necessarily a convex one. But if all locally constrained vectors at a (locally) optimal point are in the convex hull of vector tangents (this is the Guignard constraint qualification, and Monique Guignard calls that hull the 'pseudo-cone tangent'), then the Necessity Theorem holds, provided f(x) and g(x) are differentiable. The proof follows from that concerning the Abadie constraint qualification in the same way as that for weakly attainable x vectors follows from the Kuhn-Tucker proof for attainable vectors. (Guignard-1967).

Of course, every vector tangent is in the pseudo-cone tangent. Every vector in the pseudo-cone tangent is locally constrained, because the latter vectors form a convex cone, and every vector tangent is in that cone.

The Guignard qualification is weaker than any of those three previously mentioned, as can be shown by examples.

We can not continue exhibiting yet weaker qualifications, because Gould and Tolle-1971 have shown that if the Guignard qualification is not satisfied, then a function f(x) would exist for which the Necessity Theorem is not true.

References:

Abadie, J. 1967. On the Kuhn-Tucker Theorem.

Arrow, K., L. Hurwicz and H. Uzawa. 1956. Constraint qualifications in maximization problems. Naval Research Logistics Quarterly 8.

Gould, F.J. and J.W. Tolle. 1971. A necesary and sufficient qualification for constrained optimization. SIAM J. Appl. Math. 20.

Guignard, M. 1967. Conditions d'optimalité et dualité en programmation mathématique. Thèse, Lille.
Kuhn, H.W. and A.W. Tucker 1950. Nonlinear Programming. Proc. 2nd Berkeley Symp. Math. Statist. and Prob. Ed. J. Neyman.
Vajda, S. 1967. Non-linear programming and duality.
Wolfe, P. 1967. Methods of Nonlinear programming.

The papers by Abadie, Vadja, and Wolfe are in Nonlinear Programming, ed. J. Abadie, North-Holland 1967.

Mathematical Programming in Theory and Practice,
P.L. Hammer and G. Zoutendijk, (Eds.)

The Significance of Recent Developments in Mathematical Programming Systems

by

E.M.L. Beale
(Scientific Control Systems Ltd)

Abstract

This paper reviews the way that recent advances in mathematical programming methods have been applied to the solution of large-scale practical problems. The points covered are:

1) In most practical mathematical programming problems each variable occurs in only about 6 constraints on the average. This is true whether the constraints are linear or non-linear, and whether the variables are continuous or discrete. It means that problems with large numbers of variables and constraints are almost certainly very sparse; and practical solution methods must exploit this sparseness.

2) The simplex method using the product form of inverse is essentially a method of exploiting sparseness. Its practical implementations have been vastly improved in recent years.

   As Beale (1971) points out, this is largely due to the fact that a basis matrix is now expressed as a product of two triangular factors before the elementary transformations representing its inverse are created. This increases the number of elementary transformation matrices, but greatly decreases the total number of non zero coefficients required. Another important development is due to Hellerman and Rarick (1971). Forrest and Tomlin (1972) have extended the triangulation approach by updating the triangular factors after each basis change, and shown that this preserves sparseness much better than the conventional product form.

3) These and other improvements in the implementation of the simplex method, together with improvements in computer hardware, have increased the size of linear programming problem that can be solved effectively by the standard simplex method to about 10,000 rows. The have correspondingly reduced practical interest in decomposition and other techniques for specially structured linear programming problems; but this interest may revive. A partial exception to this narrowing of interest is generalized upper bounds. Facilities for generalized upper bounds are now quite widely available, and are extremely useful on problems where the majority of rows can be treated as generalized upper bounds.

4) There are great practical advantages in linking procedures for solving large mathematical programming problems, not merely with the concepts of the simplex algorithm, but with a production linear programming code. The LP code will then automatically exploit sparseness, and furthermore the whole procedure can benefit from subsequent improvements in the LP code.

5) The most dramatic improvements during the last 3 years in the capability of general mathematical programming systems have been in the field of integer programming. Integer programming is now used on practical problems with a few thousand constraints, a larger number of continuous variables and about 100 integer variables, and also on problems with a few hundred constraints and several hundred integer variables. Branch and bound search strategies have been developed to the point where they can solve many problems completely in a moderate time. But it is perhaps much more significant that they allow the search to be terminated in a reasonable time with a good solution, and often one that is almost certainly optimum, long before the work needed to guarantee optimality has been completed. The total time on the computer is then only about 5 times the time needed to solve the original problem as a linear programming problem, i.e. treating the integer variables as continuous variables. Recent developments are discussed by Benichou et al (1971) and by Forrest, Hirst and Tomlin (1974).

6) It remains true that care must be taken with the formulation of integer programming problems. Disappointing results can occur if the solution to the problem when all variables are treated as continuous variables differs radically from the solution in integer variables. A reformulation of the constraints can often help with this difficulty.

7) Many practical integer programming problems are formulated using special ordered sets, first introduced by Beale and Tomlin (1970). These are of two types: S1 sets of variables, of which at most one may be non zero in the final solution, and S2 sets of variables, of which at most two may be non zero in the final solution and where these two must be adjacent. The variables need not take integer values; but the whole procedure is regarded as part of integer programming both because it is concerned with finding global optimum solutions to non-convex problems and because it uses the indentical branch and bound mechanism as ordinary integer programming. Note that S2 sets were introduced to facilitate the finding of global optima to continuous non linear problems where the non linear functions can conveniently be separated into functions of simple arguments.

8) The use of general mathematical programming systems on more general non linear problems seems less widespread. Some recent developments of the approach first used by Griffith and Stewart (1961) are nevertheless described by Buzby (1974) and Beale (1974). The approach is to select trial values of the non linear variables and then make a local linear programming approximation to the problem, putting tolerances on the extent to which the non linear variables may depart from their current trial values. The solution to this and previous LP problems can then be used to suggest trial values and tolerances for the non linear variables in the next problem.

9) The efficient implementation of this approach is much easier on third generation computers equipped with drum or disc storage and sophisticated operating systems than it was in 1961. This primarily because there is no effective limit on the number of files that can be used to transfer information between the separate programs of the system, and no difficulty in running such separate programs sequentially as a single job on the computer. In our implementation there are four such programs in addition to the initial input. These are the Control Program, which selects trial values and tolerances for the non linear variables, the Matrix Generator, which generates the next linear approximation problem, the General Mathematical Programming System which solves the LP problem, and the Output Analyser which selects the crucial information about the solution to this LP problem.

10) We have recently used this approach on problems with over 1000 constraints and a few hundred non linear variables. Useful results were obtained after a few hours on the Univac 1108 computer, although some difficulties were encountered when we had poor initial trial values for the non linear variables. These were overcome by allowing the matrix generator to change the trial values of non linear variables in certain circumstances. We have used this procedure specifically in multi-time-period models when the trial values for later time-periods have proved incompatible with the trial values selected for earlier time-periods.

1. INTRODUCTION

This paper reviews some recent algorithmic developments in mathematical programming systems and their practical significance. Several organizations have contributed to these developments, which have made the solution of many large problems much easier. Furthermore, although these developments are based on earlier work, many represent starts along new lines of research rather than consolidations of old lines. So there are excellent prospects of further useful developments along these lines.

Section 2 of the paper discusses the crucial importance of sparsenss in large-scale mathematical programming systems, and the advances made in exploiting it in linear programming. These have enhanced the scope of other advances described in later sections: in integer programming, covered in Section 3, in the use of special ordered sets, covered in Section 4, and in large-scale non linear programming, covered in Section 5.

2. SPARSENESS

In most mathematical programming problems,each variable occurs in only about 6 constraints on the average. This is true whether the constaints are linear or non-linear, and whether the variables are continuous or discrete. It means that problems with large numbers of variables and constraints are almost certainly very sparse; and practical solution methods must exploit this sparseness.

The simplex method using the product form of inverse is essentially a method of exploiting sparseness. Its practical implementations have been vastly improved in the last 3 years. The main improvements have been in the process of inversion, sometimes called reinversion, when a new representation is found for the inverse of the current basis matrix $\underset{\sim}{B}$. These improvements are largely due to the fact that we now take the matrix $\bar{\underset{\sim}{B}}$, formed from $\underset{\sim}{B}$ by permuting the columns (explicitly) and the rows (implicitly) and write

$$\bar{\underset{\sim}{B}} = \underset{\sim}{L}\,\underset{\sim}{U},$$

where $\underset{\sim}{L}$ is lower triangular and $\underset{\sim}{U}$ is upper triangular. This can be regarded as an effective implicit representation of the inverse of $\bar{\underset{\sim}{B}}$. For example if we are given a column vector $\underset{\sim}{a}$, representing a column of the original coefficient matrix, we can find the column $\underset{\sim}{x}$, representing the corresponding column of the tableau, defined by $\underset{\sim}{x} = \underset{\sim}{B}^{-1}\underset{\sim}{a}$ as follows;

If we forget the permutations, which do not cause any computing problems, we have the equation

$$\underset{\sim}{L}\; U\; \underset{\sim}{x} = \underset{\sim}{a},$$

We can now find the column vector $\underset{\sim}{a}^*$, defined by

$$\underset{\sim}{L}\,\underset{\sim}{a}^* = \underset{\sim}{a},$$

by solving these algebraic equations sequentially. Hence we can find x, defined by

$$\underset{\sim}{U}\,\underset{\sim}{x} = \underset{\sim}{a}^*$$

by solving these algebraic equations sequentially in reverse order.

This approach is superficially quite different from the conventional product form, but in fact we can represent the columns of $\underset{\sim}{L}$ as elementary transformations identical to those used in the standard product form. These can then be followed by the columns of $\underset{\sim}{U}$ stored in reverse order but represented in the same way.

One must of course choose the permutations to produce a small number of nonzero coefficients in $\underset{\sim}{L}$ and $\underset{\sim}{U}$. We therefore start by finding the largest possible value of k such that the rows and columns of $\bar{\underset{\sim}{B}}$ have the property that, for all i = 1 ...k, either

a) $\bar{b}_{ij} = 0$ for all $j > i$, in which case we can put

$\bar{l}_{ji} = \bar{b}_{ji}$ for all $j \geq i$, and $\bar{U}_{ii} = 1$, $\bar{U}_{ij} = 0$ for $j \neq i$

and the ith row makes no contribution to $\underset{\sim}{U}$

or

b) $\bar{b}_{ji} = 0$ for all $j > i$, in which case we can put

$\bar{U}_{ij} = \bar{b}_{ij}$ for all $j \geq i$, and $\bar{l}_{ii} = 1$, $\bar{l}_{ij} = 0$ for $j \neq i$

and the ith column makes no contribution to $\underset{\sim}{L}$

The submatrix of $\bar{\underset{\sim}{B}}$ consisting of those elements with both subscripts greater than k is known as the <u>bump</u>, for historical reasons that are described for example in Beale (1971). It is in fact easy to find the largest value of k, and hence the smallest possible bump, which is unique. One simply starts with a trial solution in which $\bar{B}$ is treated as all bump. Then if there is any row with only one nonzero element in the bump, this nonzero element can be moved to the top left-hand corner of the bump and the row satisfies condition (a) above. Similarly if there is a column with only one nonzero element in the bump, this nonzero element can be moved to the top left-hand corner of the bump and the column satisfies condition (b) above. It is clear that the removal of a row and column from the bump in this way can never prevent the subsequent removal of another row or column, so the final bump must be unique, and independent of the order in which the other diagonal elements are selected.

It is less clear how the rows and columns of the bump should be ordered. But Hellerman and Rarick (1971) have made an interesting and successful proposal. This is to postpone columns in such a way that additional elements can be selected satisfying condition (b) above if entries in the postponed columns, known as spikes, are ignored.

The idea of using LU decomposition in linear algebra is not new, see for example Turing (1948), and indeed the idea of using it to invert the sparse bases arising in linear programming was specifically proposed by Markowitz (1957). But this proposal has only recently been implemented in practical mathematical programming codes. The results have been most gratifying. The number of elementary transformations has been increased, since we now need an extra transformation for each column in the bump, but the number of nonzero coefficients in the representation of the inverse has been greatly decreased, and numerical accuracy has been improved.

After inverting, one must continue with the iterative steps of the simplex method. The resulting changes of basis are then normally represented as elementary transformation matrices premultiplying the previous basis inverse. Each such matrix is formed from the coefficients in the current tableau of the variable entering the basis; in other words it uses the coefficients of the variable that one would obtain if one solved the constraints explicitly for the basic variables in terms of the nonbasic variables. Unfortunately the columns of the tableau are often quite dense, so the number of nonzero elements in the current representation of the inverse mounts rapidly with each iteration. Forrest and Tomlin (1972) have shown how this process can be avoided. If one thinks of the standard product form as related to the ordinary tableau as outlined above, then their method is similarly related to the "echelon form" of the tableau.

If the basis matrix B at any stage is written in the form

$$B = G\,U,$$

where G is a general matrix that may, but need not, be lower triangular, then if the basic variables are $X_1 \ldots . X_m$ we can write the tableau in echelon form as

$$u_{11}X_1 + u_{12}X_2 + \ldots + u_{1m}X_m = a_{10} + \sum_j a_{1j}^* (-x_j)$$
$$u_{22}X_2 + \ldots + u_2 mXm = a_{20} + \sum_j a_{2j}^* (-x_j)$$
$$u_{mm}Xm = a_{mo} + \sum_j a_{mj}^* (-x_j)$$

Here the column $\underset{\sim}{a}_j^*$ of coefficients of x_j is formed from the corresponding column $\underset{\sim}{a}_j^*$ in the original coefficient matrix by premultiplying by G^{-1}. By allowing the off-diagonal elements of $\underset{\sim}{U}$ to be nonzero, we can reduce the number of nonzero coefficients in $\underset{\sim}{G}^{-1}$, which is indeed taken as $\underset{\sim}{L}^{-1}$ immediately after an inversion. We can therefore reasonably hope that the columns $\underset{\sim}{a}_j^*$ will not be much more dense than the original columns $\underset{\sim}{a}_j$. When the basis changes, the incoming nonbasic variable x_q must change places with the outgoing basic variable X_p. If we simply exchange these columns in the equations, then the left hand side will in general no longer be triangular. But we can make it triangular again by carrying out row operations to eliminate the elements U_{pj} for $j > p$ and then permuting rows and columns. Each simplex iteration then involves premultiplying $\underset{\sim}{G}^{-1}$ by a conventional elementary matrix representing a column transformation based on the vector a_q^*, and postmultiplying U^{-1} by an elementary matrix representing a row transformation, after removing the column of $\underset{\sim}{U}$ associated with the outgoing basic variable and also any nonzero coefficients in the pivotal row of the old $\underset{\sim}{U}$. Forrest and Tomlin have found that this process greatly reduces the build-up of nonzero coefficients in the total representation of the basis inverse.

These and other improvements in the implementation of the simplex method, together with improvements in computer hardware and operating systems, have increased the size of linear programming problem that can be solved effectively by this method to about 10000 rows. Practical interest in decomposition and other techniques for large specially structured linear programming problems has correspondingly declined, but it may revive.

One special structure that is considered important, because it can easily be accommodated within a general linear programming code, is Generalized Upper Bounds. Generalized Upper Bounds, often known as GUB rows, are constraints of the form

$$\sum_j x_{jk} = b_k$$

They are subject to the restriction that no variable occurring in one GUB row may occur in any other GUB row. Nevertheless there are many important problems for which most of the rows can be treated as GUB rows, and in these circumstances the use of GUB speeds up the computations dramatically.

Non-unit positive coefficients could be allowed without any real complication of the algorithm, but it seems sensible to scale such variables so that these coefficients are all unity. Some codes allow negative coefficients in a GUB row, but this seems unnecessary in a code such as UMPIRE which allow members of GUB set to have simple upper bounds. These simple upper bounds are sometimes very useful. For example suppose we have a market in which the first A_1 units of some product can be sold at a price of £P_1 per unit, up to A_2 additional units can be sold at a lower price of £P_2 per unit, and so on. Then if x_j denotes the number of units supplied from Source j, then it may seem natural to write y_p for the number of units sold at the pth price. We could then write

$$\sum_j x_j - \sum_p y_p = 0$$

where $0 \leqslant y_p \leqslant A_p$.

This suggests that it might be useful to allow negative signs in GUB rows. But the problem can easily be transformed into one involving an ordinary GUB row, with several members having simple upper bounds, if we define z_p as the number of units that could be sold at the pth price but are not supplied. The variable y_p then does not occur in the formulation but can be computed as $A_p - z_p$. The demand row now reads

$$\sum_j x_j + \sum_p z_p = \sum_p A_p,$$

where $0 \leqslant z_p \leqslant A_p$.

The variable z_p must be assigned a unit cost of £P_p, and a constant term representing a revenue of £$\Sigma P_p A_p$ should be added to the objective function. If a problem involves several markets with this structure, then such a GUB formulation may be very helpful. The great advantage of GUB over more specialized techniques, such as those for transportation problems or optimizing flows in networks, is that the GUB formulation remains valid when other more general constraints are added to the problem. One also has all the standard facilities of a general mathematical programming system, such as the possibility of supplying a starting basis or the possibility of using parametric programming.

If we now turn away from purely linear programming to more general mathematical programming problems, we find that much research and indeed development has been made on programs not closely linked to the production linear programming codes. This is perhaps inevitable. It is generally hard for anyone to extend a production linear programming code to new problems if he has not been working on its maintenance and development; and this impediment is probably more fundamental than the other real problem of preserving commercial security in an expensive investment. Fortunately these difficulties are far less acute if the extension merely requires the repeated use of the linear programming code as a subroutine. An it remains true that there are great practical advantages in linking procedures for solving any type of large mathematical programming problem, not merely with the concepts of the simplex algorithm, but with a production linear programming code. The code will then automatically exploit sparseness, and furthermore the whole procedure can benefit automatically from any subsequent improvements in the code.

3. INTEGER PROGRAMMING

The most dramatic improvements during the last 3 years in the capability of general mathematical programming systems have been in integer programming. Integer programming is now used on practical problems with a few thousand constraints, a larger number of continuous variables, and about 100 integer variables. It is also used on problems with a few hundred constraints and several hundred or even a few thousand integer variables.

These results have been achieved using branch and bound methods. As noted below, branch and bound search strategies have now been developed to the point where they can solve many problems completely in a moderate time. But it is perhaps more significant that they allow the search to be terminated in a reasonable time with a good solution, and often one that is almost certainly optimum, long before the work needed to guarantee optimality has been completed. One can therefore be fairly confident of having a satisfactory solution after spending not more than about 5 times as long on the computer as one needs to solve the original problem as a linear programming problem, i.e. treating the integer variables as continuous variables. Many problems are solved much more quickly. Recent developments are discussed by Benichou et al (1971), and by Forrest, Hirst and Tomlin (1974). The following discussion is based on the work of both sets of authors.

The use of branch and bound methods for integer programming can be described in general terms as follows.

The problem is assumed to be a linear programming problem, with the additional requirement that certain variables are required to take integer values. In practice many or all of these variables are generally "indicator variables" which can only take the values zero or one. But the methods apply equally to integer variables that can take general nonnegative values.

We start by solving the problem as a linear programming problem, treating the indicator variables as continuous variables with a simple upper bound of one, and the general integer variables as continuous with some suitable (integer) upper bound. This bound can if necessary be made very large without significantly affecting the progress of the algorithm. If all integer variables then take integer values, then the entire problem is solved. Otherwise this continuous optimum provides an upper bound on the maximum attainable value of the objective function, with which it may be interesting to compare any feasible solution that we subsequently find. Such a feasible solution is often known as an integer solution, since it implies that all integer variables are within a suitable tolerance of integer values.

We now enter the general procedure, in which we solve a sequence in linear programming subproblems with upper and lower bounds imposed on the values of all integer variables. In general we have a list of alternative subproblems of this kind produced by previous steps of the algorithm. This list is often called the list of unexplored nodes. Initially it consists of a single subproblem, the continuous optimum. We also keep the current best known integer solution, with its value of the objective function being maximized, say x_{oc}. Until an integer solution has been found we may set x_{oc} to $-\infty$.

A general step of the algorithm is as follows.

1) See if the list is empty. If so, the whole problem is solved, i.e. the current best known integer solution is the guaranteed global optimum solution.

2) Otherwise select a subproblem from the list to be explored next, and take a variable, say X_i, that must ultimately be integer but which takes a noninteger value, say $\bar{a}_{i0}$, at the optimum solution to the subproblem. Replace the selected subproblem in the list by two new subproblems, each differing from the selected subproblem only in one bound X_i. In one new subproblem replace the upper bound on X_i by $\left[\bar{a}_{i0}\right]$ and in the other new subproblem replace the lower bound on X_i by $\left[\bar{a}_{i0}\right]$ +1.

3) Solve each new subproblem in turn. Let $\bar{a}_{oo}$ denote the optimum value of the objective function in the subproblem. If $\bar{a}_{oo} \leq x_{oc}$, then abandon the subproblem and remove it from the list. Otherwise, if all integer variables take integer values the subproblem represents a new and improved current best known integer solution, and x_{oc} can be set to $\bar{a}_{oo}$. We again remove this subproblem from the list, and also all other subproblems whose $\bar{a}_{oo}$ is less than or equal to x_{oc}. If neither condition holds we retain the subproblem in the list after solving it.

When both new subproblems have been solved the step is complete.

This procedure is often represented as a tree-search procedure. Each subproblem is represented as a node of the tree. The continuous optimum is represented as the root of the tree, and when any subproblem (or node) is explored the two subproblems generated are joined to it by branches. This tree structure clarifies the logic of the method. Fortunately it does not need to be represented explicitly inside the computer.

It will be seen that 3 important questions are left unresolved in this general description:

a) Which subproblem should be explored next?

b) Which variable should be branched on?

c) How should the new subproblems be solved?

These 3 questions are now much better understood than they were a few years ago.

It is sometimes suggested that the optimum subproblem to be explored next is always the one with the largest $\bar{a}_{oo}$. This is because this rule would minimize the number of subproblems created before completing the search if the choice of variable for branching were unaffected by the value of x_{oc}. But this rule makes no attempt to find any good integer solutions early in the search, and is subject to the following grave objections:

1) The choice of variable for branching can often be improved if we have a good value of x_{oc}, as will be shown later.

2) A good value of x_{oc} reduces the work on some subproblems, since they can be abandoned before they are fully solved if it can be shown that $\bar{a}_{oo} \leqslant x_{oc}$.

3) The list of unexplored nodes may grow very long and create computer storage problems.

4) No useful results may be obtained if the search has to be terminated before a guaranteed optimum has been found.

An alternative rule was advocated in the pioneering work on branch and bound by Little, Murty, Sweeney and Karel (1963) and used in early work on integer programming. This was called "always branch right", and may also be characterized as "Last In First Out" of the list of unexplored nodes. It is particularly convenient if one has to use mangetic tapes to store the details of the subproblems in the list, and was used quite successfully in the LP/90/94 system, see Beale (1968). Its advantages are that it minimizes the number of subproblems in the list of unexplored nodes, and it normally finds a good integer solution early which can be used to improve the rest of the search. But it has now been found inadequate on larger and more difficult problems, and has been discarded.

Practical integer programming codes now offer a bewildering variety of search options, and for a while the situation was very confused.But we can now identify a standard preferred option which will now be described.

There is not much wrong with the Last In First Out rule as long as one is continuing to branch on one of the two subproblems just created. This is particularly true if one always chooses the more promising of these subproblems. This promise can be measured entirely in terms of the objective function, but it is better to substract some measure of the integer infeasibility in the optimum solution to the linear programming subproblem. But when no further progress can be made in this way, either because one new subproblem gives an integer solution and the other gives a worse value of the objective function, or because both are infeasible, then the process of backtracking through the most recently created subproblems is often unrewarding. Our standard rule is therefore to select the more promising of the two subproblems just created, if either remains on the list after being solved, but otherwise to select the most promising of all unexplored nodes. These unexplored nodes will all have a value of $\bar{a}_{oo} \geqslant x_{oc}$ and also some measure of the sum of integer infeasibilities S. A generally effective criterion for identifying the most promising subproblem is the one that maximizes $(a_{oo} - x_{oc})/S$. And if this ratio is substantially smaller than the corresponding ratio for the continuous optimum, then on many types of problem one can fairly confident that the current best known integer solution is in fact the optimum.

We now turn to the second question, the choice of the variable to branch on.

In this connexion we first discuss the opportunities for <u>forced moves</u>, i.e. for branchings in which one of the new subproblems can be guaranteed in advance to be either infeasible or to have $\bar{a}_{oo} \leqslant x_{oc}$. Such a forced move exchanges one subproblem for a single more restricted subproblem, and is therefore a clear gain if it can be accomplished without too much computation.

The easiest opportunity for forced moves occurs with nonbasic integer variables. If any of these, say x_j, has a reduced cost $\bar{a}_{oj}$ that is greater than or equal to $\bar{a}_{oo} - x_{oc}$ then we know that x_j cannot be increased by a single unit without making the subproblem useless. We can therefore move its upper bound to its current lower bound - or vice versa if the variable is out of the basis at its upper bound and $|\bar{a}_{oj}| \geqslant \bar{a}_{oo} - x_{oc}$. This type of forced move does not exclude the current trial solution of the subproblem and so in a sense does not make immediate progress, but by preventing these nonbasic integer variables from ever taking noninteger values we may avoid unnecessary later branches.

The next opportunity for forced moves arises from the caluclation of <u>penalties</u> on the basic integer variables that take noninteger values at the current trial solution. This concept is due to Driebeck (1966). The <u>down-penalty</u> on such a variable X_i is the reduction in the objective function that would be obtained from the first iteration if one were to use the dual simplex method to solve the subproblem in which the upper bound on X_i is reduced to $\left[\bar{a}_{io}\right]$. Similarly the up-penalty is the corresponding reduction if the lower bound on X_i is increased to $\left[\bar{a}_{io}\right] + 1$. Tomlin (1970) showed how these penalties can sometimes be strengthened with a negligible computational effort when some of the nonbasic variables are integer.

If any penalty exceeds $\bar{a}_{oo} - x_{oc}$, then we can make a forced move in the other direction; and if several penalties exceed $\bar{a}_{oo} - x_{oc}$ we can make several such forced moves simultaneously.

If no forced moves are possible, then Beale (1968) recommended branching on the variable whose larger penalty is greatest. This is natural in that the choice of variable is now independent of whether or not there is a forced move, and early experience with this rule was encouraging.

Unfortunately, later experience with penalties on larger problems has often been unsatisfactory. This is because large problems often produce several nonbasic variables with small or zero reduced costs, in which case all the penalties are generally small or zero. Aonther way of seeing this is to note that in a large problem one generally needs many dual simplex iterations to solve a new subproblem and the reduction in the objective function in the first iteration may be a poor guide to the total reduction.

We therefore provide the option for the user to assign priorities to the integer variables. The program then branches on the integer variable with the highest priority, provided it differs from an integer value by more than some preassigned arbitration level. If there is more than one such integer variable, then we choose between them using either penalties, or pseudo-costs. Pseudo-costs are discussed in more detail by Benichou et al (1971) and by Forrest,Hirst and Tomlin (1974).

The third question can fortunately be answered very quickly. It has been found best to solve each new subproblem by parametric programming from the subproblem that generates it. As Benichou et al (1971) point out, parametric variation of one (or more) simple upper or lower bounds is not a standard facility in mathematical programming systems, but it can easily be implemented and has been found very useful in this context.

Various organizations, such as Scicon, now take a justifyable pride in their integer programming capability, and can quote impressive statistics. It is therefore worth emphasizing that care must still be taken with the formulation of integer programming problems. Disappointing results can occur if the solution to the problem when all variables are treated as continuous variables differ radically from the solution in integer variables. A reformulation of the constraints can sometimes help with this difficulty.

An important special case arises with constraints of the form

$$x - M\ \delta \leqslant 0,$$

where δ is a zero-one variable and M is a large number chosen to ensure that $x = 0$ if $\delta = 0$ and to avoid restricting x if $\delta = 1$. It is then well worth while taking some trouble to ensure that M is not unnecessarily large. If the model includes several such constraints one should certainly not use the same value of M in all of them just for convenience, if substantically smaller values can be justified in some cases.

In other circumstances one can usefully add linear constraints that are technically redundant when all integer variables take integer values, but which exclude the continuous optimum. I know of no general rules for this, other than to study the continuous optimum for a typical problem of the class being solved and to think how it can be excluded if it is unsatisfactory.

The other important piece of general advise about integer programming is to start with a fairly small problem of the class being solved, and to use this to check the formulation and to devise suitable priorities for the variables, before tackling a big problem.

4. SPECIAL ORDERED SETS

Special ordered sets were introduced by Beale and Tomlin (1970). They have fully lived up to their early promise, and it is appropriate that they are discussed in a recent textbook on Operational Research, see Mitchell (1972). They also illustrate the point made at the end of Section 2 about the desirability of linking new mathematical programming procedures to a production linear programming code. Hardly any of the facilities discussed in Sections 2 and 3 were in the UMPIRE system when Special Ordered Sets were first implemented: all are now available in conjunction with special ordered sets without having required any significant extra programming effort.

Special Ordered Sets are of two types: S1 sets of variables, of which at most one may be nonzero in the final solution, and S2 sets of variables, of which at most two may be nonzero in the final solution with the further constraint that if there are two they must be adjacent. The variables need not take integer values; but the whole procedure is regarded as part of integer programming both because it is concerned with finding global optimum solutions to non-convex problems and because it uses the same branch-and-bound mechanisms as ordinary linear programming.

The original motivation for the development of special ordered sets was the desire to find global optima to problems for which separable programming finds local optima, that is to say non-convex problems involving functions of single arguments in either the objective function or the constraints. When such a problem involves a nonlinear function f(z) depending on some argument z defined over a finite range of values of z, say for $a_0 < z < a_K$, then we can define an increasing sequence a_0, a_1 ....a_K of values of z, and if we can accept a piecewise linear approximation to f(z) between these values then we can define linear programming variables λ_k and write

$$\Sigma a_k \lambda_k - z = 0$$
$$\Sigma \lambda_k = 1$$

in which case f(z) is approximated by $\Sigma f(a_k)$. λ_k, provided that no more than 2 of the λ_k are nonzero and if there are as many as 2 they are adjacent.

One can find local optimum solutions to such problems using separable programming, and one can find global optimum solutions by introducing additional zero-one variables and constraints relating these variables to the λ_k. But the explicit integer formulation has proved unsatisfactory in practice and it is fortunately unnecassary. Instead we define such a set of λ_k as an S2 set. Each S2 set is assigned two "markers" in any subproblem, with the convention that the only members of the set that may take nonzero values are those between the markers. Initially the left hand marker is placed before the first member of the set and the right hand marker is placed after the last member. To branch on a set we find some suitable member λ_r and say that in the final solution <u>either</u> all members before λ_r must be zero, so that the left hand marker can be placed between λ_{r-1} and λ_r, <u>or</u> all members after λ_r must be zero, so that the right hand marker can be placed between λ_r and λ_{r+1}. These two possibilities then define the subproblems generated in Phase (2) of the general branch and bound step defined in Section 3.

Having decided to implement such a facility, we found that we could usefully extend it to cover sets of which not more than one member may be nonzero. These are now called S1 sets. The branching rule is based on the fact that in the final solution <u>either</u> all members before λ_{r+1} must be zero <u>or</u> all members after λ_r must be zero.

Further details of the algorithm are given in Beale and Tomlin (1970) or in Tomlin (1970).

S1 sets are normally used for multiple choice problems, which can be achieved by writing

$$\sum_k \lambda_k = 1,$$

but they are not restricted to such applications. For example one may have a storage facility that can be used for a number of alternative materials, but which cannot hold mixtures of materials, perhaps because they are liquids. The amounts of each material held in the store can then be treated as an S1 set, although the actual amount held is a continuous variable.

A client recently had an extension of this problem, which is of interest since it illustrates the point made about formulation at the end of Section 3. He had a set of variables x_k, each with a simple upper bound A_k, and he wanted to restrict the program to a single member of the set. He therefore defined them as an S1 set. He then found that UMPIRE does not allow bounds on members of special ordered sets, and for technical reasons it is inconvenient to provide this facility. Fortunately the problem can be formulated better without simple upper bounds, by replacing the variable x_k by the variable y_k defined as x_k/A_k. The y_k can then be treated as an S1 set subject to the generalized upper bound.

$$\Sigma y_k \leqslant 1.$$

This formulation means that if any $y_k = 1$, i.e. $x_k = A_k$, then the other members of the set must vanish even while the variables are treated as continuous. The original formulation does not have this property unless we write

$$\Sigma x_k/A_k \leqslant 1,$$

which would also avoid the need for simple upper bounds but which could not be treated as a GUB row.

5. NONLINEAR PROGRAMMING

The use of general mathematical programming systems on more general nonlinear problems than those for which separable programming or S2 sets are convenient seems less widespread. Abadie and Carpentier (1969) have developed a succesful method closely related to the simplex method but no reports have been given of the implementation of this method within a general mathematical programming system. One of the practical difficulties about nonlinear programming is the formulation of a model that is both general enough to include the true problem to be studied and also specific enough to be helpful from the point of view of data collection and algorithmic development. Theoreticians may write the problem as

$$\text{Minimize} \quad f(x_1, \ldots\ldots, x_n)$$

$$\text{subject to} \quad g_i(x_1, \ldots\ldots, x_n) \leqslant b_i \quad (i = 1, \ldots.m),$$

but this gives no hint of the problem structure. An improvement from the point of view of problems for which general mathematical programming systems may be helpful is to divide the variables into linear and nonlinear variables. If x_j $(j=1...n)$ denote the linear variables then we may write the problem as

Minimize $\Sigma a_{oj}x_j + f(y_1 \ldots y_k)$

subject to $\Sigma a_{ij}x_j + g_i(y_1 \ldots . y_k) = b_i \quad (i=1 \ldots m)$

and $0 \leqslant x_j \leqslant B_j \quad (j=1, \ldots . n).$

$L_r \leqslant y_r \leqslant U_r \quad (r=1, \ldots . k).$

But Beale (1972) suggests the alternative formulation

Maximize x_0 subject to the constraints

$$x_0 + \Sigma a_{oj}x_j = b_0$$

$$\Sigma a_{ij}x_j = b_r \quad (i=1 \ldots . m)$$

$$0 \leqslant x_j \leqslant B_j \quad (j=1 \ldots . n)$$

$$L_r \leqslant y_r \leqslant U_r \quad (r=1 \ldots . k)$$

where the coefficients a_{ij} and b_i may be either constants or functions of the nonlinear variables y_r. This formulation would be essentially the same as the preceding one if only the b_i were allowed to depend on the y_r. But by also allowing the coefficients of the nonlinear variable to depend on the nonlinear variables one can greatly reduce the number of variables that must be treated as nonlinear in many applications. This can therefore be regarded as the canonical form for a nonlinear programming problem to be solved by methods related to linear programming.

Some recent developments of an approach to such problems first used by Griffith and Stewart (1961) are described by Buzby (1974) and Beale (1974). The approach is to select trial values of the nonlinear variables and then make a local linear programming approach to the problem, putting tolerances on the extent to which the nonlinear variables may depart from their current trial values. The solution to this and previous LP problems can then be used to suggest trial values and tolerances for the nonlinear variables in the next problem.

The efficient implementation of this approach is much easier on third generation computers equipped with drum or disc storage and sophisticated operating systems than it was in 1961. This is primarily because there is no effective limit on the number of files that can be used to transfer information between the separate programs of the system; and no difficulty in running such separate programs sequentially as a single job on the computer. In our implementation there are four such programs in addition to the initial input. Each iteration of the nonlinear programming procedure involves running all four programs in the following sequence:

The Control Program, which selects trial values and tolerances for the nonlinear variables.

The Matrix Generator, which generates the linear programming problem based on these values.

The general mathematical programming system, UMPIRE, which solves the linear programming problem.

The Output Analyser, which writes a short report on the solution and records the values and reduced costs of the nonlinear variables on file for use by the control program on the next iteration.

We have recently used the implementation of this approach described by Beale (1974) on problems with over 1000 constraints, over 2000 linear variables and a few hundred nonlinear variables. Useful, though not strictly optimum, results were obtained after a few hours on the computer although some difficulties were encountered when we had poor initial trial values for the nonlinear variables. These difficulties were overcome by allowing the matrix generator to change the trial values suggested by the control program if they were obviously unsatisfactory. The details of any such procedure must be problem dependent, and one must be careful to avoid upsetting the convergence of the algorithm. But the principle may be useful on a wider range of problems than our particular applications. Specifically, it may be useful on other multi-time-period models where the trial values for later time period variables prove incompatible with the trial values for earlier time period variables.

References

J.Abadie and J.Carpenter (1969) "Generalization of the Wolfe Reduced Gradient Method to the case of Nonlinear Constraints" pp. 37 - 47 of Optimization (Ed. R.Fletcher) Academic Press London and New York.

E.M.L.Beale (1968) Mathematical Programming in Practice Pitmans London.

E.M.L.Beale (1971) "Sparseness in Linear Programming" pp. 1 - 15 of Large Sparse Sets of Linear Equations (Ed.J.K.Reid) Academic Press, London and New York.

E.M.L.Beale (1974) "A conjugate gradient method of approximation programming" to appear in Applications of Optimization Methods for Large-Scale Resource - Allocation Problems (Eds.R.W.Cottle and J.Krarup).

E.M.L.Beale and J.A.Tomlin (1970) "Special facilities in a general mathematical programming systems for non-convex problems using ordered sets of variables" pp. 447 - 454 of Proceedings of the Fifth International Conference on Operational Research (Ed.J.Lawrence) Tavistock Publications, London.

M.Benichou, J.M.Gauthier, P.Girodet, G.Hentges, G.Ribiere and D.Vincent (1971) "Experiments in Mixed-Integer Linear Programming" Mathematical Programming 1, pp. 76 - 94.

B.R.Buzby (1974) "Techniques and Experience solving really big nonlinear programs" to appear in Applications of Optimization Methods for Large-Scale Resource Allocation Problems (Eds. R.W.Cottle and J.Krarup).

N.J.Driebeck (1966) "An algorithm for the solution of mixed integer programming problems". Management Science 12, pp. 576 - 587.

J.J.H.Forrest, J.P.H.Hirst and J.A.Tomlin (1974) "Practical Solution of Large Mixed Integer Programming Problems with UMPIRE" Management Science 20, pp. 736 - 773.

J.J.H.Forrest and J.A.Tomlin (1972) "Updating the triangular factors of the basis to maintain sparsity in the product form of the simplex method" to appear in Mathematical Programming.

R.E.Griffith and R.A.Stewart (1961) "A nonlinear programming technique for the optimization of continuous processing systems" Management Science 7, pp. 379 - 392.

E.Hellerman and D.Rarick (1971) "Reinversion with the Preassigned Pivot Procedure" Mathematical Programming 1, pp. 195 - 216.

J.D.C.Little, K.C.Murty, D.W.Sweeney and C.Karel (1963) "An algorithm for the traveling salesman problem" Operations Research 11, pp. 972 - 989.

H.M.Markowitz (1957) "The elimination form of the inverse and its application to linear programming" Management Science 3, pp. 255 - 269.

G.H.Mitchell (Ed.)(1972) Operational Research: Techniques and Examples English University Press London.

J.A.Tomlin (1970) "Branch and bound methods for integer and non-convex programming" pp. 437 - 450 of Integer and Nonlinear Programming (Ed.J.Abadie).

A.M.Turing (1948) "Rounding-off errors in matrix processes" Quart J. Mech 1, pp. 287 - 308.

Mathematical Programming in Theory and Practice,
P.L. Hammer and G. Zoutendijk, (Eds.)

"Unconstrained Minimization and Extensions for Constraints"

by

M.J.D. Powell

Summary

A brief survey is given of the kinds of method that are used to minimize differentiable functions of several variables. Then a technique is described for making search directions conjugate without the calculation of derivatives. New proofs are given of the quadratic termination properties of quasi-Newton algorithms and of Dixon's (1972) theorem. The quadratic termination proof is extended to the case of minimization subject to linear equality and inequality constraints, and we note that it supports the type of algorithm developed by Goldfarb (1969). This type of algorithm is compared with the kind of method described by Murtagh and Sargent (1969), and we note that they have many features in common. Finally, for nonlinear equality and inequality constraints, a very promising idea due to Fletcher (1972a) is described for converting a constrained problem into an unconstrained one.

Theoretical Physics Division,
U.K.A.E.A. Research Group,
Atomic Energy Research Establishment,
HARWELL.

1. The use of search directions in minimization algorithms

The first five sections of this paper concern algorithms for calculating the least value of a function $F(x_1, x_2, \dots, x_n) = F(\underline{x})$, say, when there are no constraints on the variables. In Sections 6 and 7 we consider extensions for the case when there are linear constraints on $\underline{x}$, and in Section 8 a technique is described for nonlinear constraints.

This paper is not intended to provide a complete survey. Instead it describes a few recent ideas and results that seem to be particularly promising or interesting.

It is convenient to regard the objective function $F(\underline{x})$ as being defined by a computer subroutine that calculates the value of $F(\underline{x})$ for any $\underline{x}$. Further we suppose that $F(\underline{x})$ is differentiable, but for some algorithms the computer subroutine need not calculate any derivative values.

The algorithms for unconstrained minimization are iterative, and they are given a starting vector $\underline{x}^{(1)}$, say. By using calculated function values they generate a sequence of vectors $\underline{x}^{(k)}$ $(k=1,2,3\dots)$ that is intended to converge to the point at which $F(\underline{x})$ is least.

Many methods calculate the sequence of points by using search directions. At the point $\underline{x}^{(k)}$ a search direction $\underline{d}^{(k)}$ is chosen, and then the function

$$\psi(\theta) = F(\underline{x}^{(k)} + \theta \underline{d}^{(k)}) \qquad \dots(1.1)$$

is regarded as a function of the single variable θ. The vector $\underline{x}^{(k+1)}$ is defined to have the form

$$\underline{x}^{(k+1)} = \underline{x}^{(k)} + \theta \underline{d}^{(k)}, \qquad \dots(1.2)$$

the value of θ being calculated to give the inequality

$$F(\underline{x}^{(k+1)}) \leqslant F(\underline{x}^{(k)}). \qquad \dots(1.3)$$

Thus, if the set of points $\underline{x}$ satisfying the condition $F(\underline{x}) \leqslant F(\underline{x}^{(1)})$ is bounded, it follows that the sequence of points $\underline{x}^{(k)}$ $(k=1,2,3,\dots)$ is also bounded.

For example the best known method that uses search directions is the method of steepest descent. Here the direction $\underline{d}^{(k)}$ (k=1,2,3,...) is the vector $-\nabla F(\underline{x}^{(k)}) = -\underline{g}^{(k)}$, say. In the "perfect" version of this method the value of θ in equation (1.2) is calculated to minimize $F(\underline{x}^{(k+1)})$. In this case it can be proved that, if the first derivative vector of $F(\underline{x})$ is continuous, then every limit point of the sequence $\underline{x}^{(k)}$ (k=1,2,3,...) is a stationary point of $F(\underline{x})$ (Curry, 1944). Therefore in practice one expects the method to find the least value of $F(\underline{x})$.

However the steepest descent method usually converges too slowly to be useful. For example if it is used to minimise a quadratic function

$$F(\underline{x}) = a + (\underline{b},\underline{x}) + \tfrac{1}{2}(\underline{x},G\underline{x}), \qquad \ldots (1.4)$$

where G is a positive definite matrix, and if $\underline{x}^*$ is the required vector of variables, then the speed of convergence is given by the inequality

$$||\underline{x}^{(k)} - \underline{x}^*|| \leqslant C \left(\frac{M-m}{M+m}\right)^k, \qquad \ldots (1.5)$$

where C is a constant and where m and M are the smallest and largest eigenvalues of G (Kantorovich and Akilov, 1964). The slow convergence suggested by the bound (1.5) often occurs in practice, so now the steepest descent method is almost superseded by other algorithms, for instance the type of method described in Section 4.

The improvements over the steepest descent method are obtained by using techniques for choosing the search directions $\underline{d}^{(k)}$ (k=1,2,3,...) that take account of the curvature of $F(\underline{x})$. Some good techniques are considered in Sections 2-4.

Another important aspect of search direction algorithms is that one requires methods for fixing the value of θ in equation (1.2) that are suitable for numerical computation. One cannot in general find exactly the value of θ that minimizes $F(\underline{x}^{(k+1)})$ by using a few values of the objective function. Therefore a compromise has to be made between the reduction $F(\underline{x}^{(k+1)}) - F(\underline{x}^{(k)})$

and the number of times $F(\underline{x})$ is calculated.

Often this problem is not treated seriously. For instance there are many algorithms that specify that the value of θ is to minimize the function (1.1), and, on the other hand, there are some algorithms that finish adjusting θ as soon as a calculated value of the function (1.1) is less than $F(\underline{x}^{(k)})$. Fletcher (1972b) gives a nice example showing that this last strategy can ruin a good method.

However Wolfe (1969) studies some suitable finite methods for fixing the value of θ. One strategy for the case when the gradient $\underline{\nabla}F(\underline{x})$ is calculated is to satisfy the conditions

$$|(\underline{g}^{(k+1)}, \underline{d}^{(k)})| \leqslant \tfrac{1}{2} |(\underline{g}^{(k)}, \underline{d}^{(k)})| \qquad \ldots (1.6)$$

and

$$F(\underline{x}^{(k+1)}) \leqslant F(\underline{x}^{(k)}) - 0.1|\theta(\underline{g}^{(k)}, \underline{d}^{(k)})|, \qquad \ldots (1.7)$$

where again $\underline{g}^{(k)} = \underline{\nabla}F(\underline{x}^{(k)})$. Inequality (1.6) ensures that $|\theta|$ is not too small, and inequality (1.7) ensures that the iteration makes an adequate reduction in the objective function.

It is well known that useful search direction methods may find a local minimum of $F(\underline{x})$ rather than a global minimum. Therefore some techniques have been proposed (Goldstein and Price, 1971; McCormick, 1972) to direct the search for the least value of $F(\underline{x})$ away from a calculated minimum, in case this point is not the required solution. Examples show that these techniques can be successful in obtaining further reductions in the objective function. However often the first calculated minimum is the global minimum, and on other occasions there may be many local minima. Therefore these techniques are aids, rather than sure methods, for finding the global minimum of $F(\underline{x})$.

2. Methods that take account of the curvature of the objective function

In the last section we noted that the search direction of the steepest descent method, $\underline{d}^{(k)} = -\underline{g}^{(k)}$, does not give a good algorithm. Therefore now we consider better ways of choosing search directions.

When $F(\underline{x})$ is the quadratic function (1.4), the value of $F(\underline{x})$ is least at the point

$$\begin{aligned} \underline{x}^* &= -G^{-1}\underline{b} \\ &= \underline{x} - G^{-1}\underline{\nabla}F(\underline{x}) \end{aligned} \qquad \ldots (2.1)$$

for all $\underline{x}$. Therefore if for a general objective function both first and second derivatives are calculated, then the search direction

$$\underline{d}^{(k)} = -[G^{(k)}]^{-1}\underline{g}^{(k)} \qquad \ldots (2.2)$$

seems promising, where $\underline{g}^{(k)}$ is the first derivative vector and $G^{(k)}$ is the second derivative matrix of $F(\underline{x})$ at the point $\underline{x}^{(k)}$. When this direction is used for $k=1,2,3,\ldots$, and when θ is set to one in equation (1.2), the algorithm is called the "Newton-Raphson" method.

Often the Newton-Raphson method is very good, especially when the starting vector $\underline{x}^{(1)}$ is close to the required solution. Indeed, if the second derivative matrix of $F(\underline{x})$ is positive definite and Lipschitz continuous at the solution, $\underline{x}^*$ say, then the sequence of points $\underline{x}^{(k)}$ $(k=1,2,3,\ldots)$ has second order convergence to $\underline{x}^*$, provided that $\underline{x}^{(1)}$ is sufficiently close to $\underline{x}^*$ (Ortega and Rheinboldt, 1970).

However, because the Newton-Raphson method may fail to converge if $\underline{x}^{(1)}$ is too far from $\underline{x}^*$, it is helpful to check at each iteration that $\underline{x}^{(k+1)}$ satisfies inequalities (1.6) and (1.7). If either inequality fails then θ should be changed to a value that is different from one.

Even with this modification the Newton-Raphson method is not suitable for a general purpose minimization algorithm, because it may stick at a point that is not a stationary point of $F(\underline{x})$. An example is given by Powell

(1966). Here a point $\underline{x}^{(k)}$ is reached at which the direction (2.2) is well defined, but the only value of θ that satisfies condition (1.3) is $\theta=0$, because $\underline{d}^{(k)}$ is orthogonal to $\underline{g}^{(k)}$. Thus $\underline{x}^{(k+1)} = \underline{x}^{(k)}$, and no further progress is made by the iterative method.

Two techniques are used to overcome the difficulty of the last paragraph. One technique is based on the quadratic function

$$\Phi^{(k)}(\underline{\delta}) = F(\underline{x}^{(k)}) + (\underline{g}^{(k)},\underline{\delta}) + \tfrac{1}{2}(\underline{\delta},G^{(k)}\underline{\delta}), \qquad \ldots (2.3)$$

which is a good approximation to $F(\underline{x}^{(k)} + \underline{\delta})$ when $\underline{\delta}$ is small. In this case $\underline{d}^{(k)}$ is chosen to satisfy the condition

$$\Phi^{(k)}(\underline{d}^{(k)}) < F(\underline{x}^{(k)}). \qquad \ldots (2.4)$$

For example in Fletcher's (1972b) hypercube method $\underline{d}^{(k)}$ is calculated to minimize $\Phi^{(k)}(\underline{d}^{(k)})$, subject to linear constraints that limit the size of the components of $\underline{d}^{(k)}$. These constraints are adjusted automatically so that most iterations satisfy inequality (1.3) by setting $\theta=1$ in equation (1.2). Note that this calculation of $\underline{d}^{(k)}$ is a quadratic programming problem.

The other technique is to change the matrix $G^{(k)}$ if it is not positive definite, in order to force it to be positive definite. Then when $\underline{g}^{(k)}$ is non-zero the search direction (2.2) must be a direction of descent. For example a multiple of the unit matrix may be added to $G^{(k)}$ in the method proposed by Goldfeld, Quandt and Trotter (1966), and Matthews and Davies (1971) make a suitable change to a triangular factorization of $G^{(k)}$ if any negative or zero diagonal elements occur.

Fletcher's (1972b) method has an advantage over the algorithms that modify the matrix $G^{(k)}$ to make it positive definite, which is important if $\underline{x}^{(k)}$ happens to be a saddle point of $F(\underline{x})$. In this case the true second derivative matrix of $F(\underline{x})$ usually has at least one negative eigenvalue, and then from expression (2.3) a value of $\underline{d}^{(k)}$ can be found to reduce the objective function. However, if equation (2.2) is used then $\underline{d}^{(k)}$ is zero.

As a rule computer users prefer not to calculate second derivatives, so now we consider some algorithms that take account of the curvature of $F(\underline{x})$, although they require the calculation of only values and first derivatives of the objective function. First we consider some "quasi-Newton methods", and then "conjugate direction methods" are mentioned briefly.

We define a quasi-Newton method to be one that is of the type described already in this section, except that the second derivative matrix $G^{(k)}$ is estimated from calculated first derivatives. Therefore the main new problem is to estimate this matrix. Some methods of estimation give symmetric positive definite matrices automatically, and then it is usual to define each search direction by equation (2.2).

The most straightforward method of estimation is to use differences of gradient vectors along the coordinate directions. Specifically for $i=1,2,\ldots,n$ we let the ith column of $G^{(k)}$ be the vector

$$\{\underline{g}(\underline{x}^{(k)} + h_i\underline{e}_i) - \underline{g}^{(k)}\}/h_i, \qquad \ldots (2.5)$$

where $\underline{e}_i$ is the ith coordinate vector and where h_i is a suitable step-length. This method may be extended to give a symmetric matrix by replacing $G^{(k)}$ by $\frac{1}{2}\{G^{(k)} + G^{(k)T}\}$.

However most quasi-Newton algorithms do not calculate gradient vectors just for the purpose of estimating second derivatives. Instead they make use of the gradients that are calculated at the points $\underline{x}^{(k)}$ $(k=1,2,3,\ldots)$. For example in the rank one method (Murtagh and Sargent, 1970) we have a symmetric matrix $G^{(k)}$ at the beginning of the kth iteration, and $\underline{x}^{(k+1)}$ is calculated from equations (1.2) and (2.2). Then for the next iteration the second derivative approximation is the matrix

$$G^{(k+1)} = G^{(k)} + \frac{(\underline{\gamma} - G^{(k)}\underline{\delta})(\underline{\gamma} - G^{(k)}\underline{\delta})^T}{(\underline{\gamma} - G^{(k)}\underline{\delta},\ \underline{\delta})}, \qquad \ldots (2.6)$$

where $\underline{\delta}$ and $\underline{\gamma}$ are the vectors

$$\left.\begin{aligned} \underline{\delta} &= \underline{x}^{(k+1)} - \underline{x}^{(k)} \\ \underline{\gamma} &= \underline{g}^{(k+1)} - \underline{g}^{(k)} \end{aligned}\right\}, \qquad \ldots (2.7)$$

and where the superscript "T" distinguishes a row vector from a column vector. The formula (2.6) has the remarkable property that, if $F(\underline{x})$ is a quadratic function, and if no zero denominators occur in equation (2.6), then the matrix $G^{(n+1)}$ is the exact second derivative matrix of $F(\underline{x})$, whatever the choice of the initial symmetric matrix $G^{(1)}$ and the search directions $\underline{d}^{(k)}$ $(k=1,2,\ldots,n)$ (Fiacco and McCormick, 1968).

Note that the formula (2.6) satisfies the equation

$$G^{(k+1)}\underline{\delta} = \underline{\gamma} . \qquad \ldots (2.8)$$

This is good because, if $F(\underline{x})$ is a quadratic function whose second derivative matrix is G, then $G\underline{\delta} = \underline{\gamma}$. The equation (2.8) is satisfied by many quasi-Newton algorithms, in particular by the methods that define $G^{(k+1)}$ by the formula

$$G^{(k+1)} = G^{(k)} - \frac{G^{(k)}\underline{\delta}\,\underline{\delta}^T G^{(k)}}{(\underline{\delta}, G^{(k)}\underline{\delta})} + \frac{\underline{\gamma}\underline{\gamma}^T}{(\underline{\delta},\underline{\gamma})} + \alpha\underline{u}\underline{u}^T, \qquad \ldots (2.9)$$

where $\underline{u}$ is the vector

$$\underline{u} = G^{(k)}\underline{\delta} - \frac{(\underline{\delta}, G^{(k)}\underline{\delta})}{(\underline{\delta},\underline{\gamma})}\,\underline{\gamma} , \qquad \ldots (2.10)$$

and where α is a parameter. Different choices of α give different algorithms, and this class of methods includes the rank one method, the DFP algorithm and the BFS algorithm (Broyden, 1970; Fletcher, 1970). Sections 4 to 7 study these methods in more detail, so for the present we note only that these quasi-Newton algorithms are used frequently and successfully to minimize functions of up to about two hundred variables.

Next we consider conjugate direction methods. The term conjugacy is properly defined only when the objective function is quadratic, and in this case we say that the directions $\underline{d}_i$ and $\underline{d}_j$ are conjugate if the equation

$$(\underline{d}_i, G\underline{d}_j) = 0 \qquad \ldots (2.11)$$

is satisfied, where G is the second derivative matrix of $F(\underline{x})$. Conjugate directions are useful because, if $F(\underline{x})$ is a positive definite quadratic function, and if we have n mutually conjugate search directions $\underline{d}_i$ $(i=1,2,\ldots,n)$, then the following method finds the least value of $F(\underline{x})$. Let $\underline{x}^{(1)}$ be any starting vector. For $k=1,2,\ldots,n$, define $\underline{x}^{(k+1)}$ by equation (1.2), where $\underline{d}^{(k)} = \underline{d}_k$ and where the value of θ is calculated to minimize $F(\underline{x}^{(k+1)})$. Then $\underline{x}^{(n+1)}$ is the point at which $F(\underline{x})$ is least (Kowalik and Osborne, 1968).

In fact in the quadratic case many quasi-Newton methods generate search directions that happen to be conjugate, so some explanation is needed to distinguish quasi-Newton algorithms from conjugate direction algorithms. We say that a method is a quasi-Newton method if it tries to approximate a second derivative matrix, and we say that a method is a conjugate direction method if it generates search directions in order to take advantage of the construction of the last paragraph. One takes advantage of this construction when $F(\underline{x})$ is not a quadratic function by using search directions that tend to be conjugate with respect to the second derivative matrix at the limit of the sequence $\underline{x}^{(k)}$ $(k=1,2,3,\ldots)$, when this limit exists.

There are some conjugate direction algorithms for the case when the gradient of $F(\underline{x})$ is available but second derivatives are not calculated. For this case we note that equations (1.2) and (1.4) imply the equation

$$\theta G\underline{d}^{(k)} = \underline{g}^{(k+1)} - \underline{g}^{(k)}. \qquad \ldots (2.12)$$

It follows that a direction is conjugate to $\underline{d}^{(k)}$ if and only if it is orthogonal to $\{\underline{g}^{(k+1)} - \underline{g}^{(k)}\}$. Thus conjugacy conditions can be expressed in terms of first derivative vectors (Zoutendijk, 1960). At this conference Wolfe (1972) will describe some conjugate gradient methods that do not require second derivatives to be calculated, and that are particularly useful because they do not require computer storage for any matrices.

When only values of $F(\underline{x})$ are calculated, first derivatives can be approximated by differences in quasi-Newton methods. For instance Stewart (1967) extends the DFP algorithm (Fletcher and Powell, 1963) in this way, and his method gives good results.

Alternatively conjugate direction methods can be found that do not require any derivative calculations. Some of these methods are especially interesting because they do not approximate any derivatives directly. For example the following construction generates a pair of conjugate directions.

Let $\underline{d}$ be a search direction, and let $\underline{x}_1$ and $\underline{x}_2$ be any two points in the space of the variables such that $(\underline{x}_2 - \underline{x}_1)$ is not a multiple of $\underline{d}$. From the point $\underline{x}_1$ we search for the least value of $F(\underline{x})$ in the direction $\underline{d}$ and reach the point $\underline{y}_1$, say. Similarly we reach the point $\underline{y}_2$ by searching from $\underline{x}_2$ in the direction $\underline{d}$. Then, if $F(\underline{x})$ is a quadratic function, the direction $(\underline{y}_2 - \underline{y}_1)$ is conjugate to $\underline{d}$.

An extension of this construction has been developed into a minimization algorithm by Powell (1964). This algorithm is used often, although on average Stewart's (1967) method seems to be slightly faster. Possibly this is because Powell's method has to include some precautions to ensure that the search directions span the full space of the variables.

Therefore in the next section a very recent technique for constructing conjugate directions is described (Powell, 1972a), that automatically keeps linear independence in the search directions. This technique has not yet been developed into an algorithm.

3. The construction of conjugate directions without derivatives

In this section a recent method (Powell, 1972a) is described for changing a set of search directions ($\underline{d}_1, \underline{d}_2, \ldots, \underline{d}_n$) to improve their conjugacy properties. This method does not require the calculation of any derivatives, but it does require the estimation of a few second derivatives. Specifically when the function (1.1) is considered the derivative $\psi''(\theta)$ has to be estimated, and therefore the second derivative approximations seldom require extra values of $F(\underline{x})$ to be calculated.

In order to be able to state that a new set of search directions has better conjugacy properties than the set ($\underline{d}_1, \underline{d}_2, \ldots, \underline{d}_n$), we require a criterion for measuring the conjugacy of a set of directions. To define conjugacy properly we suppose that $F(\underline{x})$ is the positive definite quadratic function (1.4), and in this case we let $\Delta(\underline{d}_1, \underline{d}_2, \ldots, \underline{d}_n)$ be the quantity

$$\Delta = \left|\det\{\underline{d}_1, \underline{d}_2, \ldots, \underline{d}_n\}\right| / \prod_{i=1}^{n} (\underline{d}_i, G\underline{d}_i)^{\frac{1}{2}}, \qquad \ldots (3.1)$$

where $\{\underline{d}_1, \underline{d}_2, \ldots, \underline{d}_n\}$ is the matrix whose i^{th} column is $\underline{d}_i$ $(i=1,2,\ldots,n)$. Then we say that $(\underline{d}_1^*, \underline{d}_2^*, \ldots, \underline{d}_n^*)$ is more conjugate than $(\underline{d}_1, \underline{d}_2, \ldots, \underline{d}_n)$ if the inequality

$$\Delta(\underline{d}_1^*, \underline{d}_2^*, \ldots, \underline{d}_n^*) > \Delta(\underline{d}_1, \underline{d}_2, \ldots, \underline{d}_n) \qquad \ldots (3.2)$$

is satisfied. The following theorem shows that this criterion is suitable.

<u>Theorem 3.1.</u> The value of $\Delta(\underline{d}_1, \underline{d}_2, \ldots, \underline{d}_n)$ is greatest if and only if the directions satisfy the conjugacy conditions

$$(\underline{d}_i, G\underline{d}_j) = 0, \quad i \neq j. \qquad \ldots (3.3)$$

<u>Proof.</u> Suppose that $(\underline{d}_i, G\underline{d}_j)$ is different from zero, $i \neq j$. Then if $\underline{d}_i$ is replaced by $\underline{d}_i + \lambda \underline{d}_j$, and if the other directions are unchanged, the value of Δ is multiplied by the factor

$$\frac{(\underline{d}_i, G\underline{d}_i)^{\frac{1}{2}}}{\{(\underline{d}_i, G\underline{d}_i) + 2\lambda(\underline{d}_i, G\underline{d}_j) + \lambda^2(\underline{d}_j, G\underline{d}_j)\}^{\frac{1}{2}}} \cdot \qquad \ldots (3.4)$$

This factor is greater than one if λ is sufficiently small and $\lambda(\underline{d}_i, G\underline{d}_j)$ is negative. It follows that $\Delta(\underline{d}_1, \underline{d}_2, \ldots, \underline{d}_n)$ is greatest only if condition (3.3) holds for all i not equal to j.

If we choose a suitable normalization of the directions, for instance $(\underline{d}_i, G\underline{d}_i) = 1$, $i=1,2,\ldots,n$, then the function Δ is a continuous and bounded function of the search directions. Therefore its maximum value is attained. Say it occurs when the search directions are $(\underline{d}_1^*, \underline{d}_2^*, \ldots, \underline{d}_n^*)$. We have proved already that these directions are mutually conjugate, and to complete the proof of the theorem we must show that if $(\underline{d}_1, \underline{d}_2, \ldots, \underline{d}_n)$ is any other set of mutually conjugate directions, then the equation

$$\Delta(\underline{d}_1, \underline{d}_2, \ldots, \underline{d}_n) = \Delta(\underline{d}_1^*, \underline{d}_2^*, \ldots, \underline{d}_n^*) \qquad \ldots (3.5)$$

is satisfied.

Because the definition (3.1) is independent of the scaling of the directions, we assume, without loss of generality, that the normalization conditions

$$\left.\begin{array}{l}(\underline{d}_i, G\underline{d}_i) = 1 \\ (\underline{d}_i^*, G\underline{d}_i^*) = 1\end{array}\right\}, \quad i=1,2,\ldots,n, \qquad \ldots (3.6)$$

are obtained.

There exists a matrix, P say, satisfying the equation

$$\underline{d}_i = \sum_{j=1}^{n} P_{ij}\, \underline{d}_j^*, \quad i=1,2,\ldots,n, \qquad \ldots (3.7)$$

and from expressions (3.1) and (3.6) we obtain the identity

$$\Delta(\underline{d}_1, \underline{d}_2, \ldots, \underline{d}_n) = |\det P|\, \Delta(\underline{d}_1^*, \underline{d}_2^*, \ldots, \underline{d}_n^*). \qquad \ldots (3.8)$$

Further, by using the conjugacy properties of the directions and expression (3.6)again, we deduce the relation

$$\delta_{ij} = (\underline{d}_i, G\underline{d}_j)$$

$$= \sum_{k\ell} P_{ik}\, P_{j\ell}\, (\underline{d}_k^*, G\underline{d}_\ell^*)$$

$$= \sum_k P_{ik} P_{jk} , \qquad \ldots (3.9)$$

where δ_{ij} is the Kronecker delta. It follows that the matrix P is orthogonal, and therefore equation (3.8) shows that the expression (3.5) is true. Theorem 3.1 is proved.

This theorem shows that it is good to change the search directions by methods that increase the value of $\Delta(\underline{d}_1,\underline{d}_2,\ldots,\underline{d}_n)$. The method to be described is suggested by the next theorem.

Theorem 3.2. Let $(\underline{d}_1,\underline{d}_2,\ldots,\underline{d}_n)$ be a set of independent directions satisfying the condition

$$(\underline{d}_i,G\underline{d}_i) = 1, \; i=1,2,\ldots,n, \qquad \ldots (3.10)$$

and let Ω be any orthogonal matrix. Then the set of directions $(\underline{d}_1^*,\underline{d}_2^*,\ldots,\underline{d}_n^*)$, defined by the equation

$$\underline{d}_i^* = \sum_{j=1}^{n} \Omega_{ij}\, \underline{d}_j , \qquad \ldots (3.11)$$

satisfies the inequality

$$\Delta(\underline{d}_1^*,\underline{d}_2^*,\ldots,\underline{d}_n^*) \geq \Delta(\underline{d}_1,\underline{d}_2,\ldots,\underline{d}_n) . \qquad \ldots (3.12)$$

Proof. Because the definition (3.11) and the orthogonality of Ω imply the equation

$$|\det\{\underline{d}_1^*,\underline{d}_2^*,\ldots,\underline{d}_n^*\}| = |\det\{\underline{d}_1,\underline{d}_2,\ldots,\underline{d}_n\}|, \qquad \ldots (3.13)$$

it follows from expressions (3.1) and (3.10) that we just have to prove that the inequality

$$\prod_{i=1}^{n} (\underline{d}_i^*,G\underline{d}_i^*) \leq 1 \qquad \ldots (3.14)$$

is satisfied. This condition is deduced from the geometric/arithmetic mean inequality, the definition (3.11), the orthogonality of Ω, and equation (3.10), in the following way:

$$\{\prod_{i=1}^{n} (\underline{d}_1^*,G\underline{d}_i^*)\}^{1/n} \leq \frac{1}{n}\sum_i (\underline{d}_i^*,G\underline{d}_i^*)$$

$$= \frac{1}{n} \sum_{ijk} \Omega_{ij} \Omega_{ik} (\underline{d}_j, G\underline{d}_k)$$

$$= \frac{1}{n} \sum_{j} (\underline{d}_j, G\underline{d}_j)$$

$$= 1. \qquad \ldots (3.15)$$

Theorem 3.2 is proved.

Therefore Powell (1972a) suggests the following general method for minimizing a function without calculating any derivatives.

Choose a starting point $\underline{x}^{(1)}$ and n independent search directions $(\underline{d}_1, \underline{d}_2, \ldots, \underline{d}_n)$, for example the coordinate directions. Use the search directions in sequence. During each search, estimate the second derivative of the objective function along the direction, and normalize the search direction in accordance with equation (3.10). When the directions have all been used once, choose a suitable orthogonal matrix Ω. Then for i = 1,2,..., n, replace $\underline{d}_i$ by the vector (3.11). Replace $\underline{x}^{(1)}$ by the point that has given the least calculated value of $F(\underline{x})$. Go back to the operation of using the search directions in sequence etc. Note that this method can be applied to general objective functions.

In order to develop this method into an algorithm, it is necessary to specify the choice of Ω, but this question has not been studied yet, except in the trivial case when n=2. In this case it is good to let $\underline{d}_1^*$ and $\underline{d}_2^*$ be the vectors

$$\left.\begin{aligned} \underline{d}_1^* &= (\underline{d}_1 + \underline{d}_2)/\sqrt{2} \\ \underline{d}_2^* &= (-\underline{d}_1 + \underline{d}_2)/\sqrt{2} \end{aligned}\right\}, \qquad \ldots (3.16)$$

because then the normalization condition (3.10) and the symmetry of G imply that the conjugacy equation

$$(\underline{d}_1^*, G\underline{d}_2^*) = 0 \qquad \ldots (3.17)$$

is satisfied. However, for n>2, it is not yet known whether it is better to fix the matrices Ω in advance, or whether it is better for their definition to depend on the progress of the iterative method.

4. A family of quasi-Newton algorithms

We consider the class of quasi-Newton algorithms that use the formula (2.9) to revise symmetric second derivative approximations, and that use equation (2.2) to generate search directions. We are interested in this class because it includes many very useful methods.

In the early algorithms of this class the step-length θ of equation (1.2) in theory has the value that minimizes $F(\underline{x}^{(k+1)})$, so for these methods the problem of obtaining a suitable value of θ using few function evaluations is left to the programmer. However some good algorithms have been proposed recently that specify θ in a practical way, for example the method described by Fletcher (1970). Because these algorithms are derived mainly by good sense and numerical experimentation, the theory of the methods lags behind our practical experience. Indeed many important theoretical questions are still unanswered for the case when, for each value of k, the quantity θ in equation (1.2) is determined by an exact line search. And for the case when θ is defined in a practical way there are hardly any theorems.

In this section and the next one we dwell on two theoretical points that are important to the understanding of present methods and to future developments. This section gives a new proof of the quadratic termination properties of the class of quasi-Newton algorithms, because the method of proof can be extended to the case when there are linear constraints on the variables. The extension is made in Section 7, and it helps our understanding of Goldfarb's (1969) algorithm. In the next section a new proof of the following statement is given. If every iteration obtains the value of θ by an exact linear search, then, if a single critical value of α is avoided in equation (2.9), the choice of α does not influence the sequence of points $\underline{x}^{(k)}$ $(k=1,2,3,\ldots)$. Huang (1970) proves this theorem in the case when $F(\underline{x})$ is a quadratic function, and Dixon (1972) gives the surprising result that the theorem holds for quite general objective functions.

Because the matrix $[G^{(k)}]^{-1}$ occurs in equation (2.2), it is usual to work with the matrix

$$[G^{(k)}]^{-1} = H^{(k)}, \qquad \ldots (4.1)$$

say, instead of with $G^{(k)}$. Therefore expression (2.2) is replaced by the equation

$$\underline{d}^{(k)} = -H^{(k)}\underline{g}^{(k)}. \qquad \ldots (4.2)$$

Further by some straightforward algebra it can be shown that equations (2.9) and (4.1) give the formula

$$\begin{aligned} H^{(k+1)} &= [G^{(k+1)}]^{-1} \\ &= H_{BFS}^{(k+1)} + \beta\,\underline{\omega\omega}^T \qquad \ldots (4.3) \end{aligned}$$

for calculating $H^{(k+1)}$, where $H_{BFS}^{(k+1)}$ is the matrix

$$H_{BFS}^{(k+1)} = \left(I - \frac{\underline{\delta\gamma}^T}{(\underline{\delta},\underline{\gamma})}\right) H^{(k)} \left(I - \frac{\underline{\gamma\delta}^T}{(\underline{\delta},\underline{\gamma})}\right) + \frac{\underline{\delta\delta}^T}{(\underline{\delta},\underline{\gamma})}, \qquad \ldots (4.4)$$

where $\underline{\omega}$ is the vector

$$\underline{\omega} = H^{(k)}\underline{\gamma} - \frac{(\underline{\gamma},H^{(k)}\underline{\gamma})}{(\underline{\delta},\underline{\gamma})}\,\underline{\delta}, \qquad \ldots (4.5)$$

and where β is a parameter that is related to the parameter α of expression (2.9) by the equation

$$\beta = \frac{-\alpha(\underline{\delta},G^{(k)}\underline{\delta})^2}{(\underline{\delta},\underline{\gamma})^2 + \alpha(\underline{\delta},G^{(k)}\underline{\delta})\{(\underline{\delta},G^{(k)}\underline{\delta})(\underline{\gamma},H^{(k)}\underline{\gamma}) - (\underline{\delta},\underline{\gamma})^2\}}. \qquad \ldots (4.6)$$

The subscript "BFS" occurs in equation (4.3) because the value of β is zero in the Broyden-Fletcher-Shanno algorithm (Broyden, 1970).

In order to derive the quadratic termination properties of the class of algorithms, we suppose that $F(\underline{x})$ is the positive definite quadratic function (1.4). Further we define $S(\underline{g},H)$ to be the linear space of vectors $\underline{z}$ that satisfy the two conditions

$$\left.\begin{aligned} G H \underline{z} &= \underline{z} \\ (\underline{z}, H\underline{g}) &= 0 \end{aligned}\right\} . \qquad \ldots (4.7)$$

Then the quadratic termination properties are deduced from the following theorem.

Theorem 4.1. If the vector $\theta \underline{d}^{(k)}$ that occurs in equation (1.2) is non-zero, then the inequality

$$\dim\{S(\underline{g}^{(k+1)}, H^{(k+1)})\} \geqslant \dim\{S(\underline{g}^{(k)}, H^{(k)})\} \qquad \ldots (4.8)$$

is satisfied, where $\dim\{S\}$ denoted the dimension of the space S. Further, if the value of θ is calculated to minimize the function (1.1), then the inequality

$$\dim\{S(\underline{g}^{(k+1)}, H^{(k+1)})\} \geqslant \dim\{S(\underline{g}^{(k)}, H^{(k)})\} + 1 \qquad \ldots (4.9)$$

holds.

Proof. To prove that inequality (4.8) holds we show that $S(\underline{g}^{(k)}, H^{(k)})$ is a subspace of $S(\underline{g}^{(k+1)}, H^{(k+1)})$. Therefore let $\underline{z}$ be any vector that satisfies the conditions

$$\left.\begin{aligned} G H^{(k)} \underline{z} &= \underline{z} \\ (\underline{z}, H^{(k)} \underline{g}^{(k)}) &= 0 \end{aligned}\right\} . \qquad \ldots (4.10)$$

From equations (1.2), (2.7) and (4.2) and from the second line of expression (4.10) we obtain the identity

$$(\underline{z}, \underline{\delta}) = \theta(\underline{z}, \underline{d}^{(k)}) = -\theta(\underline{z}, H^{(k)} \underline{g}^{(k)}) = 0 . \qquad \ldots (4.11)$$

Further, by using equations (1.4) and (2.7), then by using the first line of expression (4.10), and then by using equation (4.11), we obtain the identity

$$(\underline{z}, H^{(k)} \underline{\gamma}) = (\underline{z}, H^{(k)} G \underline{\delta}) = (G H^{(k)} \underline{z}, \underline{\delta}) = (\underline{z}, \underline{\delta}) = 0 . \qquad \ldots (4.12)$$

It follows from expressions (4.3), (4.4), (4.5), (4.11) and (4.12) that the equation

$$H^{(k+1)} \underline{z} = H^{(k)} \underline{z} \qquad \ldots (4.13)$$

is satisfied. Therefore from equations (2.7), (4.10) and (4.12) we deduce that the conditions

$$GH^{(k+1)}\underline{z} = \underline{z} \qquad \dots (4.14)$$

and

$$(\underline{z}, H^{(k+1)}\underline{g}^{(k+1)}) = (\underline{z}, H^{(k)}\underline{g}^{(k+1)})$$
$$= (\underline{z}, H^{(k)}\underline{g}^{(k)}) + (\underline{z}, H^{(k)}\underline{\gamma})$$
$$= 0 \qquad \dots (4.15)$$

hold, which show that $\underline{z}$ is in $S(\underline{g}^{(k+1)}, H^{(k+1)})$. Thus it is proved that $S(\underline{g}^{(k)}, H^{(k)})$ is a subspace of $S(\underline{g}^{(k+1)}, H^{(k+1)})$, so inequality (4.8) is true.

In the case that the value of θ is calculated to minimize the function (1.1), we prove that the dimension of $S(\underline{g}^{(k+1)}, H^{(k+1)})$ is greater than the dimension of $S(\underline{g}^{(k)}, H^{(k)})$, by proving that the vector $\underline{\gamma}$ is in $S(\underline{g}^{(k+1)}, H^{(k+1)})$ but is not in $S(\underline{g}^{(k)}, H^{(k)})$.

Indeed, by using equations (4.3), (4.4) and (4.5), and then equations (1.4) and (2.7), we obtain the identity

$$GH^{(k+1)}\underline{\gamma} = G\underline{\delta} = \underline{\gamma}. \qquad \dots (4.16)$$

Further, by using the symmetry of $H^{(k+1)}$, then equations (4.3), (4.4) and (4.5) again, and then the fact that θ is calculated to minimize the function (1.1), we obtain the identity

$$(\underline{\gamma}, H^{(k+1)}\underline{g}^{(k+1)}) = (H^{(k+1)}\underline{\gamma}, \underline{g}^{(k+1)}) = (\underline{\delta}, \underline{g}^{(k+1)}) = 0 . \qquad \dots (4.17)$$

It follows that $\underline{\gamma}$ is in the space $S(\underline{g}^{(k+1)}, H^{(k+1)})$.

However, because by hypothesis the vector $\theta\underline{d}^{(k)}$ is non-zero, and because equations (4.2), (1.2), (2.7) and (1.4) give the relation

$$\theta(\underline{\gamma}, H^{(k)}\underline{g}^{(k)}) = -(\underline{\gamma}, \underline{\delta}) = -(G\underline{\delta}, \underline{\delta}), \qquad \dots (4.18)$$

we deduce from the positive definiteness of G that the scalar product

$(\underline{\gamma}, H^{(k)}\underline{g}^{(k)})$ is different from zero, which implies that $\underline{\gamma}$ is not in the space $S(\underline{g}^{(k)}, H^{(k)})$. Therefore $S(\underline{g}^{(k)}, H^{(k)})$ is a proper subspace of $S(\underline{g}^{(k+1)}, H^{(k+1)})$, so inequality (4.9) is true. The proof of Theorem 4.1 is complete.

This theorem is useful because the dimension of $S(\underline{g}, H)$ cannot exceed n, the number of variables, and because inequality (4.8) does not require the value of θ to be calculated by an exact line search. Thus we obtain the following quadratic termination result.

Theorem 4.2. If one of the class of variable metric algorithms is applied to minimize a positive definite quadratic function, if the algorithm terminates when the search direction (4.2) is identically zero, if on some iterations the value of θ is calculated by a perfect line search to minimize the function (1.1), and if every iteration sets θ to a value that is different from zero, then not more than n perfect line searches are used.

Note that if the matrix $H^{(k)}$ has both positive and negative eigenvalues, then it may happen that the search direction (4.2) is non-zero, but that the value of θ that minimizes the function (1.1) is zero. Partly to avoid this case, most of the algorithms from the class that are available as computer programs ensure that every matrix $H^{(k)}$ $(k=1,2,3...)$ is positive definite, except that sometimes positive definiteness is lost due to computer rounding errors.

The possibility of negative eigenvalues does complicate the theory of quadratic termination, because, for example, we should have made some remarks in the statement of Theorem 4.2 to cover the special case when an exact line search gives the value $\theta = 0$. To treat similar special cases properly in Sections 5-7 would lengthen and complicate this paper, so we ignore them. Therefore, if the reader is not familiar with the kind of provisos that have to be made to take account of these special cases, then he is advised to suppose that no negative eigenvalues occur in the matrices $H^{(k)}$.

In this case results that are stronger than Theorems 4.1 and 4.2 are given by Powell (1972b). He proves that, if every matrix $H^{(k)}$ is positive definite or positive semi-definite, then the dimension of the space $S(\underline{g}^{(k+1)}, H^{(k+1)})$ is either $\dim\{S(\underline{g}^{(k)},H^{(k)})\}$ or $\dim\{S(\underline{g}^{(k)},H^{(k)})\}+1$. Also he relates these alternative values to the parameters θ and β of equations (1.2) and (4.3).

In Section 7 we use Theorem 4.1 to prove a rather strong quadratic termination theorem for a family of algorithms that minimize $F(\underline{x})$ subject to linear equality and inequality constraints on the variables.

5. A new proof of Dixon's (1972) theorem

Dixon's (1972) theorem concerns the points $\underline{x}^{(k)}$ $(k=1,2,3,\ldots)$ that are calculated by the algorithms of the family described in Section 4, when $F(\underline{x})$ is a quite general differentiable function, and when on every iteration the value of θ is calculated to minimize the function (1.1). To ensure that such a value of θ exists for each search direction, we impose the condition that the set $\{\underline{x}\,|\,F(\underline{x}) \leqslant F(\underline{x}^{(1)})\}$ is bounded. Then the theorem states that if any ambiguity in the definition of θ is resolved in a consistent way, and if in equation (4.3) the value

$$\beta = -1/(\underline{g}^{(k+1)}, H^{(k)}\underline{g}^{(k+1)}) \qquad \ldots (5.1)$$

is never used, then the sequence of points $\underline{x}^{(k)}$ $(k=1,2,3\ldots)$ depends only on the starting point $\underline{x}^{(1)}$ and the starting matrix $H^{(1)}$. In particular the sequence is independent of the value of β that is used in equation (4.3) on each iteration.

Before proving the main theorem, we prove a weaker result that is due to Shanno and Kettler (1970).

Theorem 5.1. The direction of the search vector $\underline{d}^{(k+1)}$ is independent of the value of β that is used in equation (4.3) to define the matrix $H^{(k+1)}$, provided that the value of β is not equal to $-1/(\underline{g}^{(k+1)}, H^{(k)}\underline{g}^{(k+1)})$.

Proof. Equations (4.2) and (4.3) imply that $\underline{d}^{(k+1)}$ is the vector

$$\underline{d}^{(k+1)} = -H_{BFS}^{(k+1)}\,\underline{g}^{(k+1)} - \beta\,\underline{\omega}(\underline{\omega},\underline{g}^{(k+1)}). \qquad \ldots (5.2)$$

Therefore to prove the theorem we must show that the vector $\underline{\omega}$ is a multiple of $H_{BFS}^{(k+1)}\underline{g}^{(k+1)}$.

The fact that θ is calculated to minimize the function (1.1) gives the equation

$$(\underline{\delta},\underline{g}^{(k+1)}) = 0. \qquad \ldots (5.3)$$

Therefore from equation (4.4) we obtain the relation

$$H_{BFS}^{(k+1)}\underline{g}^{(k+1)} = H^{(k)}\underline{g}^{(k+1)} - \frac{(\underline{\gamma},H^{(k)}\underline{g}^{(k+1)})\underline{\delta}}{(\underline{\delta},\underline{\gamma})}$$

$$= H^{(k)}\underline{\gamma} + H^{(k)}\underline{g}^{(k)} - \frac{(\underline{\gamma},H^{(k)}\underline{g}^{(k+1)})\underline{\delta}}{(\underline{\delta},\underline{\gamma})}, \quad \ldots (5.4)$$

the last line being a consequence of the definition (2.7). Now equations (1.2), (2.7) and (4.2) show that $\underline{\delta}$ is a multiple of $H^{(k)}\underline{g}^{(k)}$, so equation (5.4) gives the identity

$$H_{BFS}^{(k+1)}\underline{g}^{(k+1)} = H^{(k)}\underline{\gamma} + \underline{\delta}\left[\frac{(\underline{\gamma},H^{(k)}\underline{g}^{(k)}) - (\underline{\gamma},H^{(k)}\underline{g}^{(k+1)})}{(\underline{\delta},\underline{\gamma})}\right]$$

$$= H^{(k)}\underline{\gamma} - (\underline{\gamma},H^{(k)}\underline{\gamma})\,\underline{\delta}/(\underline{\delta},\underline{\gamma})$$

$$= \underline{\omega}, \quad \ldots (5.5)$$

from the definition (4.5). Thus the direction of the vector (5.2) is independent of β, provided that this vector is not identically zero.

Expression (5.2) is zero if β satisfies the equation

$$1 + \beta(\underline{\omega},\underline{g}^{(k+1)}) = 0, \quad \ldots (5.6)$$

and, because equation (5.5) implies that $\underline{\omega}$ is equal to the first line of expression (5.4), we deduce from equation (5.3) the identity

$$(\underline{\omega},\underline{g}^{(k+1)}) = (\underline{g}^{(k+1)},H^{(k)}\underline{g}^{(k+1)}). \quad \ldots (5.7)$$

Therefore the value (5.1) must not be used. The proof of the theorem is now complete.

For the remainder of this section it is convenient to indicate that there is a parameter β for each iteration. Specifically, for k=1,2,3,..., we let $\beta^{(k)}$ denote the parameter value that is used in the definition (4.3) of $H^{(k+1)}$. Therefore, for example, Theorem 5.1 states that the direction of $\underline{d}^{(k+1)}$ is independent of $\beta^{(k)}$, and Dixon's theorem is the much stronger statement that the direction of $\underline{d}^{(k+1)}$ is independent of $\beta^{(j)}$ for j=1,2,...,k.

The new feature of our proof of Dixon's theorem is that simultaneously it is shown that the matrix $H_{BFS}^{(k+1)}$, defined by equation (4.4), is independent of $\beta^{(j)}$ $(j=1,2,...,k-1)$, although equation (4.3) shows that $H^{(k)}$ depends on $\beta^{(k-1)}$. Indeed we prove the following theorem.

Theorem 5.2. If the method described in Section 4 is used to minimize a differentiable function $F(\underline{x})$, if each iteration calculates θ by a perfect line search to minimize the function (1.1), and if this specification of θ is unambiguous, then the sequence of points $\underline{x}^{(k)}$ and the sequence of matrices $H_{BFS}^{(k+1)}$ $(k=1,2,3,...)$, defined by equation (4.4), are independent of the parameter values $\beta^{(k)}$ $(k=1,2,3,...)$, provided that no iteration sets $\beta^{(k)}$ to the value (5.1).

Proof. Since the matrix $H^{(1)}$ is given, the search direction $\underline{d}^{(1)}$, defined by equation (4.2), is independent of the parameters $\beta^{(k)}$ $(k=1,2,3,...)$. Therefore $\underline{x}^{(2)}$ is independent of the parameters, and it follows from the definition (4.4) that the matrix $H_{BFS}^{(2)}$ is also independent of the parameters.

We complete the proof by induction. Specifically for $k=2,3,4,...$ we suppose that $\underline{x}^{(k)}$ and $H_{BFS}^{(k)}$ are independent of the parameters, and we deduce that $\underline{x}^{(k+1)}$ and $H_{BFS}^{(k+1)}$ are also independent of the parameters.

Theorem 5.1 states that the search direction $\underline{d}^{(k)}$ that is actually used is a multiple of the search direction that would be used if $\beta^{(k-1)}$ were zero, which, from the definitions (4.2) and (4.3), is the vector $-H_{BFS}^{(k)}\underline{g}^{(k)}$. Therefore the inductive hypothesis implies that $\underline{d}^{(k)}$ is a multiple of a vector that is independent of the parameters. It follows that $\underline{x}^{(k+1)}$ is independent of the parameters. Consequently the vectors $\underline{\delta}$ and $\underline{\gamma}$ of the kth iteration are also independent of the parameters.

To complete the proof we show that the inductive hypothesis also implies that $H_{BFS}^{(k+1)}$ is independent of the parameters. Equations (4.3) and (5.5) give the equation

$$H^{(k)} = H_{BFS}^{(k)} + \beta^{(k-1)}\underline{\omega}^{(k-1)}\underline{\omega}^{(k-1)T}$$

$$= H_{BFS}^{(k)} + \beta^{(k-1)} H_{BFS}^{(k)} \underline{g}^{(k)} \underline{g}^{(k)T} H_{BFS}^{(k)}. \quad \ldots (5.8)$$

Therefore the definition (4.4) shows that $H_{BFS}^{(k+1)}$ depends on the parameters if and only if the term

$$\beta^{(k-1)} \left(I - \frac{\underline{\delta}\, \underline{\gamma}^T}{(\underline{\delta},\underline{\gamma})} \right) H_{BFS}^{(k)} \underline{g}^{(k)} \underline{g}^{(k)T} H_{BFS}^{(k)} \left(I - \frac{\underline{\gamma}\, \underline{\delta}^T}{(\underline{\delta},\underline{\gamma})} \right) \quad \ldots (5.9)$$

can be non-zero. Now we noted in the previous paragraph that $\underline{\delta}$ is a multiple of $H_{BFS}^{(k)} \underline{g}^{(k)}$. It follows that expression (5.9) is zero, so the inductive hypothesis implies that $H_{BFS}^{(k+1)}$ is independent of $\beta^{(j)}$ $(j=1,2,3,\ldots)$. Theorem 5.2 is proved.

This theorem shows that the BFS formula (4.4) has the following special property. If the choice of $\beta^{(j)}$ is arbitrary for $j=1,2,\ldots,k-1$, but if $\beta^{(k)}$ is set to zero, then the matrix $H^{(k+1)}$ is independent of $\beta^{(j)}$ $(j=1,2,\ldots,k-1)$. I believe that this property has not been noted before. It may be important in practice, because it implies that any bad features in $H^{(k)}$ that are due to the choice of $\beta^{(j)}$ $(j=1,2,\ldots,k-1)$ are suppressed in the definition of $H^{(k+1)}$.

However the results of this section depend on exact line searches, and we have noted that in practice exact line searches are not feasible. Therefore the given theorems are only an aid to the understanding of practical quasi-Newton algorithms.

6. The minimization of a quadratic function subject to linear equality constraints

From now on we consider extensions of methods for unconstrained minimization to take account of constraints on the variables. We consider two types of extensions. In one type the constraints are satisfied during the calculation, and in the other type the calculation leads to a point that satisfies the constraints, but constraint violations are allowed during the calculation. The first type of method is practicable when the constraints are linear, but it can hardly be applied for nonlinear equality constraints. Moreover a linear constraint is welcome in the first type of method, because it can be used to reduce the number of free variables. Therefore as a general rule the first type of method should be used for linear constraints, and this topic is considered in Sections 6 and 7. Then in Section 8 we consider using the second type of method to take account of constraints that are usually non-linear.

We require the algorithms for minimization subject to linear constraints to work well when the objective function is quadratic, and when the objective function is not quadratic the quasi-Newton algorithms are based on quadratic approximations to $F(\underline{x})$. Therefore much of the theory of Sections 6 and 7 concerns the quadratic case. It suggests algorithms, and we indicate how the algorithms are applied to general functions $F(\underline{x})$.

In this section we consider the case when the only constraints on the variables are the equality conditions

$$c_i(\underline{x}) \equiv (\underline{n}_i, \underline{x}) - d_i = 0, \quad i=1,2,\ldots,p. \qquad \ldots (6.1)$$

Here $\underline{n}_i$ is the gradient vector of the linear constraint function $c_i(\underline{x})$ and d_i is a constant. In practice the equality constraint case is very important because, to solve the more interesting case of inequality constraints, we use an "active set strategy" (Fletcher, 1971). The active set is a list of inequality constraints that are regarded as equalities for the present, and then during the calculation the members of the active set are changed automatically. Thus the inequality problem is regarded as a sequence of equality problems.

Equation (2.1) gives the vector of variables that minimizes a quadratic function when there are no constraints on the variables. The following theorem extends this result to the case when the equality constraints (6.1) are present.

Theorem 6.1. If $F(\underline{x})$ is the quadratic function (1.4), if the matrix G is positive definite, and if the matrix, N say, whose columns are $\{\underline{n}_1, \underline{n}_2, \ldots \underline{n}_p\}$ has rank p, then the vectors of variables, $\underline{x}^*$ say, that minimizes $F(\underline{x})$ subject to the constraints (6.1), is defined by the equation

$$\underline{x}^* = \underline{x} - \{G^{-1} - G^{-1}N(N^TG^{-1}N)^{-1}N^TG^{-1}\} \underline{\nabla}F(\underline{x}), \qquad \ldots (6.2)$$

where $\underline{x}$ is any vector that satisfies the conditions (6.1).

Proof. We require the stationary point of the Lagrangian function

$$\begin{aligned} \Phi(\underline{x},\underline{\lambda}) &= F(\underline{x}) - \Sigma\lambda_i c_i(\underline{x}) \\ &= F(\underline{x}) - (\underline{\lambda}, N^T\underline{x} - \underline{d}), \end{aligned} \qquad \ldots (6.3)$$

where the last scalar product is between two vectors that each have p components. We define $\underline{x}^*$ and $\underline{\lambda}^*$ to be the parameter values at the stationary point of this Lagrangian function, and therefore the equations

$$\left.\begin{aligned} N^T\underline{x}^* - \underline{d} &= 0 \\ \underline{b} + G\underline{x}^* - N\underline{\lambda}^* &= 0 \end{aligned}\right\} \qquad \ldots (6.4)$$

are satisfied, the second one being obtained from the condition that at $\underline{x}^*$ the derivative of $\Phi(\underline{x},\underline{\lambda}^*)$ with respect to $\underline{x}$ is zero. To eliminate $\underline{b}$ and $\underline{d}$ from these equations, we note that if $\underline{x}$ is any vector that satisfies the conditions (6.1), then expression (6.4) is equivalent to the equations

$$\left.\begin{aligned} N^T(\underline{x}^* - \underline{x}) &= 0 \\ \underline{\nabla}F(\underline{x}) + G(\underline{x}^* - \underline{x}) - N\underline{\lambda}^* &= 0 \end{aligned}\right\}. \qquad \ldots (6.5)$$

From these equations we require an expression for $\underline{x}^*$ that is independent of $\underline{\lambda}^*$.

Because G is non-singular, the second of the equations (6.5) gives the identity

$$(\underline{x}^* - \underline{x}) = G^{-1}\{N\underline{\lambda}^* - \underline{\nabla}F(\underline{x})\}, \qquad \dots (6.6)$$

which is substituted in the first of the equations (6.5) to give the relation

$$(N^T G^{-1} N)\underline{\lambda}^* = N^T G^{-1} \underline{\nabla}F(\underline{x}). \qquad \dots (6.7)$$

Now $(N^T G^{-1} N)$ is a square matrix of dimension p, and the conditions in the statement of the theorem imply that this matrix is non-singular. Therefore $\underline{\lambda}^*$ is the vector

$$\underline{\lambda}^* = (N^T G^{-1} N)^{-1} N^T G^{-1} \underline{\nabla}F(\underline{x}). \qquad \dots (6.8)$$

By substituting this expression in equation (6.6) we obtain the required result (6.2). The proof of the theorem is complete.

Equation (6.2) shows one way in which a quasi-Newton method can be extended to take account of linear equality constraints. If $H^{(k)}$ is an estimate of the matrix G^{-1}, obtained for example by one of the family of methods considered in Section 4, then in place of the search direction (4.2) we use the vector

$$\underline{d}^{(k)} = -\{H^{(k)} - H^{(k)}N[N^T H^{(k)} N]^{-1} N^T H^{(k)}\}\underline{g}^{(k)}, \qquad \dots (6.9)$$

where the point $\underline{x}^{(k)}$ has to satisfy the constraints (6.1). In particular this technique is used by Murtagh and Sargent (1969). Note that if the matrix $[N^T H^{(k)} N]^{-1}$ is available and if $H^{(k+1)}$ is defined by equation (4.3), then $[N^T H^{(k+1)} N]^{-1}$ can be calculated in of order np operations by using the Sherman/ M rrison formula (Householder, 1964). Thus algorithms based on equation (6.9) require only of order n^2 operations per iteration. Later in this section it is proved that they have good quadratic termination properties.

A technique for linear equality constraints that seems to be quite different is described by Goldfarb (1969). He lets the initial matrix $H^{(1)}$

be any positive semi-definite matrix that annihilates just the vectors that are in the linear space spanned by the constraint normals $\underline{n}_i$ $(i=1,2,\ldots,p)$, and the starting point $\underline{x}^{(1)}$ is chosen to satisfy the constraints. Then one of the methods of Section 4 is applied directly, except that the convergence criterion must not be based on a small value of $\underline{\nabla}F(\underline{x})$, because when constraints are present $\underline{\nabla}F(\underline{x})$ usually tends to a non-zero value, which is a linear combination of the normals $\underline{n}_i$ $(i=1,2,\ldots,p)$.

Goldfarb's method is particularly easy to understand when the constraints (6.1) are the conditions

$$x_i = d_i \quad (i=1,2,\ldots,p). \qquad \ldots (6.10)$$

In this case the matrix $H^{(1)}$ has the form

$$H^{(1)} = \left(\begin{array}{c|c} 0 & 0 \\ \hline 0 & \bar{H} \end{array}\right), \qquad \ldots (6.11)$$

where $\bar{H}$ is a symmetric non-singular matrix of dimension $(n-p)$. Further $\underline{x}^{(1)}$ must be chosen so that its first p components are d_i $(i=1,2,\ldots,p)$. Then it follows from equations (4.2) and (4.3) that every vector $\underline{\delta}$ has its first p components equal to zero, that every vector $\underline{x}^{(k)}$ satisfies the constraints (6.10), and that every matrix $H^{(k)}$ has the form (6.11). Thus we have really eliminated the first p variables from the calculation, and we are applying the method of Section 4 to a function of $(n-p)$ variables, whose values are unconstrained. Therefore the theorems of Sections 4 and 5 hold. In particular Theorem 4.2 shows that quadratic termination requires not more than $(n-p)$ exact line searches.

Now the quasi-Newton methods that use equations (4.2) and (4.3) are not affected by linear transformations of the variables, provided that an appropriate transformation is made to the matrix $H^{(1)}$. This point is explained for the DFP algorithm in Lemma 7 of Powell (1971). Therefore, provided that the matrix N has rank p, there is no loss of generality in supposing that a linear transformation is applied to the variables to make the constraints (6.1) become the constraints (6.10). It follows that the conclusions of the last paragraph, in particular the quadratic termination property, hold generally for Goldfarb's algorithm.

However it is not clear that Theorem 4.2 is valid for algorithms that use equations (6.9) and (4.3) instead of equations (4.2) and (4.3), because the proof of Theorem 4.1 depends on the fact that the search direction is the vector (4.2). Therefore we now show that the use of equation (6.9) does not spoil the quadratic termination, by proving the remarkable result that equation (6.9) gives an algorithm that is virtually identical to Goldfarb's method, for general objective functions $F(\underline{x})$. I believe that this result is new.

For the proof we require some notation. We use the convention that $H^{(k)}$ is always the matrix that occurs in equation (6.9), and therefore it is non-singular. We let $P\{H\}$ be the matrix

$$P\{H\} = H - HN[N^T HN]^{-1} N^T H. \qquad \ldots (6.12)$$

Because its rank is (n-p) and because it satisfies the equation

$$P\{H\}N = 0, \qquad \ldots (6.13)$$

it is a suitable initial matrix for Goldfarb's algorithm. Moreover we let $U_\beta\{H\}$ be the matrix

$$U_\beta\{H^{(k)}\} = H_{BFS}^{(k+1)} + \beta \underline{\omega}\, \underline{\omega}^T, \qquad \ldots (6.14)$$

corresponding to formula (4.3). First we treat the case $\beta = 0$ by using the following theorem.

<u>Theorem 6.2</u>. The equation

$$P[U_o\{H\}] = U_o[P\{H\}] \qquad \ldots (6.15)$$

is true, if $\underline{\delta}$ is a multiple of the vector (6.9).

<u>Proof</u>. By definition $U_o\{H\}$ is the matrix

$$U_o\{H\} = \left(I - \frac{\underline{\delta}\,\underline{\gamma}^T}{(\underline{\delta},\underline{\gamma})}\right) H \left(I - \frac{\underline{\gamma}\,\underline{\delta}^T}{(\underline{\delta},\underline{\gamma})}\right) + \frac{\underline{\delta}\,\underline{\delta}^T}{(\underline{\delta},\underline{\gamma})}. \qquad \dots (6.16)$$

Moreover, because $\underline{\delta}$ is a multiple of the vector (6.9), it satisfies the equation

$$N^T\underline{\delta} = 0. \qquad \dots (6.17)$$

From these two equations and the definition (6.12) we obtain the result

$$\begin{aligned} P[U_o\{H\}] &= U_o\{H\} - \left(I - \frac{\underline{\delta}\,\underline{\gamma}^T}{(\underline{\delta},\underline{\gamma})}\right) HN[N^THN]^{-1}N^TH \left(I - \frac{\underline{\gamma}\,\underline{\delta}^T}{(\underline{\delta},\underline{\gamma})}\right) \\ &= \left(I - \frac{\underline{\delta}\,\underline{\gamma}^T}{(\underline{\delta},\underline{\gamma})}\right)\{H - HN[N^THN]^{-1}N^TH\} \left(I - \frac{\underline{\gamma}\,\underline{\delta}^T}{(\underline{\delta},\underline{\gamma})}\right) + \frac{\underline{\delta}\,\underline{\delta}^T}{(\underline{\delta},\underline{\gamma})} \\ &= U_o[P\{H\}], \end{aligned} \qquad \dots (6.18)$$

which proves the theorem.

Now, in the algorithm based on equation (6.9) and the BFS formula, the search directions (6.9) are the vectors $-P\{H^{(k)}\}\underline{g}^{(k)}$ $(k=1,2,3,\dots)$, where the sequence of matrices $H^{(k)}$ is calculated from the formula

$$H^{(k+1)} = U_o\{H^{(k)}\},\ k=1,2,3,\dots\ . \qquad \dots (6.19)$$

However Theorem 6.2 shows that these search directions are also the vectors $-\bar{H}^{(k)}\underline{g}^{(k)}$ $(k=1,2,3,\dots)$, where $\bar{H}^{(1)} = P\{H^{(1)}\}$, and where the sequence of matrices $\bar{H}^{(k)}$ is calculated from the formula

$$\bar{H}^{(k+1)} = U_o\{\bar{H}^{(k)}\}. \qquad \dots (6.20)$$

This alternative method of calculation is really Goldfarb's algorithm. Therefore we have proved the equivalence of the two methods in the case that the BFS formula is used on every iteration.

To study the case when β is different from zero we require a lemma.

Lemma 6.1. If $\underline{u}$ is any vector such that the matrix $\{H+\beta H\underline{u}\underline{u}^TH\}$ is non-singular, then the equation

$$P\{H + \beta H\underline{u}\underline{u}^T H\} = P\{H\} + \frac{\beta P\{H\}\underline{u}\underline{u}^T P\{H\}}{1 + \beta\underline{u}^T HN[N^T HN]^{-1} N^T H\underline{u}} \qquad \ldots (6.21)$$

is true.

Proof. The Sherman/Morrison formula gives the identity

$$[N^T\{H + \beta H\underline{u}\underline{u}^T H\}N]^{-1} = X - \frac{\beta XN^T H\underline{u}\underline{u}^T HNX}{1 + \beta\underline{u}^T HNXN^T H\underline{u}}, \qquad \ldots (6.22)$$

where X is the matrix $[N^T HN]^{-1}$. By substituting this expression in the definition (6.12) and by combining terms we obtain the equation

$$P\{H + \beta H\underline{u}\underline{u}^T H\} = H - HNXN^T H$$
$$+ \frac{\beta\{H\underline{u} - HNXN^T H\underline{u}\}\{H\underline{u} - HNXN^T H\underline{u}\}^T}{1 + \beta\underline{u}^T HNXN^T H\underline{u}}. \qquad \ldots (6.23)$$

The truth of the lemma follows from equation (6.12).

We now extend Theorem 6.2 in a way which allows us to show that the algorithm based on equation (6.9) is equivalent to Goldfarb's method, when the value of β in expression (4.3) is general.

Theorem 6.3. The equation

$$P[U_\beta\{H\}] = U_{\beta'}[P\{H\}] \qquad \ldots (6.24)$$

is true, where β' has the value

$$\beta' = \frac{\beta}{1 + \beta\underline{\gamma}^T HN[N^T HN]^{-1} N^T H\underline{\gamma}}. \qquad \ldots (6.25)$$

Proof. Because the vector (4.5) can be expressed in the form

$$\underline{\omega} = \left(1 - \frac{\underline{\delta}\,\underline{\gamma}^T}{(\underline{\delta},\underline{\gamma})}\right) H^{(k)}\underline{\gamma}, \qquad \ldots (6.26)$$

expressions (4.4) and (6.14) imply the identity

$$U_\beta\{H\} = U_o\{H + \beta H\underline{\gamma}\underline{\gamma}^T H\}. \qquad \ldots (6.27)$$

Therefore, by using equation (6.27), then Theorem 6.2, then Lemma 6.1, and then equation (6.27) again, we obtain the result

$$\begin{aligned} P[U_\beta\{H\}] &= P[U_0\{H+\beta H\underline{\gamma}\underline{\gamma}^T H\}] \\ &= U_0[P\{H+\beta H\underline{\gamma}\underline{\gamma}^T H\}] \\ &= U_0[P\{H\} + \beta' P\{H\}\underline{\gamma}\underline{\gamma}^T P\{H\}] \\ &= U_{\beta'}[P\{H\}]. \qquad \ldots (6.28) \end{aligned}$$

Theorem 6.3 is proved.

We now use an argument that is similar to the one following Theorem 6.2. We note that the search directions of the method based on equation (6.9) are $-P\{H^{(k)}\}\underline{g}^{(k)}$ $(k=1,2,3,\ldots)$, where the sequence of matrices $H^{(k)}$ is calculated from the equation

$$H^{(k+1)} = U_\beta\{H^{(k)}\}, \; k=1,2,3,\ldots, \qquad \ldots (6.29)$$

and we deduce from Theorem 6.3 that these search directions are also the vectors $-\bar{H}^{(k)}\underline{g}^{(k)}$ $(k=1,2,3,\ldots)$, where $\bar{H}^{(1)} = P\{H^{(1)}\}$, and where the sequence of matrices $\bar{H}^{(k)}$ is defined by the formula

$$\bar{H}^{(k+1)} = U_{\beta'}\{\bar{H}^{(k)}\}, \qquad \ldots (6.30)$$

provided that the value of β' is related to the value of β in equation (6.29) by expression (6.25) for each value of k. Again this alternative method of defining the search directions is really Goldfarb's algorithm. Thus the equivalence of the two methods is proved in the case that the method based on equation (6.9) uses a general value of β. To emphasise this interesting result it is stated as a theorem.

Theorem 6.4. Let the algorithm based on equations (6.9) and (4.3) be applied, starting with the vector $\underline{x}^{(1)}$ and the non-singular matrix $H^{(1)}$, where on every iteration the parameters β and θ of equations (4.3) and (1.2) are arbitrary. Also let Goldfarb's algorithm be applied, starting with the same vector $\underline{x}^{(1)}$ and the matrix $P\{H^{(1)}\}$, where on each iteration the value of θ is the same as that used by the first algorithm, and the value of β is is related to the one used by the first algorithm by equation (6.25). Then the two methods generate the same sequence of points $\underline{x}^{(k)}$ (k=1,2,3,...). Moreover, if the sequence of matrices used by the first algorithm is $H^{(k)}$ (k=1,2,3,...), then the sequence of matrices used by Goldfarb's algorithm is $P\{H^{(k)}\}$ (k=1,2,3,...), where P is defined by equation(6.12).

It follows from this theorem and from Theorem 5.2 that, if on every iteration θ is calculated to minimize the function (1.1), then the sequence of points $\underline{x}^{(k)}$ (k=1,2,3,...) is independent of the choice of β on each iteration, even if the relation (6.25) is ignored.

Because equation (6.25) shows that β' is zero when $\beta=0$, we have deduced that the BFS formula in the algorithm based on equation (6.9) corresponds to the BFS formula in Goldfarb's algorithm, but, for any other value of β, β' is different from β. However it does not follow that we have found another property that is special to the BFS formula, because in some well-known formulae the value of β depends on H. In particular equation (6.29) is the DFP formula (Fletcher and Powell, 1963) if and only if β has the value

$$\beta = -1/(\underline{\gamma},H^{(k)}\underline{\gamma}). \qquad \dots (6.31)$$

In this case expression (6.25) is the quantity

$$\begin{aligned}\beta' &= -1/\{(\underline{\gamma},H^{(k)}\underline{\gamma}) - \underline{\gamma}^T H^{(k)} N[N^T H^{(k)} N]^{-1} N^T H^{(k)}\underline{\gamma}\} \\ &= -1/(\underline{\gamma},P\{H^{(k)}\}\underline{\gamma}), \qquad \dots (6.32)\end{aligned}$$

the last line being a consequence of the definition (6.12). Since the matrix $P\{H^{(k)}\}$ is equal to the matrix $\bar{H}^{(k)}$ of equation (6.30), we deduce that the value (6.32) causes equation (6.30) to be the DFP formula. Thus the use of the DFP formula in the method based on equation (6.9) is equivalent to the use of the DFP formula in Goldfarb's algorithm. A similar conclusion holds for the rank-one correction formula.

The actual arithmetic operations of the two methods of Theorem 6.4 are different. Goldfarb's algorithm requires less work, because equation (4.2) is used instead of equation (6.9). The saving of computer arithmetic is particularly great if a change of variables is made so that the constraints have the form (6.10), for then $H^{(k)}$ has the zero elements shown in expression (6.11). A change of variables of this type is in the spirit of Wolfe's (1967) reduced gradient method and McCormick's (1970) variable reduction method.

In the next section we find that, when the linear constraints are inequalities, then there are some advantages in using methods based on the formula (6.9) instead of Goldfarb's algorithm.

Note that in Theorem 6.1 the condition that G is to be positive definite can be weakened (see Fletcher, 1971, for instance). It is important only that $(\underline{x}, G\underline{x})$ be positive for all non-zero vectors $\underline{x}$ that satisfy the condition $N^T\underline{x} = 0$.

7. Extensions for linear inequality constraints

Many methods have been proposed for minimizing a general function subject to linear inequality constraints. The paper by Rosen (1960) pioneered much of the research on this subject, but his actual algorithm is no longer attractive because it has the disadvantages of the steepest descent method described in Section 2. A recent discussion of current methods is given in the excellent review paper by Fletcher (1971). This review includes the algorithms that are obtained by extending the methods of Section 6 to inequality constraints (Goldfarb, 1969; Murtagh and Sargent, 1969), and it is noted that these algorithms are good. They are described in this section, and in particular we consider the quadratic termination properties of Goldfarb's algorithm, because they were shown to be very satisfactory (Powell, 1972b) after Fletcher wrote his review.

Already we have remarked that the Goldfarb (1969) and Murtagh and Sargent (1969) algorithms treat inequality constraints by an active set strategy (Fletcher, 1971). This means that at any stage some constraints are treated as equalities, and the other constraints do not affect the calculation, unless they are violated during the search for the least value of $F(\underline{x})$. For example at the beginning of the kth iteration we may have a feasible point $\underline{x}^{(k)}$ that satisfies the conditions (6.1), where these conditions occur through treating all the constraints in the active set as equalities. Then one of the methods described in Section 6 is used to calculate $\underline{x}^{(k+1)}$, unless this choice of $\underline{x}^{(k+1)}$ conflicts with the inequality constraints that are not in the active set. If a conflict occurs, then the value of θ in equation (1.2) is decreased in magnitude, until all the inequality constraints are satisfied at $\underline{x}^{(k+1)}$, and one of the constraints that is not in the active set is satisfied as an equality. This constraint is added to the active set before $\underline{x}^{(k+2)}$ is calculated. However, if $\underline{x}^{(k+1)}$ does not have to be modified to satisfy the constraints that are not in the active set, then a constraint may be deleted from the active set before $\underline{x}^{(k+2)}$ is calculated, depending on a criterion that is mentioned in Section 8.

If the linear constraint

$$c_q(\underline{x}) \equiv (\underline{n}_q,\underline{x}) - d_q \geqslant 0 \qquad \ldots (7.1)$$

is added to the active set, then the matrix N of equation (6.2) gains an extra column, namely the vector $\underline{n}_q$, and if the constraint

$$c_p(\underline{x}) \equiv (\underline{n}_p,\underline{x}) - d_p \geqslant 0 \qquad \ldots (7.2)$$

is deleted from the active set, then N loses the column $\underline{n}_p$. Thus the methods based on equation (6.9) are extended to take account of changes in the active set. Note in particular that for these methods the matrix $H^{(k)}$ of equation (6.9) is not changed because the active set is changed. However in Goldfarb's algorithm the matrix $H^{(k)}$ must be altered, because it is important that this matrix has the property that it annihilates just the vectors that are in the column space of N.

Therefore Goldfarb (1969) proposes the following changes to the matrix $H^{(k)}$ to take account of changes to the active set. If the constraint (7.1) is added to the active set on the kth iteration, then the matrix

$$H^{(k+1)} = H^{(k)} - \frac{H^{(k)}\underline{n}_q\underline{n}_q^T H^{(k)}}{(\underline{n}_q, H^{(k)}\underline{n}_q)}, \qquad \ldots (7.3)$$

is used instead of expression (4.3). Moreover when the constraint (7.2) is deleted from the active set, then $H^{(k)}$ is replaced by the matrix

$$H^{(k)} + \frac{\bar{P}\underline{n}_p\underline{n}_p^T\bar{P}}{(\underline{n}_p,\bar{P}\underline{n}_p)} = H^{(k)}_{new}, \qquad \ldots (7.4)$$

say, where $\bar{P}$ is the projection operator

$$\bar{P} = I - \bar{N}(\bar{N}^T\bar{N})^{-1}\bar{N}^T, \qquad \ldots (7.5)$$

$\bar{N}$ being the matrix whose columns are $\{\underline{n}_1,\underline{n}_2,\ldots,\underline{n}_{p-1}\}$. It is straightforward to verify that the matrices (7.3) and (7.4) have the required properties.

If the matrix $(\bar{N}^T\bar{N})$ of equation (7.5) or the matrix $[N^T H^{(k)} N]$ of equation (6.9) is singular, then it has no inverse. Therefore it is important to ensure that the normals of the constraints in the active set do not become linearly dependent. Fortunately the method that has been described for changing the active set cannot cause this to happen.

To prove this statement we note that linear dependence can be introduced only when a constraint, say inequality (7.1), is added to the active set. If this constraint becomes active on the kth iteration, then $c_q(\underline{x}^{(k)}) \geqslant 0$ and $c_q(\underline{x}^{(k+1)}) < 0$, where $\underline{x}^{(k+1)}$ is calculated by one of the methods of Section 6. Therefore the scaler product $(\underline{n}_q, \underline{x}^{(k+1)} - \underline{x}^{(k)})$ is different from zero, but the methods of Section 6 are such that the equations

$$(\underline{n}_i, \underline{x}^{(k+1)} - \underline{x}^{(k)}) = 0, \; i=1,2,\ldots,p, \qquad \ldots (7.6)$$

are satisfied. It follows that $\underline{n}_q$ is not in the subspace spanned by the vectors $\{\underline{n}_1, \underline{n}_2, \ldots, \underline{n}_p\}$. Therefore, if the vectors $\{\underline{n}_1, \underline{n}_2, \ldots, \underline{n}_p\}$ are independent, then the new constraint normals are also linearly independent.

Next we consider quadratic termination properties. Because the matrices (7.3) and (7.4) differ from $H^{(k)}$ by a matrix of rank one, the following theorem is relevant to the quadratic termination properties of Goldfarb's algorithm.

<u>Theorem 7.1</u>. Let $S(\underline{g}, H)$ be the linear space of vectors $\underline{z}$ that satisfy the conditions (4.7), and let the matrix H be changed by the formula

$$H_{new} = H_{old} \pm \underline{v}\underline{v}^T, \qquad \ldots (7.7)$$

where $\underline{v}$ is some vector. Then the inequality

$$\dim\{S(\underline{g}-\theta G H_{old}\underline{g}, H_{new})\} \geqslant \dim\{S(\underline{g}, H_{old})\} - 1 \qquad \ldots (7.8)$$

is true for all θ, including $\theta = 0$.

<u>Proof</u>. The inequality (7.8) is trivial if $\dim\{S(\underline{g}, H_{old})\}$ is not greater than one. Therefore we suppose that the dimension of $S(\underline{g}, H_{old})$ exceeds one, and we let $\bar{S}$ be the linear space of vectors $\underline{z}$ that satisfy the conditions

$$\left.\begin{aligned} GH_{old}\underline{z} &= \underline{z} \\ (\underline{z},H_{old}\underline{g}) &= 0 \\ (\underline{z},\underline{v}) &= 0 \end{aligned}\right\} . \qquad \ldots (7.9)$$

It follows that the dimension of $\bar{S}$ is at least $\dim\{S(\underline{g},H_{old})\} - 1$, and that $\bar{S}$ is a subspace of $S(\underline{g}-\theta GH_{old}\underline{g},H_{new})$. Therefore inequality (7.8) is satisfied. Theorem 7.1 is proved.

It is now straightforward to prove the following quadratic termination theorem (Powell, 1972b) for Goldfarb's algorithm.

Theorem 7.2. If Goldfarb's (1969) algorithm is applied to minimize a positive definite quadratic function of n variables, subject to linear constraints on the variables, if during the calculation the active constraint set is changed ℓ times, and if some of the values of θ in equation (1.2) are calculated to minimize the function (1.1), then not more than $(n+\ell)$ of these exact line searches are required to achieve quadratic termination.

Proof. Let $\{\underline{g}^{(1)},H^{(1)}\}$ be the initial values of $\underline{g}$ and H, let $\{\underline{g}^{(f)},H^{(f)}\}$ be the final values of these quantities, and let s be the number of iterations on which the value of θ is calculated to minimize the function (1.1). Let $\theta = \theta^{(k)}$ in Theorem 7.1 when the matrix (7.3) is used and let $\theta = 0$ when the matrix (7.4) is used. Then Theorems 4.1 and 7.1 give the bound

$$\dim\{S(\underline{g}^{(f)},H^{(f)})\} \geqslant \dim\{S(\underline{g}^{(1)},H^{(1)})\} + s-\ell .$$

Thus we obtain the inequality

$$\begin{aligned} s &\leqslant \dim\{S(\underline{g}^{(f)},H^{(f)})\} - \dim\{S(\underline{g}^{(1)},H^{(1)})\} + \ell \\ &\leqslant n+\ell , \end{aligned} \qquad \ldots (7.10)$$

which proves the theorem.

Note that this theorem is much stronger than the one given by Goldfarb on quadratic termination, for he proves only that quadratic termination exists, using a method that suggests that the value of s, defined above, may be of the order of the product of n and ℓ. This is unfortunate, because I think it has caused the power of Goldfarb's algorithm to be underestimated.

To show that in the case of linear inequality constraints the methods based on equation (6.9) also have a quadratic termination property, we extend the part of Section 6 that compares the methods based on equation (6.9) with Goldfarb's technique. We recall that this comparison depends on the remark that, if the non-singular matrices $H^{(k)}$ $(k=1,2,3,\ldots)$ occur in a method based on equation (6.9), then the matrices $P\{H^{(k)}\}$ $(k=1,2,3,\ldots)$ can be regarded as matrices $H^{(k)}$ that are generated naturally by Goldfarb's algorithm, where P is defined by equation (6.12). Therefore, to extend the comparison to inequality constraints, we consider the change to $P\{H^{(k)}\}$ that occurs when a method based on equation (6.9) is in use and when the active constraint set is changed.

If the constraint (7.1) is added to the active set, then the matrix N is replaced by the matrix

$$(N|\underline{n}_q) = N_+, \qquad \ldots (7.11)$$

say. Corresponding to equation (6.12), we let $P_+\{H\}$ be the matrix

$$P_+\{H\} = H - HN_+[N_+^T HN_+]^{-1} N_+^T H. \qquad \ldots (7.12)$$

We use Rosen's (1960) formula for the inverse of a partitioned matrix to express $[N_+^T HN_+]^{-1}$ in terms of $N, \underline{n}_q$ and H. Thus from equations (6.12), (7.11) and (7.12) we deduce the identity

$$P_+\{H\} = P\{H\} - \frac{P\{H\}\underline{n}_q \underline{n}_q^T P\{H\}}{(\underline{n}_q, P\{H\}\underline{n}_q)} . \qquad \ldots (7.13)$$

It follows that if $P\{H^{(k)}\}$ is regarded as the matrix $H^{(k)}$ of Goldfarb's algorithm, then the matrix $P_+\{H^{(k)}\}$ is equal to the matrix (7.3) of Goldfarb's algorithm. Therefore the equivalence between a method based on equation (6.9) and Goldfarb's algorithm is preserved when constraints are added to the active set.

If the constraint (7.2) is deleted from the active set, then the new constraint matrix, $\bar{N}$ say, is related to N by the equation

$$(\bar{N}|\underline{n}_p) = N. \qquad \ldots (7.14)$$

We let $P_-\{H\}$ be the matrix

$$P_-\{H\} = H - H\bar{N}[\bar{N}^T H\bar{N}]^{-1}\bar{N}^T H. \qquad \ldots (7.15)$$

Because equations (6.12), (7.11) and (7.12) give the identity (7.13), it follows by analogy that equations (6.12), (7.14) and (7.15) give the identity

$$P_-\{H\} = P\{H\} + \frac{P_-\{H\}\underline{n}_p\underline{n}_p^T P_-\{H\}}{(\underline{n}_p, P_-\{H\}\underline{n}_p)}. \qquad \ldots (7.16)$$

Therefore, if the matrix $P\{H^{(k)}\}$ is regarded as the matrix $H^{(k)}$ of Goldfarb's algorithm, then the change from $P\{H^{(k)}\}$ to $P_-\{H^{(k)}\}$ in the method based on equation (6.9) is the same as the change from $H^{(k)}$ to the matrix (7.4) in Goldfarb's algorithm, if and only if the vector $P_-\{H^{(k)}\}\underline{n}_p$ is equal to $\bar{P}\underline{n}_p$. Usually this is not the case. However, because the change (7.4) is somewhat arbitrary, the spirit of Goldfarb's algorithm is preserved if we generalize the matrix (7.4) to the form

$$H^{(k)} + \underline{v}\underline{v}^T, \qquad \ldots (7.17)$$

where $\underline{v}$ is any vector that satisfies the conditions

$$\left.\begin{aligned} (\underline{v},\underline{n}_i) &= 0, \quad i=1,2,\ldots,p-1 \\ (\underline{v},\underline{n}_p) &\neq 0 \end{aligned}\right\}. \qquad \ldots (7.18)$$

In particular by letting $\underline{v}$ be the vector $P_-\{H^{(k)}\}\underline{n}_p/(\underline{n}_p, P_-\{H^{(k)}\}\underline{n}_p)^{\frac{1}{2}}$, it follows that the method based on equation (6.9) is equivalent to a generalization of Goldfarb's algorithm when constraints are deleted from the active set.

The proof of Theorem 7.2 remains valid when Goldfarb's algorithm is generalized in the way described in the previous paragraph. Therefore the methods based on equation (6.9) for minimization subject to linear inequality constraints do have the quadratic termination property that is stated in Theorem 7.2.

One advantage in using a method based on equation (6.9) instead of Goldfarb's algorithm is that in practical computations the matrix (7.4) may not be satisfactory. Indeed it is straightforward to show that the Euclidean norm of the matrix $\bar{P}\underline{n}_p\underline{n}_p^T\bar{P}/(\underline{n}_p,\bar{P}\underline{n}_p)$ is equal to one, and therefore the difference $[H_{new}^{(k)} - H^{(k)}]$ is negligible in practice if, for example, the elements of $H^{(k)}$ happen to be of order 10^{20}. In other words there is no natural scaling present in equation (7.4), so the kind of difficulty that is discussed by Bard (1968) may occur. However there is no similar difficulty in the methods based on equation (6.9), because the matrix $H^{(k)}$ is not altered to take account of changes in the active constraint set.

Moreover replacing $H^{(k)}$ by the matrix (7.3) has the undesirable property that the loss of rank in the matrix we are working with may imply the loss of useful information, which is another good reason for using a method based on equation (6.9). Therefore it is possible that the quadratic termination properties of the method based on equation (6.9) are stronger than those we have given in this section.

However E.M.L. Beale pointed out to me that there is one situation when Goldfarb's algorithm is much superior to the methods based on equation (6.9). It occurs when n is large and when the number of constraints in each active set is close to n. In this case it is well worthwhile to transform the variables so that each matrix $H^{(k)}$ has the form (6.11), for thus we reduce considerably the dimension of the part of $H^{(k)}$ that has to be computed. Suitable transformations of variables correspond to using the contraints in the active set to eliminate variables. Therefore the application of these transformations has been studied already, in the reduced gradient method (Wolfe, 1967). Note that when eliminating variables one can make good use of any sparsity in the matrix of coefficients of the linear constraints. Thus this version of Goldfarb's algorithm provides an efficient extension of linear programming methods to the case when the objective function is general.

This technique is also proposed by McCormick (1970). However McCormick prefers to set the matrix $\bar{H}$ of equation (6.11) to the unit matrix when the active constraint set is changed, which is unsatisfactory, because it destroys the nice quadratic termination properties of Theorem 7.2.

8. Some Lagrange multiplier techniques for constraints

So far in this paper we have discussed subjects in which the author has researched recently. Therefore we have considered only a few of the many important aspects of unconstrained minimization and its extensions for constraints. Some other topics are mentioned briefly in this section, including criteria for dropping constraints from active sets, and Fletcher's (1972a) recent work on minimization subject to nonlinear inequality constraints.

A good discussion of the techniques for dropping constraints from active sets is given by Fletcher (1971). Most of them are based on estimates of Lagrange multipliers, for the sign of a Lagrange multiplier of an active constraint indicates whether the constraint function becomes positive or negative if the constraint is dropped from the active set. Because expression (6.8) gives the Lagrange multipliers at the solution of the quadratic programming problem stated in Theorem 6.1, the estimate

$$\underline{\lambda}^* \approx [N^T H^{(k)} N]^{-1} N^T H^{(k)} \underline{g}^{(k)} \qquad \ldots (8.1)$$

is used in the algorithms based on equation (6.9). However in Goldfarb's algorithm we have to use the estimate

$$\underline{\lambda}^* \approx (N^T N)^{-1} N^T \underline{g}^{(k)}, \qquad \ldots (8.2)$$

because the matrix $N^T H^{(k)}$ is zero.

Note that if $\underline{g}^{(k)}$ is equal to $N\underline{\lambda}$ for some vector $\underline{\lambda}$, then expressions (8.1) and (8.2) are both equal to $\underline{\lambda}$. Therefore the estimates (8.1) and (8.2) are accurate at the solution to the constrained problem derived from the active constraint set. However away from the solution the estimate (8.2) is at a disadvantage because it does not take account of any second derivatives of the objective function.

The simplest technique for dropping constraints from the active set is to drop one constraint immediately a Lagrange parameter estimate indicates that this should be beneficial. However this technique may cause "zig-zagging" when the estimate (8.2) is employed (Zoutendijk, 1970, for instance). Therefore a number of authors, including Rosen (1960) and Goldfarb (1969), recommend methods that change the active constraint set less readily. However it seems that inefficient zig-zagging should not occur if $F(\underline{x})$ has continuous second derivatives, and if the estimate (8.1) is used, provided that the formula for calculating the matrices $H^{(k)}$ $(k=1,2,3,\ldots)$ absorbs second derivative information well.

In Fletcher's (1972a) method for minimization subject to general constraints, it is also necessary to estimate Lagrange multipliers and active constraints, but a quite different method is used. First we describe this method in the case that all the constraints are linear, and then we give the extension for nonlinear constraints. An important feature of the method is that the estimates depend on the value of $\underline{x}$ and on a parameter q, but, in contrast to the active set strategy, they do not depend on any previous calculations.

We let the set of linear constraints be R, which may include both equality and inequality conditions, and we let $\underline{g}$ be the gradient vector of $F(\underline{x})$ at $\underline{x}$. Then to estimate Lagrange multipliers and active constraints we calculate the vector $\underline{\delta}$ that minimizes the quadratic function

$$Q(\underline{\delta}) = (\underline{g},\underline{\delta}) + \tfrac{1}{2}q\,||\underline{\delta}||^2, \qquad \ldots (8.3)$$

subject to the condition

$$\underline{x} + \underline{\delta} \in R. \qquad \ldots (8.4)$$

We estimate that the active constraints are the ones that are active at the solution to this problem, and we estimate that the corresponding Lagrange multipliers are the ones that are present at the solution to this quadratic programming problem.

For example if there is one variable, if the objective function is $F(x) \equiv x$, if R is the single condition $x-1 \geqslant 0$, and if $q = 1$, then the quadratic programming problem is to minimize $Q(\delta) = \delta + \frac{1}{2}\delta^2$, subject to the condition $x + \delta - 1 \geqslant 0$. The solution is the quantity

$$\delta = \begin{cases} -1, & x \geqslant 2 \\ 1-x, & x \leqslant 2. \end{cases} \qquad \dots (8.5)$$

Therefore the constraint is active if $x \leqslant 2$, and the value of the Lagrange parameter is

$$\lambda = \begin{cases} 0, & x \geqslant 2 \\ 2-x, & x \leqslant 2, \end{cases} \qquad \dots (8.6)$$

where for convenience we set the Lagrange parameter to zero it the constraint is inactive. Note that λ is a continuous function of $\underline{x}$, and that there is not an exact correspondence between the estimate that the constraint is active and the condition $\underline{x} \varepsilon R$.

This method is extended to nonlinear constraints by approximating each constraint function by a linear function. The linear approximations are the first-order Taylor series expansions about $\underline{x}$. For example the constraint $c(\underline{x} + \underline{\delta}) \geqslant 0$ would be replaced by the inequality

$$c(\underline{x}) + \{\underline{\delta}, \underline{\nabla} c(\underline{x})\} \geqslant 0 \qquad \dots (8.7)$$

in the set of conditions (8.4). Then $\underline{\delta}$ is calculated as before, and thus we estimate Lagrange multipliers and active constraints in the general non-linear programming problem. We let $\underline{\lambda}(\underline{x})$ denote the vector of Lagrange multipliers that is estimated from first derivatives calculated at $\underline{x}$, and, as in expression (8.6), we set a component of $\underline{\lambda}(\underline{x})$ to zero whenever the corresponding constraint is predicted to be inactive.

This method gives a very promising technique for solving the general constrained problem: minimize $F(\underline{x})$ subject to the conditions $\underline{c}(\underline{x}) \geqslant 0$ (Fletcher, 1972a). Indeed if we let $\underline{x}^*$ be the solution to this problem, and if we define $\underline{\lambda}(\underline{x})$ in the way described above, then $\underline{x}^*$ is a stationary point of the function

$$\Phi(\underline{x}) = F(\underline{x}) - \{\underline{\lambda}(\underline{x}), \underline{c}(\underline{x})\}. \qquad \dots (8.8)$$

Further, if the value of q in equation (8.3) is sufficiently large, then $\underline{x}^*$ is a local minimum of $\Phi(\underline{x})$. Therefore many general nonlinear programming problems can be solved by applying an algorithm for unconstrained minimization to the function (8.8).

In the example where $F(x) \equiv x$ and there is one constraint, $x-1 \geq 0$, it follows from equation (8.6) that expression (8.8) is the function

$$\Phi(x) = \begin{cases} x, & x \geq 2 \\ x - (2-x)(x-1), & x \leq 2. \end{cases} \quad \ldots (8.9)$$

This function is least when $x = 1$, which is the solution to the constrained problem. Note that, although the function (8.9) is continuous, its first derivative has a discontinuity, at the point where the estimate of the active constraint set changes.

Fletcher (1972a) gives a number of theorems on the function (8.8), and he discusses the problem of finding an algorithm that is suitable for minimizing $\Phi(\underline{x})$. There are two main difficulties. One is that the definition of $\Phi(\underline{x})$ depends on first derivatives of $F(\underline{x})$ and the constraint functions. The other is that, although $\Phi(\underline{x})$ is continuous, it has discontinuous first derivatives. Fortunately quite mild conditions ensure that the first derivative of $\Phi(\underline{x})$ is continuous at the required solution $\underline{x}^*$.

I have chosen to describe some of Fletcher's work at this conference, because minimizing the function (8.8) is the most promising idea I have encountered for treating nonlinear inequality constraints.

References

Bard, Y. (1968) "On a numerical instability of Davidon-like methods", Maths. Comp., Vol. 22, pp 665-666.

Broyden, C.G. (1970) "The convergence of a class of double-rank minimization algorithms 2. The new algorithm", J. Inst. Maths. Applics., Vol. 6, pp 222-231.

Curry, H.B. (1944). "The method of steepest descent for nonlinear minimization problems", Quart. Appl. Maths., Vol. 2, pp 258-261.

Dixon, L.C.W. (1972) "Quasi-Newton algorithms generate identical points", Math. Prog., Vol. 2, pp 383-387.

Fiacco, A.V. and McCormick, G.P. (1968) "Nonlinear programming: sequential unconstrained minimization techniques", John Wiley Inc. (New York).

Fletcher, R. (1970) "A new approach to variable metric algorithms", Computer Journal, Vol. 13, pp 317-322.

Fletcher, R. (1971) "Minimizing general functions subject to nonlinear constraints", Report T.P. 453, A.E.R.E., Harwell.

Fletcher, R. (1972a) "An exact penalty function for nonlinear programming with inequalities", Report T.P. 478, A.E.R.E., Harwell.

Fletcher, R. (1972b). "An algorithm for solving linearly constrained optimization problems", Math. Prog., Vol. 2, pp 133-165.

Fletcher, R. and Powell, M.J.D. (1963) "A rapidly convergent descent method for minimization", Computer Journal, Vol. 6, pp 163-168.

Goldfarb, D. (1969) "Extension of Davidon's variable metric algorithm to maximization under linear inequality and equality constraints", SIAM J. Appl. Maths., Vol. 17, pp 739-764.

Goldfeld, S.M., Quandt, R.E. and Trotter, H.F. (1966) "Maximization by quadratic hill-climbing", Econometrica, Vol. 34, pp 541-551.

Goldstein, A.A. and Price, J.F. (1971) "On descent from local minima", Maths. Comp., Vol. 25, pp 569-574.

Householder, A.S. (1964) "The theory of matrices in numerical analysis", Blaisdell Publishing Co. (New York).

Huang, H.Y. (1970) "Unified approach to quadratically convergent algorithms for function minimization", J.O.T.A., Vol. 5, pp 405-423.

Kantorovich, L.V. and Akilov, G.P. (1964) "Functional analysis in normed spaces", Pergamon Press (Oxford).

Kowalik, J. and Osborne, M.R. (1968) "Methods for unconstrained optimization problems", Elsevier Publishing Co. Inc. (New York).

McCormick, G.P. (1970) "The variable reduction method for nonlinear programming", Management Science, Vol. 17, pp 146-160.

McCormick, G.P. (1972) "Attempts to calculate global solutions of problems that may have local minima" in "Numerical methods for nonlinear optimization", ed. F.A. Lootsma, Academic Press (London).

Matthews, A. and Davies, D (1971) "A comparison of modified Newton methods for unconstrained optimization", Computer Journal, Vol. 14, pp 293-294.

Murtagh, B.A. and Sargent, R.W.H. (1969) "A constrained minimization method with quadratic convergence" in "Optimization", ed. R. Fletcher, Academic Press (London).

Murtagh, B.A. and Sargent, R.W.H. (1970) "Computational experience with quadratically convergent minimization methods", Computer Journal, Vol. 13, pp 185-194.

Ortega, J.M. and Rheinboldt, W.C. (1970) "Iterative solution of nonlinear equations in several variables", Academic Press (New York).

Powell, M.J.D. (1964) "An efficient method for finding the minimum of a function of several variables without calculating derivatives", Computer Journal, Vol. 7, pp 155-162.

Powell, M.J.D. (1966) "Minimization of functions of several variables" in "Numerical analysis: an introduction", ed. J. Walsh, Academic Press (London).

Powell, M.J.D. (1971) "On the convergence of the variable metric algorithm", J. Inst. Maths. Applics., Vol. 7, pp 21-36.

Powell, M.J.D. (1972a) "Unconstrained minimization algorithms without computation of derivatives", Report T.P. 483, A.E.R.E., Harwell.

Powell, M.J.D. (1972b) "Quadratic termination properties of a class of double-rank minimization algorithms", Report T.P. 471, A.E.R.E., Harwell.

Rosen, J.B. (1960) "The gradient projection method for nonlinear programming, Part 1. Linear constraints", SIAM Journal, Vol. 8, pp 181-217.

Shanno, D.F. and Kettler, P.C. (1970) "Optimal conditioning of quasi-Newton methods", Maths. Comp., Vol. 24, pp 657-664.

Stewart, G.W. (1967) "A modification of Davidon's minimization method to accept difference approximations of derivatives", J. Assoc. Comput. Mach., Vol. 14, pp 72-83.

Wolfe, P. (1967) "Methods of nonlinear programming" in "Nonlinear programming", ed. J. Abadie, North Holland Publishing Co. (Amsterdam).

Wolfe, P. (1969) "Convergence conditions for ascent methods", SIAM Review, Vol. 11, pp 226-235.

Wolfe, P. (1972) "The method of conjugate gradients", presented at this conference.

Zoutendijk, G. (1960) "Methods of feasible directions", Elsevier Publishing Co. (Amsterdam).

Zoutendijk, G. (1970) "Nonlinear programming, computational methods" in "Integer and nonlinear programming", ed. J. Abadie, North Holland Publishing Co. (Amsterdam).

Mathematical Programming in Theory and Practice,
P.L. Hammer and G. Zoutendijk, (Eds.)

On linearly constrained nonlinear programming and some extensions

by G. Zoutendijk, University of Leiden, The Netherlands

1. Introduction

The general nonlinear programming problem can be defined as the problem

$$\text{Max} \{ f(x) \mid x \in R \}, \tag{1.1}$$

in which $f(x)$ is a continuous - usually differentiable - function of the vector $x \in E^n$, while R is a non-empty connected set in n-space satisfying a number of regularity conditions which will not be further explained in this paper.

Methods of solution for (1.1) will usually converge to a local stationary point. In most cases this will be a local maximum. There is no guarantee that a global maximum will be found unless of course we know beforehand that any local maximum will be a global one like in the case of $f(x)$ pseudoconcave and R convex.

For any $x \in R$ we can introduce a cone of feasible directions $S(x)$ by:

$$s \in S(x) \iff \exists \lambda > 0 \; \forall \mu, 0 \leq \mu \leq \lambda : x + \mu s \in R.$$

A direction $s \in S(x)$ will be usable if a λ exists such that for all μ, $0 < \mu \leq \lambda$, $f(x + \mu s) > f(x)$ will hold. For a differentiable function this entails $\nabla f(x)^T s > 0$ with $\nabla f(x)$ being the gradient vector in x.

A method of feasible directions to solve (1) is a method with the following characteristics:

1. $x^0 \in R$,
   (the starting point should be feasible).

2. Suppose $x^0, x^1, \ldots, x^k$ have already been determined. Then the next point will be obtained by the following operations:

   a. first a usable feasible direction is to be found:

   $$s^k \in S(x^k) \cup \{ s \mid \nabla f(x^k)^T s > 0 \};$$

   b. next the steplength λ_k will be determined in such a way that

$x^k + \lambda_k s^k \in R$ and $f(x^k + \lambda_k s^k) > f(x^k)$ will hold.

c. $x^{k+1} = x^k + \lambda_k s^k$

3. If $\bar{x}$ is a point of accumulation of the sequence x^k, so that

$f(\bar{x}) = \lim f(x^k)$, then $\bar{x}$ is a local stationary point.

If the set R is defined by

$$R = \{x \mid f_i(x) \leq b_i, i = 1, \ldots, m\} \tag{1.2}$$

with $f_i(x)$ being differentiable functions of $x \in E^n$, then a point of accumulation $\bar{x}$ will satisfy the optimality conditions (Kuhn-Tucker conditions):

$$\exists\, \bar{u} \geq 0, \bar{v} \geq 0 : \nabla f(\bar{x}) = \sum_{i=1}^{m} \bar{u}_i \nabla f_i(\bar{x}) - \bar{v}, \quad \bar{u}^T \bar{y} = 0, \quad \bar{v}^T \bar{x} = 0,$$

in which $\bar{y}_i = b_i - f_i(\bar{x})$ is the slack variable belonging to the i-th constraint.

In chapter 2 the principles underlying methods of feasible directions will be further investigated. In chapters 3, 4 and 5 these methods will be applied to linear programming, unconstrained optimization and linearly constrained nonlinear programming, respectively. In chapter 6 some extensions to general nonlinear programming will be considered. In particular, some recent work carried out by Dr. J. D. Buys will be reported.

Methods of feasible directions have been described in many books and papers (see e.g. Zoutendijk (1960), (1970a) and 1970b)). Their convergence properties have been studied by Topkis and Veinnott (1967) and Pironneau and Polak (1973).

2. Principles of methods of feasible directions

As we have seen in chapter 1 a feasible starting point x^0 should be available. If this is not the case the problem can be modified in such a way that any x^0 can be used as starting point: If the problem is defined by (1.1) and (1.2), and if $f_i(x^0) > b_i$ for $i \in I_0$, then solve the modified problem

$$\operatorname{Max} \{f(x) - \mu \xi \mid f_i(x) - \delta_i \xi \leq b_i, i = 1, \ldots, m;\ x \geq 0, \xi \geq 0\}, \tag{2.1}$$

in which $\delta_i = 1$ if $i \in I_c$ and $= 0$ if $i \notin I_c$, while $\mu > 0$ should be chosen so large that in the final solution $\xi = 0$ will hold. It is not difficult to prove that for non-empty regions R there exists a $\mu_c > 0$ such that for $\mu \geq \mu_c$ problems (1.1), (1.2) and (2.1) will have the same local stationary points.

Methods of feasible directions differ in the way the directions are being chosen. In all methods a direction-finding problem is explicitly or implicitly solved, i.e. a problem is solved in which a direction has to be determined satisfying the following requirements:

(1) $s \in S(\tilde{x})$,
($\tilde{x}$ being the current solution),

(2) $\nabla f(\tilde{x})^T s > 0$, (2.2)

(3) additional requirements.

For reasons of simplicity we will assume that R is a convex polyhedron:

$$R = \{x \mid Ax \leq b, x \geq 0\}, \tag{2.3}$$

with A an m by n matrix with rows a_i, and $b \in E^m$.
If $\tilde{x} \in R$, then

$$I(\tilde{x}) = \{i \mid a_i^T \tilde{x} = b_i\}, \tag{2.4}$$

$$J(\tilde{x}) = \{j \mid \tilde{x}_j = 0\}, \tag{2.5}$$

$$S(\tilde{x}) = \{s \mid a_i^T s \leq 0, i \in I(\tilde{x}) ; s_j \geq 0, j \in J(\tilde{x})\} \tag{2.6}$$

so that the requirement $s \in S(\tilde{x})$ results in a set of homegeneous linear relations.

In the case of nonlinear contraints the situation is slightly more complicated due to the fact that directions in tangent planes need no longer be feasible. If R is defined by (1.2) with $f_i(x)$ strictly convex, $\tilde{x} \in R$, then

$$S(\tilde{x}) = \{s \mid \nabla f_i(\tilde{x})^T s < 0, i \in I(\tilde{x})\}.$$

There are at least two ways to overcome this difficulty of a strict inequality requirement in such a way that linear programming or related techniques can be used to solve the direction finding problems. Disregarding the additional requirements (3) in (2.2) we can either introduce an additional variable s_0 and replace (1) and (2) in (2.2) by

(1) $\nabla f_i(\tilde{x})^T s + s_0 \leqq 0 \,, \; i \in I(\tilde{x}) \,,$

(2) $-\nabla f(\tilde{x})^T s + s_0 \leqq 0 \,, \; s_0 > 0$

or we can try to find a solution of the non-homogeneous system:

(1) $\nabla f_i(\tilde{x})^T s \leqq -1 \,, \; i \in I(\tilde{x}) \,,$

(2) $\nabla f(\tilde{x})^T s = 1 \,.$

Both possibilities have been discussed in earlier papers.
The additional requirements will be added to the direction-finding problem in order to quarantee or speed-up convergence. They will also consist of linear homogeneous inequalities or equalities, so that a direction-finding method - to be called direction generator - can easily cope with them. Different direction generators will result in completely different methods of feasible directions. These methods will not differ in the way the step-length determination is being carried out. This can be formulated as a one dimensional maximization problem:

$$\operatorname{Max} \{ f(\tilde{x} + \lambda \tilde{s}) \mid x = \tilde{x} + \lambda \tilde{s} \in R \} \,, \tag{2.7}$$

where it is usually accepted that the first maximum will be taken.
The maximum will be either determined by the boundary or an interior stop will take place. In the unconstrained case the latter will always take place; in the case of a linear objective function the former possibility will always occur. A boundary stop can easily be calculated in the case of linear constraints. This is another reason that it is computationally much easier to apply a method of feasible directions to a linearly constrained nonlinear programming problem than to one with nonlinear constraints. Highly sophisticated methods have been developed to solve the so-called linear search problem $\max f(\tilde{x} + \lambda \tilde{s})$.

Most of these methods are based on interpolation techniques; some of them have been described by Kowalik and Osborne (1969). In practice the one-dimensional maximization problem is only roughly approximated, at least as long as we are far from the optimum. Especially in unconstrained optimization it has moreover been tried to get rid of linear searches since from the computational point of view they are rather expensive due to the number of function and gradient evaluations involved. See further Powell (1971 and 1974).
We shall distinguish three classes of direction generators:

A. Direct methods

B. Optimization methods

C. Feasibility methods.

Without trying to be complete we will now describe some methods from each class. On purpose we have omitted those methods which - according to our experience - have no advantage over other methods. We will assume that R is a convex polyhedron, defined by (2.3).

A. Direct methods

In these methods a direction is implicitly determined; the direction problems are in the original variables x_j rather than in the direction variables s_j. This is for instance being done in the linear approximation method originally suggested by Frank and Wolfe (1956) for linearly constrained problems. In this method at each step a linear subproblem is solved of the form

$$\mathrm{Max}\left\{\nabla f(\tilde{x})^T x \;\middle|\; Ax \leqq b,\, x \geqq 0,\, x \in K\right\}, \tag{2.8}$$

in which the requirements denoted by $x \in K$ are also linear; they are added to avoid infinite solutions or to speed-up convergence. The constraint set is supposed to be linear. It is also possible, however, that the linear constraints are linearizations of non-linear ones like in the modified feasible direction method (Zoutendijk (1966)). In that case the point $\tilde{x}$ should be interior with respect to the nonlinear constraints, i.e. $f_i(\tilde{x}) < b_i$ for f_i nonlinear.
If the solution of (2.8) is $\bar{x}$, $\nabla f(\tilde{x})^T(\bar{x}-\tilde{x}) = 0$ and the relations $x \in K$ are not binding, then the maximum of the nonlinear problem has obviously been arrived at. If $\nabla f(\tilde{x})^T(\bar{x}-\tilde{x}) > 0$, then $\tilde{s} = \bar{x}-\tilde{x}$ will be a usable feasible direction. In the linear approximation method we shall then carry out a constrained linear search resulting in a new point $\tilde{x}$.

Instead of solving (2.4) we could also use the simplex method to determine a vertex $\bar{x}$, not necessarily the optimum one, with $\nabla f(\hat{x})^T(\bar{x}-\hat{x}) > 0$.
This might have advantages in large practical problems since the increase in objective function value is usually small during the last part of the iterations. We could even stop at the first vertex for which the inequality holds. This variant has been suggested by Zangwill (1969). If applied to a linearly constrained problem with steplength determination through constrained linear searches and with a special way to improve convergence it is called the convex simplex method.

B. Optimization Methods

The direction-finding problem is now of the form:

$$\text{Max}\ \{\nabla f(\hat{x})^T s \mid s \in S(\hat{x}),\ s \in T,\ s \in N\}, \tag{2. 9}$$

in which $S(\hat{x})$ is defined by (2.6), T is defined by a set of linear homogeneous (in)equalities introduced to speed up or guarantee convergence and N is defined by a set of relations introduced to prevent infinite solutions.
We could for instance take

$$N = \{s \mid \|s\| \leq 1\}, \tag{2.10}$$

leading to direction generators of the general form

$$\text{Max}\ \{p^T s \mid A_1 s \leq 0,\ A_2 s = 0,\ \|s\| \leq 1\}, \tag{2.11}$$

in which p, A_1 and A_2 have been introduced to simplify writing.
Examples:

a. $\sum |s_j| \leq 1$ (L_1 norm)
b. $s^T s \leq 1$ (L_2 or Euclidean norm)
c. $\forall_j : -1 \leq s_j \leq 1$ (L_∞ or Chebyshev norm)
d. $s^T P s \leq 1$ for P symmetric and positive definite (metricized norm)

Direction generators Ba and Bc will result in linear programming problems of a special type for which special methods based upon the simplex method can be developed. For Bc this has been shown by Zoutendijk (1960, 1970a), for Ba it makes little sense to report the details since we have not found any advantages in this direction generator compared to other ones. Methods Bb and Bd will result in quadratic programming problems of a very special type (Bb being a special case of Bd with $P = I$, for which a special method of solution can be developed. Hereto we consider the problem:

$$\operatorname{Max} \{ p^T x \mid Ax \leqq 0 , x^T P x \leqq 1 \} \tag{2.12}$$

From the optimality conditions it follows that in the maximum we shall have

$$p = A^T u + \beta P x , u \geqq 0 , \beta \geqq 0 , u^T A x = 0 .$$

Multiplying from the left by $-AP^{-1}$ and writing $v = -\beta Ax$ we obtain:

$$-AP^{-1}A^T u + v = -AP^{-1} p , u \geqq 0 , v \geqq 0 , u^T v = 0 , \tag{2.13}$$

$$\beta x = P^{-1} p - P^{-1} A^T u . \tag{2.14}$$

The system (2.13) can be solved by complimentary pivoting. The matrix $-AP^{-1}A^T$ is negative semi-definite. If diagonal elements are only used as pivot elements and if after a number of changes of basis (interchanges of u_i and the corresponding v_i) the transformed system is being written as:

$$-Qw + z = q \quad (w_i = u_i \text{ or } v_i , z_i = v_i \text{ or } u_i) ,$$

then it is easy to prove that

(1) $q_{ii} \geqq 0$ for all i ;

(2) if $q_{ii} = 0$, then $q_i = 0$.

This justifies the following method of solution:

1. choose $q_i < 0$ (in practice take the most negative one or the one for which $q_i / \sqrt{q_{ii}}$ is most negative);

2. increase w_i until
   a. either $z_i = 0$ holds,
   b. or $z_r = 0$ holds for $r \neq i$, while $q_r > 0$
   (whatever comes first)

3. in case 2a, interchange w_i and z_i and go to 1;
   in case 2b, interchange w_r and z_r and go to 2.

In case 2b a bottleneck is being removed, so that a further increase of w_i becomes possible (disregarding degeneracy which can be easily coped with). A validity and finiteness proof for this method is left to the reader. If we denote the active rows of the matrix A by A_1, so that $A_1 x = 0$ will hold for the solution of (2.12), then this solution can be explicitly written as

$$x = P^{-1}p - P^{-1}A_1^T(A_1 P^{-1} A_1^T)^{-1} A_1 P^{-1} p = Q_{A_1} p.$$

The matrix Q_{A_1} may also be calculated using the recurrence relation

$$Q_0 = P^{-1},$$
$$Q_k = Q_{k-1} - \frac{Q_{k-1} a_{k.} a_{k.}^T Q_{k-1}}{a_{k.}^T Q_{k-1} a_{k.}}, \quad k = 1, \dots, m_1, \tag{2.15}$$

where we have assumed that the matrix A_1 consists of the rows $a_{1.}, \dots, a_{m_1.}$. A special case is $P = I$. In that case $x = (I - A_1^T(A_1 A_1^T)^{-1}A_1)p$, the projection of p onto the cone $Ax \leq 0$ or onto the linear subspace $A_1 x = 0$.

Instead of using a norm we could bound the directions in another way, for instance $c \leqq s \leqq d$

With $d_j = 1$ if $p_j > 0$ and $= \infty$ if $p_j \leqq 0$, $c_j = -1$ if $p_j < 0$ and $-\infty$ if $p_j \geqq 0$ we would have a method, which, if applied to the dual of a linear programming problem would be equivalent to the so-called primal-dual method for linear programming applied to the primal problem.

Another class of optimization methods would be obtained by considering direction generators of the form

$$\text{Min}\left\{ \|p - s\| \;\middle|\; s \in S(\bar{x}) \right\}. \tag{2.16}$$

In the case of the Euclidean norm this method will be computationally equivalent to method Bb as can be easily verified by considering the optimality conditions (Kuhn-Tucker conditions) for both problems. In the case of the L_1 or Chebyshev norm a linear programming problem would result. For the Chebyshev norm this would be:

$$\text{Min}\left\{ s_0 \;\middle|\; -s_0 \leqq p_j - s_j \leqq s_0,\ j = 1, \dots, n;\ s \in S(\bar{x}) \right\}. \tag{2.17}$$

A special method could easily be developed for this problem. From the computational point there do not seem to be any advantages, however.

C. Feasibility Methods

Instead of trying to find the (locally) best direction according to some optimality criterion like is being done in the methods of class B we are now satisfied with just a direction which only satisfies all the requirements, hence for which holds:

(1) $s \in S(\tilde{x})$, $s \in T$;

(2) $\nabla f(\tilde{x})^T s > 0$, e.g. $= 1$.

The methods in this class differ in the way the system of linear inequalities and equalities is being solved. We will mention three ways to do this:

a. The simplex method for linear programming could be applied. We could maximize $\nabla f(\tilde{x})^T s$ under the linear (homogeneous) constraints $s \in S(\tilde{x})$ and $s \in T$.
   This being a homogeneous problem an unbounded solution will be the result (at least if (1) and (2) are not inconsistent). The extreme ray leading to infinity will then indicate the solution of (1) and (2).
   To further explain this we will assume that the programming problem to be solved is linear and of the form:

$$Max \{p^T x \mid Ax = b, x \geqq 0\}, \tag{2.18}$$

   so that the directions will always have to satisfy:

$$\begin{aligned} &As = 0, \; s_j \geqq 0, j \in J(\tilde{x}); \\ &p^T s > 0. \end{aligned} \tag{2.19}$$

   This is no restriction.
   Using the simplex method to maximize $p^T s$ the matrix A will be subdivided in an m by m submatrix B and a remaining part D, so that we can write:

$$\begin{aligned} &B s_B + D s_D = 0, \\ &p_B^T s_B + p_D^T s_D = 1 \end{aligned} \tag{2.20}$$

or

$$
\begin{aligned}
&s_B + B^{-1} D \, s_D = 0, \\
&-(p_B^T B^{-1} D - p_D^T) \, s_D = 1.
\end{aligned}
\tag{2.21}
$$

For reasons of simplicity we will assume that there is no degeneracy; hence at least m components of any feasible $\tilde{x}$ will be strictly positive. The matrix B could always be chosen in such a way that $(\tilde{x}_B)_i > 0$ will hold. Suppose that the infinite solution is found when $(s_D)_k$ wants to enter the basis.
Introducing

$$v^T = p_B^T B^{-1} D - p_D^T$$

we can distinguish the following possibilities:

a. $(\tilde{x}_B)_i > 0$: the sign of $(B^{-1}D)_{ik}$ does not matter;

b. $(\tilde{x}_B)_i = 0$: then $(B^{-1}D)_{ik} \leqq 0$ when $v_k < 0$ ($(s_D)_k$ wants to enter at positive value);

$(B^{-1}D)_{ik} \geqq 0$ when $v_k > 0$ ($(s_D)_k$ wants to enter at negative value).

The direction will be:

$$
\begin{aligned}
(\tilde{s}_D)_k &= 1 \quad \text{if} \quad v_k < 0, \\
&= -1 \quad \text{if} \quad v_k > 0 \quad \text{(this is only possible when } (\tilde{x}_D)_k > 0\text{)}, \\
(\tilde{s}_D)_j &= 0, \; j \neq k, \\
\tilde{s}_B &= -(B^{-1}D)_{.k} \, (\tilde{s}_D)_k .
\end{aligned}
$$

If one wants a normalization can be carried out afterwards, so that $p^T s = 1$ will hold but there is no reason to do this.

b. Instead of only giving a non-zero value to one non-basic variable we could also give non-zero values to all non-basis variables according to the corresponding components of the "reduced gradient" vector v, i.e. we could take

$$(\tilde{s}_D)_j = -v_j \quad \text{if } (\tilde{x}_D)_j \neq 0 \quad \text{or } (\tilde{x}_D)_j = 0 \quad \text{and } v_j \leqq 0 ;$$

$$= 0 \quad \text{if } (\tilde{x}_D)_j = 0 \quad \text{and } v_j > 0 \qquad \text{(to maintain feasibility)}$$

$$\tilde{s}_B = -B^{-1} D \tilde{s}_D .$$

This direction is being called the reduced gradient. It is clear that for feasibility reasons we have to require now $(\tilde{x}_B)_i > 0$.
In the case of non-degeneracy this is always possible, if necessary by making a further change of basis.

c. Another way to find a solution of the system defined by (1) and (2) is by using successive projection.
Let us first assume that the problem is given by

$$A s = 0 , p^T s > 0 ,$$

and that the rows of A are linearly independent. Then the projection of p onto the linear subspace defined by $A s = 0$ is given by

$$s = P_A p \tag{2.22}$$

with

$$P_A = \left(I - A^T (A A^T)^{-1} A\right) \tag{2.23}$$

A recurrence relation to calculate P_A can be easily derived:

$$Q_0 = I ,$$

$$Q_k = Q_{k-1} - \frac{Q_{k-1} a_{k.} a_{k.}^T Q_{k-1}}{a_{k.}^T Q_{k-1} a_{k.}} , k = 1, \ldots, m , \tag{2.24}$$

$$P_A = Q_m$$

(assuming A has m rows).
In this way successive projection can be carried out.
In the case of inequalities,

$$A s \leqq 0 , p^T s > 0$$

with the rows $a_{k.}$ coming either from the constraints $a_{i.}^T s \leqq 0$ or from $s_j \geqq 0$ the same procedure can be applied with three additions:

(1) As soon as $AQ_k p \leqq 0$ holds the procedure can be stopped, $\tilde{s} = Q_k p$ will be a usable feasible direction.

(2) The rows of A should not be selected in their natural order but according to some reasonable criterion like

$$\frac{a_{i.}^T Q_k p}{|a_{i.}|} \quad \text{largest.}$$

(3) If, for some k, $Q_k p = 0$ will hold it should not be concluded that there is no usable feasible direction since one of the choices made during the row selection process could have been a wrong one. Denoting by A_k the matrix consisting of those rows of A that have been used by the projection procedure we must investigate whether $p^T s \leqq 0$ holds for all s satisfying $A_k s \leqq 0$. If that were true, then we derive from Farkas' theorem that u, $u \geqq 0$ exists such that $p = A_k^T u$. It follows

$$u = (A_k A_k^T)^{-1} A_k p . \tag{2.25}$$

If there are negative components in this vector, then the most negative one should be selected and the corresponding row, say $a_{t.}$ should be dropped from A_k.
If A_k is derived from A_{k-1} by dropping the row $a_{t.}$, then it is not difficult to calculate $(A_{k-1} A_{k-1}^T)^{-1}$ from $(A_k A_k^T)^{-1}$. If this matrix is being stored, then (2.23), (2.22) and (2.25) can always be calculated. It then becomes clear that this way of gradient projection which was suggested by Rosen (1960) is closely related but not equivalent to the optimization method Bb. The difference is that instead of solving (2.13) by complimentary pivoting the following procedure is applied:

1. A $v_i < 0$ is chosen; u_i and v_i are interchanged.
2. This is continued until $v_i \geqq 0$ for all i.
3. If $x = p - A^T u \neq 0$, x is accepted as usable feasible direction; otherwise the most negative u_i is chosen, u_i and v_i will be interchanged and we return to 1.

It is clear that this procedure does in general not result in the solution of (2.13). As far as the choice of $v_i < 0$ is concerned the criterion

$\frac{a_{i\cdot}^T Q_k p}{|a_{i\cdot}|}$ largest differs from the criterion mentioned in rule 1 of the complimentary pivoting procedure to solve (2.13). The criterion

$q_i / \sqrt{q_{ii}}$ most negative appears to be equivalent - in the case of a variable v_i leaving the basis - with $a_{i\cdot}^T Q_k p / (a_{i\cdot}^T Q_k a_{i\cdot})^{\frac{1}{2}}$ largest. This criterion seems to be more appropriate.

3. Application to Linear Programming

A feasible starting point, a direction generator like one of those discussed in chapter 2 and steplength determination by constrained linear search (which in view of the linearity of the objective function will always result in a boundary stop) will lead to a method of solution for the linear programming problem

$$Max \{ p^T x \mid Ax \leq b, x \geq 0 \}.$$

The direct method of class A and method Ca - if started in a vertex - will actually reduce to the simplex method. The other direction generators do not have any specific advantage over the simplex method, except for special problems. Due to the generality of the direction generation process no general finiteness proof can be given. Such a proof is not difficult for generators Bb and Bd (if the directions are normalized such that $s^T s = 1$ or $s^T P s = 1$ then $p^T s$ will decrease monotonically, so that the same sets $I(\tilde{x})$ and $J(\tilde{x})$ will never return in a direction problem). A general convergence proof - holding for any direction generator - will be easy if we add an anti-zigzagging requirement like:

require $a_{i\cdot}^T s = 0$ instead of ≤ 0 or $s_j = 0$ instead of ≥ 0 when we arrive in a certain hyperplane $a_{i\cdot}^T x = b_i$ or $x_j = 0$. Relax at least one of these additional requirements if no usable feasible vector can be found anymore.

With the anti-zigzagging requirement we are sure that after a finite number of steps a maximum of a restricted problem will be obtained, the restriction being that we have to be in a certain combination of hyperplanes. As soon as we have relaxed some of the additional requirements, so that we can move again we are sure that we will never return in this combination of hyperplanes. From this the finiteness follows. When we have to relax some of the

additional requirements we can do that in several ways: just as few as necessary to obtain a usable direction or all of them at the same time or something in between. This gives rise to several variants. If the variant is chosen that just one is relaxed, say the oldest one, then after a finite number of steps we will be obliged to follow a vertex to vertex path just like in the simplex method. From that moment on there is only one degree of freedom and it does not matter anymore which direction generator is chosen. If the choice of the equality to be changed into an inequality is made according to their dual values, then the vertex to vertex path will be exactly the same as in the simplex method.
In all direction generators we are working with linear subproblems. It is clear that the final information from one tableau (final matrix or inverse of the basis) should be used for the next one, so that this subproblem is solved after only a few iterations. This means that we have to add or delete rows to or from a final matrix. Formulae for this can be easily derived in most cases.
Most practical linear programming problems are large and structured. The matrix has very few non-zero elements and will show a special network type structure which will to a large extent be maintained during the iterations. It is clear that those direction generators will then be preferable which maintain this structure as much as the simplex method does. This holds for generators of class A, for Ba, Bc, Ca and Cb but not for Bb, Bd and Cc. In the latter cases the operation AA^T or $AP^{-1}A^T$ destroys most of the structure. Moreover the simple non-negativity requirements cannot be handled in as simple a way as in the other methods. Nonlinear programming problems which are linear to a very large extent (only few and simple non-linearities) and for which the same remarks about structure will apply will be called of the LP type. They may be encountered in production planning and distribution problems. For these problems the use of a direction generator of class A, Bc, Ca or Cb is to be recommended (we have not found any advantages in using Ba).

4. Application to Unconstrained Optimization

Our problem now is

$$\max f(x), \tag{4.1}$$

so that for any $\tilde{x}$ $S(\tilde{x}) = E^n$ will hold. In this chapter it is not our primal intention to develop methods which will succesfully compete with the existing methods (for this see the survey articles by Powell (1971), (1974) and the recent book, edited by Murray (1972)), although this might be one of the results. It is our primary intention to show convergence can be speeded up by using the (near)- conjugacy principle in such a way that extension to linearly constrained nonlinear programming becomes easy.

We will write

$$g^k = \nabla f(x^k),$$
$$\Delta g^k = g^{k+1} - g^k.$$

A general iterative method could then be as follows:

1. x^0 arbitrary, steplength λ_k by linesearching (one-dimensional maximization);

2. in x^0 require for s^0 : $(g^0)^T s > 0$;

   in x' require for s' : $(\Delta g^0)^T s = 0$,
   $(g')^T s > 0$;

   in x^k require for s^k : $(\Delta g^k)^T s = 0$, $k = 0, 1, \ldots, k-1$; $k \leq n-1$;
   $(g^k)^T s > 0$

3. for $k \geq n$ omit one or more of the "conjugacy" relations so that a usable direction can be obtained (see further below).
   When $f(x)$ is quadratic, $f(x) = p^T x - \frac{1}{2} x^T C x$ with C symmetric, positive definite, then

$$g(x) = p - Cx,$$
$$g^{k+1} = g^k - \lambda_k C s^k,$$
$$(\Delta g^k)^T s = -\lambda_k (s^k)^T C s,$$

so that $(\Delta g^k)^T s = 0$ entails that s should be conjugate to s^k. The general iterative method will then result in a sequence of mutually conjugate feasible directions, so that the maximum will be obtained in at most n steps. For a more general objective function it is to be expected that requirements of the type $(\Delta g^k)^T s = 0$ will speed-up convergence. This happens to be the case in many practical problems. The methods developed this way all have the so-called quadratic termination property. For non-quadratic functions the maximum will not be obtained after n steps. However, after n steps a tableau of n linearly independent

relations of the type $(\Delta g^k)^T s = 0$ will be obtained, so that no non-trivial solution will be found anymore.
We may then proceed in one of the following ways:

a. Start again, hence omit all conjugacy relations and take x^0(new) = x^n(old).

b. Omit the oldest relation and add a new one; this is being called the moving tableau approach (in exceptional cases more than one of the oldest relations has to be omitted due to linear dependence).

c. Work with a tableau of size $m \leq n$, where m is gradually growing, dependent on the progress made.

d. When the n (near)-conjugate directions have been stored, then it is possible to improve method 3a considerably by making one additional quasi-Newton step before starting afresh.
Let

$$H = -\sum_{k=0}^{n-1} \lambda_k \frac{s^k (s^k)^T}{(\Delta g^k)^T s^k},$$

then for a quadratic function $p^T x - \frac{1}{2} x^T C x$ we will have $H = C^{-1}$.
It therefore makes sense to choose

$$s^n = H g^n$$

and to make a final step in the direction s^n before starting afresh. The matrix H need not even be calculated. If the dual variables of the final tableau consisting of the transformed relations $(\Delta g^k)^T s = 0$, $k = 0, 1, \ldots, n-1$ are denoted by u_k, then it is not difficult to show that for a quadratic function

$$C^{-1} g^n = \sum_{k=0}^{n-1} u_k \tilde{s}^k$$

with $\tilde{s}^k = -\lambda_k s^k$, so that for a more general function the n-th step can be taken in the direction $s^n = \sum_{k=0}^{n-1} u_k \tilde{s}^k$.

This means that $\tilde{s}^k = -\lambda_k s^k = -(x^{k+1} - x^k)$ should be stored.

e. As Dr. M. Best pointed out to me recently it is also possible to alternate between an acceleration step (like the one in 3d) and a moving tableau step. If s^n is determined by the acceleration procedure, then s^{n+1} will satisfy $(\Delta g^k)^T s = 0$, $k = 1, \ldots, n-1$.

The relation to be added to the tableau will then be

$$(\Delta g^{n+1})^T s = 0.$$

A new acceleration step will then be made to determine s^{n+2}.
s^{n+2} will then have to satisfy $(\Delta g^k)^T s = 0$, $k = 2, 3, \ldots, n-1, n+1$, etc. In the general iterative method the relations to be considered are linear. They have already been indicated in chapter 2 by writing $s \in T$. It is clear that any direction generator can be used to solve the direction problems. If for the continuation after n steps variant 3b or 3e is chosen, then it does not matter anymore which direction generator is being used since there is only one degree of freedom at each step. We can then choose the one that is organizationally easiest and computationally fastest, i.e. method Ca. With variants 3a and 3d the choice of the direction generator is of importance, however. Computational experience shows that for most unconstrained problems 3b is to be preferred over 3a; the number of steps, hence the number of function evaluations is so much lower in 3b compared to 3a that the additional amount of work to solve the n by n direction problem (about three times more on the average) is more than compensated in most cases. This is due to the fact that a line search entails several function evaluations and gradient calculations, which are usually expensive operations. There is reason to assume that 3d is much better than 3 a and that 3e is the most promising method for unconstrained optimization. Methods 3d and 3e have the additional advantage that the accuracy of the steplength calculation is of less importance (the relation $H = C^{-1}$ for a quadratic function remains valid for arbitrary λ's).
If a class A direction generator is used, then $(\Delta g^k)^T s = 0$ should be replaced by $(\Delta g^k)^T (x - x^{k+1}) = 0$.
In the case of direction generators Bb or Cc an explicit formula can be derived:

$$s^k = Q_k g^k \tag{4.2}$$

with

$$Q_0 = I, \tag{4.3}$$

$$Q_{k+1} = Q_k - \frac{Q_k \Delta g^k (\Delta g^k)^T Q_k}{(\Delta g^k)^T Q_k \Delta g^k}. \tag{4.4}$$

In this case the gradient g^k is being projected onto the linear subspace defined by the relations $(\Delta g^k)^T s = 0$. After n steps we will have $Q_n = 0$, so that when applying the explicit formula we have to start afresh after every n steps (the acceleration step can be made, however). An explicit formula is also possible when using Bd. Formulae (4.2) and (4.4) remain the same but (4.3) is replaced by

$$Q_0 = P^{-1}. \tag{4.5}$$

In the case of a quadratic objective function and direction generator Bb the gradients at different steps will be mutually orthogonal, $(g^i)^T g^j = 0,\ i \neq j$. It can be proved that we will then obtain the updating formula:

$$s^0 = \frac{g^0}{(g^0)^T g^0},$$
$$s^{k+1} = s^k + \frac{g^{k+1}}{(g^{k+1})^T g^{k+1}}, \quad k = 0, 1, 2, \dots . \tag{4.6}$$

This formula together with linear searches, leads to the conjugate gradient method for unconstrained quadratic maximization. The method has the advantage that no matrix has to be stored and updated; it is a vector method. The same formulae can, of course, be applied to a general objective function; this has been suggested by Fletcher and Reeves (1964). After n steps we can then continue using the formula, although from the computational point of view the advantage of this compared with a restart after n steps is subject to doubt.

Most well-known for the unconstrained optimization problem are the variable metric methods like the method of Davidson, Fletcher and Powell, in which

$$s^k = H_k g^k, \tag{4.7}$$

$$H_0 = I, \tag{4.8}$$

$$H_{k+1} = H_k - \lambda_k \frac{s^k (s^k)^T}{(\Delta g^k)^T s^k} - \frac{H_k \Delta g^k (\Delta g^k)^T H_k}{(\Delta g^k)^T H_k \Delta g^k}, \tag{4.9}$$

while line searching is also applied. In this method the matrix H_k is an approximation to the inverse of the matrix of second partial derivatives. For a quadratic function $f(x) = p^T x - \frac{1}{2} x^T C x$ we will have:

(1) $H_n = C^{-1}$,

(2) $(s^i)^T C s^j = 0$, $i \neq j$,

so that the directions will be mutually conjugate.
This method and other variable metric methods are behaving quite well in practice, although a periodic restart might be worthwhile in some cases. This observation has stimulated us to consider the following method which is based on direction generator Bd:

1. Choose x^0 arbitrary, $P_0 = I$, $\ell = 0$.

2. For $k = 0, 1, \ldots, n-1$:

   a. $s^k := \max \{ (g^k)^T s \mid (\Delta g^k)^T s = 0, k = 0, 1, \ldots, k-1 ; s^T P_\ell s \leq 1 \}$,

   b. $\lambda_k := \max f(x^k + \lambda s^k)$,

   c. $x^{k+1} = x^k + \lambda_k s^k$, calculate g^{k+1}.

3. Perform a (near) optimality test; if not passed:

4. $P_{\ell+1}^{-1} = - \sum_{k=0}^{n-1} \lambda_k \frac{s^k (s^k)^T}{(\Delta g^k)^T s^k}$; $x^0(new) = x^n(old)$; $\ell = \ell + 1$, go to 2.

(In 2a it is understood that for $k = 0$ there are no conjugacy relations).
Since for a quadratic function and n mutually conjugate directions we have

$$C^{-1} = - \sum_{k=0}^{n-1} \lambda_k \frac{s^k (s^k)^T}{(\Delta g^k)^T s^k} \tag{4.10}$$

the first step of each cycle of n steps will always be a quasi-Newton step (equivalent to the acceleration step in the general method 3d).
In the direction problems we have a variable metricized norm given by the approximation to the matrix of second partial derivations as calculated from the previous cycle. This metricized norm method has been developed independently by Hestenes (1969). To calculate s^k we could also use formulae (4.2), (4.4) and (4.5).

The method is then:

1. x^0, $Q_0 = I$, $\ell = 0$,

2. For $k = 0, 1, \ldots, n-1$:

   a. $s^k = Q_k g^k$,

   b. $\lambda_k := \max f(x^k + \lambda s^k)$,

   c. $x^{k+1} = x^k + \lambda_k s^k$, g^{k+1},

   d. $Q_{k+1} = Q_k - \dfrac{Q_k \Delta g^k (\Delta g^k)^T Q_k}{(\Delta g^k)^T Q_k \Delta g^k}$.

3. Near-optimality test.

4. $Q_0 = -\sum_{k=0}^{n-1} \lambda_k \dfrac{s^k (s^k)^T}{(\Delta g^k)^T s^k}$; $x^0(\text{new}) = x^n(\text{old})$; $\ell = \ell + 1$, go to 2.

In most unconstrained optimization problems the line searches will be rather time-consuming. Much research is therefore being undertaken to develop methods which avoid linear searches. For this the reader is again referred to Powell's survey papers (1971), (1974). Our objective here is only to outline a very simple way to avoid extensive linear searches which is moreover easily applied in linearly constrained problems as well. Let $\tilde{x}$ be the current solution with gradient $\tilde{g}$ and let $\tilde{s}$ be a usable direction, hence $\tilde{g}^T \tilde{s} > 0$. Make a reasonable guess $\lambda' > 0$ of the steplength and calculate $x' = \tilde{x} + \lambda' \tilde{s}$. Let $x = \tilde{x} + \alpha(x' - \tilde{x})$ indicate the one-dimensional maximum sought. Assuming linearity of the gradient (a valid assumption in case of a quadratic function) we have

$$0 = g(x)^T \tilde{s} = (1-\alpha)\, \tilde{g}^T \tilde{s} + \alpha (g')^T \tilde{s}, \text{ so that } \quad \alpha = \frac{-\tilde{g}^T \tilde{s}}{(g' - \tilde{g})^T \tilde{s}}.$$

In this way an approximation $\bar{x}$ to $\tilde{x}$ can be calculated, together with $\bar{g}$; the relation $(\bar{g} - \tilde{g})^T s = 0$ can then be added to the tableau. With this modification all methods described still have the quadratic termination property. The choice of λ' should be made in an intelligent way. If the metricized norm method is applied to a quadratic maximization problem formula (4.10) is independent of the choice of the λ's. As long as the s^k

are conjugate the expression holds. This indicates that even with an inaccurate choice of the λ's the direction generation procedure in this method is gradually improved which seems to be a favourable property of this method: it seems to be less crucially dependent on accurate linear searches (the same holds for the general methods 3d and 3e).

It has been proved earlier (Zoutendijk 1970a) that all methods for which $\sum \theta_k^2 = \infty$ with $\theta_k = (g^k)^T s^k / |g^k| \, |s^k|$ are convergent. Most of the methods mentioned in this chapter either satisfy this criterion immediately or satisfy it when a periodic restart is applied or they can be modified, so that the criterion is satisfied.

For some of the methods outlined superlinear convergence has been proved; for others this is still a conjecture.

5. Linearly Constrained Nonlinear Programming

The problem to be considered in this chapter is

$$\mathrm{Max}\,\{ f(x) \mid Ax \leqq b,\ x \geqq 0 \}\,. \tag{5.1}$$

Two important special problems, viz. linear programming and unconstrained optimization, have already been studied. They may be considered at the two pillars on which linearly constrained nonlinear programming is built. Any method for the latter problem will contain elements borrowed from linear programming and from unconstrained optimization. It cannot be expected that a method will be developed which is best for all problems. It can rather be expected that different methods will be used for

(1) problems of the LP type (large, structured matrices, few non-linearities),

(2) problems of the UO type (few constraints, highly nonlinear objective function).

As a first example of a method for the LP type problem we mention the linear approximation method suggested by Frank and Wolfe (1956). In its original form the method proceeds as follows (one step is described starting in $\tilde{x} \in R$);

1. $\bar{x} := \max\,\{ \nabla f(\tilde{x})^T x \mid Ax \leqq b,\ x \geqq 0,\ x \in K \}$;

2. $\lambda' := \max\,\{ f(\tilde{x} + \lambda(\bar{x} - \tilde{x})) \mid x = \tilde{x} + \lambda(\bar{x} - \tilde{x}) \in R \}$;

3. $\tilde{x}\,(new) = \tilde{x} + \lambda'(\bar{x} - \tilde{x})$.

Here additional requirements $x \in K$ have only been added to avoid infinite solutions. K should be such that the final maximum is in its interior. It is not difficult to give examples in which convergence of this method is extremely bad. This is usually so when the maximum is not in a vertex. An important exception to this bad convergence is formed by the class of functions satisfying

$$\forall x^1, x^2 \in R : \left(\nabla f(x^1)^T (x^2 - x^1) > 0 \implies \nabla f(x^2)^T (x^2 - x^1) \geqq 0 \right).$$

Examples are

- linear programming;
- problems with a linear fractional objective function;
- problems with a (pseudo) convex objective function.

In these cases an interior stop will never occur; each $\tilde{x}$ (with the possible exception of the starting point) will be a vertex, so that the method will be finite. In the case of a linear fractional objective function:

$$\operatorname{Max} \left\{ \frac{p_0 + p^T x}{q_0 + q^T x} \,\middle|\, Ax \leqq b, \, x \geqq 0 \right\},$$

where we assume $q_0 + q^T x > 0$ for all $x \in R$ (if $= 0$ an infinite solution would result, the sign assumption is no restriction), it might be worth mentioning that a simple change of variables will transform the problem into a linear programming problem, viz.

$x^0 = q_0 + q^T x, \; w_0 = \frac{1}{x_0}, \; w = \frac{x}{x_0}$ We obtain:

$$\operatorname{Max} \{ p_0 w_0 + p^T w \mid Aw - b w_0 \leqq 0, \, w \geqq 0, \, w_0 \geqq 0, \, q_0 w_0 + q^T w = 1 \}.$$

The solution $\hat{x}$ can then be calculated when $\hat{w}, \hat{w}_0$ have been obtained. In the general case the convergence of the linear approximation method (or the convex simplex method) may be improved by adding the conjugacy relation

$\{ \nabla f(x^k) - \nabla f(x^{k-1}) \}^T (x - x^k) = 0$ when there is an interior stop (unconstrained one-dimensional maximum). These relations have to be removed after a while since in general they will not contain the maximum of (5.1). There are several ways to do this, so that a number of variants of the method can be distinguished. The relaxation policies will be discussed later in this chapter.

The above mentioned direct methods are special cases of the more general methods of conjugate feasible directions which work along the following lines:

1. A feasible starting point should be selected.

2. A sequence of feasible points will be determined

   $x^{k+1} = x^k + \lambda_k s^k$, where

   a. s^k will be the solution of a direction problem in which any direction generator can be used;

   b. λ_k is the solution of the constrained linear search problem.

3. If 2b results in an interior stop, then the conjugacy relation $(\Delta g^k)^T s = 0$ is added to the direction problem;
   If 2b results in hitting a new hyperplane, then the relation $a_i^T s = 0$ or $s_j = 0$ (observe the equality instead of the inequality sign) is added to the tableau.

4. A relaxation policy is applied, so that requirements which have been added to improve convergence but which are too strong will be gradually omitted.

The relaxation policy can be established in several different ways, so that a number of variants can be distinguished. We only mention:

V1: a. omit all "conjugacy" relations when a new hyperplane is hit;

   b. when no progress can be made anymore consider the oldest conjugacy or anti-zigzagging relation in the tableau:
   - if it is a conjugacy relation, remove it;
   - if it is an anti-zigzagging requirement, relax it.

   (this to be repeated until progress can be made again).

V2: a. omit all "conjugacy" relations when a new hyperplane is hit;

   b. when no progress can be made anymore and the last n' directions $s^{k-n'}, s^{k-n'+1}, \ldots, s^{k-1}$ have led to an interior stop, then make one additional step in the direction $s^k = Hg^k$ with

   $$H = -\sum_{k=k-n'}^{k-1} \lambda_k \frac{s^k (s^k)^T}{(\Delta g^k)^T s^k}.$$

   After this start afresh

   with $x^0(\text{new}) = x^{k+1}(\text{old})$

   (hence all conjugacy and anti-zigzagging requirements will be omitted).

V3: a. omit all "conjugacy" relations when a new hyperplane is hit;

   b. apply method 3e of chapter 4 (alternately an acceleration and a moving tableau step) until a new hyperplane is hit.

V4: like V1 but without a. (hence a "conjugacy" relation is kept in the tableau until it becomes the oldest one; although there is little theoretical justification for this it has been successfully applied in some special methods, like the one suggested by Ritter (1973) which is essentially our general method with relaxation policy V4).

In all cases the quadratic termination property will hold, i.e. the method will be finite for a quadratic programming problem. The rationale behind a. in V1-V3 is that for a quadratic objective function the old conjugacy relations will no longer contain the constrained maximum if a new hyperplane has to be added to the set of active constraints; the old relations therefore do not make sense anymore. Relaxation policies V1 and V3 probably are the best ones to be applied in practice for the nearly linear problems. In that case it does not matter anymore after the initial steps (in number $\leq n$) which direction generator is being taken, so that the computationally simplest one, i.e. Ca, is to be preferred. For the initial steps another direction generator could be chosen. For problems with a highly nonlinear objective function variant V2 might also be considered. The special cases of linear programming and unconstrained optimization are included in this general method. Linear equality contraints in (5.1) can be easily coped with (always require $a_i^T s = 0$); upperbounds, $x_j \leq c_j$, can be dealt with in the same way as the non-negativity requirements unrestricted variables do not present any problems either.
The quadratic termination is a direct consequence of the following lemma: let A be m by n with full row rank m and let $s^0, s^1, \ldots, s^{n-m+1}$ be mutually conjugate directions with respect to the symmetric, positive semi-definite n by n matrix C, satisfying $As^i = 0$, then the solution of the problem

$$\text{Max}\ \{p^T x - \tfrac{1}{2} x^T C x \mid Ax = b\} \quad \text{can be written as} \quad \hat{x} = x^0 + \sum_{i=0}^{n-m} \lambda_i s^i$$

$$\text{with } Ax^0 = b,\ \lambda_i = \frac{(g^i)^T s^i}{(s^i)^T C s^i} \quad \text{and} \quad g^{i+1} = g^i - \lambda_i C s^i.$$

The proof of the lemma is simple. Due to this lemma and the way relations are added and removed from the tableau we shall obtain the maximum of a restricted problem after a finite number of steps, the restriction being that we are forced to be in certain hyperplanes. A continuation after relaxation entails that we will never return in this combination of hyperplanes. From this the finiteness of the quadratic programming method follows.
The acceleration steps in V2 and V3 do not make sense for a quadratic objective function (unless the perfect line search has been replaced by a less accurate steplength determination). For more general functions it may be an important improvement, however.

For general objective functions convergence properties have been studied by Zoutendijk (1960), Topkis and Veinnott (1967) and Pironneau and Polak (1972). The anti-zigzagging requirements are required to guarantee or improve convergence. Without them convergence cannot even be proved in the linear programming case. This is due to the fact that the direction generation procedure is quite general. If this is being done in a specific way, then an interesting problem arises for which class of functions convergence is guaranteed without anti-zigzagging precaution. Wolfe (1972) has given an example that even with direction generator Bb (gradient projection) convergence to a non-stationary point may take place. For a quadratic objective function it is not difficult to give an example that without anti-zigzagging precaution the method is no longer finite, although still convergent.

For the linearly constrained nonlinear programming problems of the UO type (highly nonlinear, few constraints, no matrix structure) it could also be considered to adapt a variable metric method to linear contraints. This has first been tried by Goldfarb (1969).
One of the ways to do this is the following:

1. Start with x^0 and calculate $Q_{A_1(x^0)}$, in which $A(x^0)$ is the set of active constraints in x^0, $A_1(x^0)$ the set of those determining the projection when applying generator Bb and $Q_{A_1(x^0)}$ the corresponding projection matrix. Let $H_0 = Q_{A_1(x^0)}$.
2. At step k a matrix H_k will be available. Determine

   a. $s^k = H_k g^k$,

   b. λ_k by constrained linear search

   c. x^{k+1} and g^{k+1},

   d1. if during the search an interior stop is made, then

$$H_{k+1} = H_k - \lambda_k \frac{s^k (s^k)^T}{(\Delta g^k)^T s^k} - \frac{H_k \Delta g^k (\Delta g^k)^T H_k}{(\Delta g^k)^T H_k \Delta g^k} ;$$

   d2. if during the search a new hyperplane is hit, say the plane $a_{i.}^T x = b_i$, then

$$H_{k+1} = H_k - \frac{H_k a_{i.} a_{i.}^T H_k}{a_{i.}^T H_k a_{i.}} .$$

This procedure should not be continued indefinitely since we might iterate in the wrong combination of hyperplanes. The best time to stop is after $n-\ell$ consequent iterates of type d1 (interior stops), where ℓ is the dimension of the active set of hyperplanes. If a plane has to be added to this set (case d2) the counting is started afresh. After a stop we continue with a complete restart, hence a new x^0 and new $Q_{A,(x^0)}$.

Any variable metric updating formula can be used instead of the one mentioned in 2d1. For a quadratic objective function the maximum will again be arrived at after a finite number of steps. The method outlined here has the advantage over Goldfarb's method that the danger of making many small steps in the wrong combination of hyperplanes is evaded.

The special method for unconstrained optimization with a metricized norm which has been described at the end of chapter 4 is not easily generalized to linearly constrained problems, at least not when there are inequalities involved. However, this method is easily adapted to equality constrained problems, hence problems of the type

$$\text{Max}\,\{f(x)\mid Ax=b\}\,. \tag{5.2}$$

The method is then (assuming full row rank for A, no real restriction):

1. x^0 such that $Ax^0=b$, $Q_0=I-A^T(AA^T)^{-1}A$.

2. For $k=0,1,\dots,n-m-1$:

   a. $s^k=Q_k g^k$,

   b. $\lambda_k := \max f(x^k+\lambda s^k)$,

   c. x^{k+1}, g^{k+1},

   d. $Q_{k+1}=Q_k-\dfrac{Q_k\,\Delta g^k(\Delta g^k)^T Q_k}{(\Delta g^k)^T Q_k\,\Delta g^k}$.

3. Near-optimality test.

4. $Q_0 = -\sum_{k=0}^{n-m-1}\lambda_k\dfrac{s^k(s^k)^T}{(\Delta g^k)^T s^k}$; $x^0(\text{new})=x^{n-m}(\text{old})$; go to 2.

The only differences with the unconstrained method are another starting matrix at the very beginning and a lower dimensionality of the linear space we are in.

6. Extensions to general nonlinear programming

In this chapter some methods will be reviewed for the general nonlinear programming problem:

$$Max \{ f(x) \mid f_i(x) \leqq b_i \ , i \in I_1 \ ; f_i(x) = b_i \ , i \in I_2 \ ; x \in L \} , \qquad (6.1)$$

in which f and the f_i are nonlinear but continuous and differentiable functions of $x \in E^n$, while linear constraints, if any, are represented separately by $x \in L$, a convex polyhedron. Many methods have been proposed. They can be distinguished into four classes:

1. Direct methods.
2. Barrier function and penalty function methods.
3. Primal-dual methods.
4. Special methods for special problems.

Methods belonging to the first class are direct extensions of methods, originally developed for linearly constrained problems. There are three possibilities:

- At each $\bar{x} \in R$ a usable feasible direction is determined in the way described in chapter 2, i.e. by requiring $\nabla f_i(\bar{x})^T s < 0$ for $i \in I_1(\bar{x})$; together with a steplength procedure and a number of additional requirements to guarantee convergence we then obtain a general method. Nonlinear equalities cannot easily be dealt with, hence we assume $I_2 = \emptyset$. This approach has been investigated by Zoutendijk (1960).

- Given an interior point with respect to the nonlinear constraints ($f_i(\bar{x}) < b_i \ , i \in I_1$; again $I_2 = \emptyset$ should hold), a linearized problem is solved giving a solution $\bar{y}$. If the region is convex $\bar{y}$ will not be feasible. Constrained linear search on the line $x = \bar{x} + \lambda(\bar{y} - \bar{x})$ will either give a new interior point, or a boundary point $\bar{z}$. In the latter case the tangent plane in $\bar{z}$ to the bounding constraint is added to the set of linear and linearized constraints and a new interior point is chosen somewhere on the line connecting $\bar{x}$ and $\bar{z}$.
Repetition of this procedure gives a method that has been suggested by Zoutendijk (1966).
It can be extended to non-convex problems.

- At $\bar{x} \in R$ a direction is determined which is usable and feasible with

respect to the linearizations of the constraints active in $\tilde{x}$; in other terms the direction is allowed to lie in one or more tangent planes, so that in the convex case we shall leave R by making a small step. After this a new and better point in R is determined by solving a set of nonlinear equations. This approach is called hemstitching. With Cb as direction generator (reduced gradient) it has been successfully implemented in a computer program and is being called the generalized reduced gradient method (GRG, see Abadie and Carpenter, 1969).
In principle the same approach can be followed with nonlinear equalities. These need not be excluded therefore.

These direct methods will probably find their major application for large, structural, nearly linear problems.
For smaller, highly nonlinear problems the methods of class 2 may be more appropriate. They can be distinguished into:

- Barrier function methods or interior point methods in which a sequence of unconstrained or better linearly constrained problems is to be solved with an objective function constructed in such a way that once in the feasible region we will not be able to leave it. Since we need an interior point we must require $I_2 = \phi$. These methods have been thoroughly investigated by Fiacco and McCormick (1968) and Huard (1967).

- Penalty function methods or exterior point methods in which a sequence of unconstrained or better linearly constrained problems is to be solved with an objective function having a penalty term for each constraint which is violated by the current solution. By gradually increasing the penalties feasibility is being forced but this is being done in such a way that optimality is obtained at the same time. An example is the method suggested by Zangwill (1967). Nonlinear equalities do not give any additional problem.

- Mixed methods, say a barrier function method with respect to the nonlinear inequalities and a penalty function method with respect to the nonlinear equalities. See Fiacco and McCormick (1968).

Special methods have for instance been devised for convex programming problems, for separable programming problems and for geometric programming problems. They will not be discussed here. We will rather concentrate in the remaining part of this paper on what could be called primal-dual methods since in this class there might be scope for further developments. To simplify the discussion as well as the writing we will assume that our problem is:

$$\text{Max}\,\{f(x)\mid f_i(x) \leq b_i\,,\, i=1,\ldots,m\}. \tag{6.2}$$

The Lagrangean function will be defined by

$$L(x,u) = f(x) - \sum_{i=1}^{m} u_i \{f_i(x) - b_i\}, \quad u_i \geq 0. \tag{6.3}$$

We can then distinguish

- the primal function $\varphi_1(x) = \min_u \{L(x,u) \mid u \geq 0\}$

  and

- the dual function $\varphi_2(u) = \max_x L(x,u)$,

so that we will have:

$$\forall x \; \forall u \geq 0 : \; \varphi_1(x) \leq L(x,u) \leq \varphi_2(u).$$

There is a saddlepoint $\hat{x}, \hat{u} \geq 0$, hence $\varphi_1(\hat{x}) = \varphi_2(\hat{u})$ if $\hat{x} \in R$, $\hat{u} \geq 0$ satisfy the optimality conditions, i.e. $\nabla_x L(\hat{x}, \hat{u}) = 0$ and if $-f$ and f_i are convex functions. This is well-known and will not be proved here. In that case it makes sense to introduce

- the primal problem $\max_x \varphi_1(x)$

  and

- the dual problem $\min_u \{\varphi_2(u) \mid u \geq 0\}$

since both problems will have the same optimal value, while moreover

$$\varphi_1(x) = -\infty \quad \text{for } x \notin R,$$

$$= f(x) \quad \text{for } x \in R,$$

so that the maximization of $\varphi_1(x)$ is equivalent to solving the original nonlinear programming problem (6.2), while as a consequence of the just mentioned duality the dual problem $\min\{\varphi_2(u) \mid u \geq 0\}$ could be solved instead. This dual problem can be reformulated as

$$\text{Min} \left\{ f(x) - \sum_{i=1}^{m} u_i (f_i(x) - b_i) \;\middle|\; \nabla f(x) - \sum u_i \nabla f_i(x) = 0, \; u \geq 0 \right\},$$

a formulation which is due to Wolfe (1961). Since in general the original variables x_j are still present in the dual problem it is not easier to solve the dual instead of the primal. An important exception is in geometric programming, where it can be shown that the dual problem reduces to a linearly constrained nonlinear programming problem, so that one of the methods

of chapter 5 can be applied. See for instance Zangwill (1969), pp. 9-11 and 68-76.
For non-convex functions Roode (1968) has tried to develop a theory for generalized Lagrangean functions. These are functions $\psi(x,u)$ satisfying the following requirements:

(1) $\psi(x,u) \geqq f(x)$ for all $x \in R$ and $u \in D$,

(2) $\psi(x,0) = f(x)$,

(3) if $x \notin R$, then for any $\mu > 0$ there is a $u(x)$ such that
$\psi(x, u(x)) < -\mu$
Here D is a convex subset of E^p for some $p \geqq 1$, $0 \in D$.
The Lagrangean function $L(x,u)$ satisfies these conditions. In that case
$D = \{u \in E^m \mid u \geqq 0\}$.
Another example is

$$g(x,\rho) = f(x) - \rho \sum_{i \in I}\{y_i^2 \mid y_i < 0\}, \rho \geqq 0,$$

where $y_i = b_i - f_i(x)$. In this case $p = 1$, $D = \{\rho \mid \rho \geqq 0\}$. This is the penalty function used by Zangwill (1967).
Again we can introduce
(1) a primal problem, equivalent to (6.2):

$$\text{Max}\ \{\psi(x,u) \mid \psi(x,u) = \inf_\eta \{\psi(x,\eta) \mid \eta \in D\},$$

(2) a dual problem

$$\text{Min}\ \{\psi(x,u) \mid u \in D,\ \psi(x,u) = \sup_\xi \psi(\xi,u)\}.$$

These problems will have the same solution $(\hat{x},\hat{u})$, so that a duality theorem holds, if

a. ψ is upper semi-continuous and strictly quasi-concave in x for each $u \in D$;

b. ψ is lower semi-continuous and strictly quasi-convex on D for each x;

c. if $(\hat{x},\hat{u})$ is the solution of the dual problem, then $\psi(x,\hat{u})$ is monotonic strictly quasi-concave in some neighbourhood of $\hat{x}$.

The proof will be found in Roode's book.
If some of the nonlinear inequalities are equalities, then the same theory will apply. In the Lagrangean function the corresponding u_i will then be unrestricted. If there are also linear constraints, to be indicated by $x \in L$,

then $x \in L$ should be added in all maximizations, so that the unconstrained maximizations in x become linearly constrained maximizations. This separate treatment of linear constraints has computational advantages compared to including these contraints in the set of nonlinear constraints.

As a relatively simple first example we shall consider the problem

$$\operatorname{Max} \{ f(x) \mid f_i(x) = 0, \; i = 1, \ldots, m \}, \tag{6.3}$$

in which we assume the gradient vectors $\nabla f_i(x)$ to be linearly independent. As auxiliary function (generalized Lagrangean) we take:

$$\varphi(x,u) = f(x) - \sum_{i=1}^{m} u_i f_i(x) - \tfrac{1}{2}\beta \sum_{i=1}^{m} f_i(x)^2, \; \beta > 0, \tag{6.4}$$

in which β is fixed. It can be proved that for β fixed but sufficiently large the duality still holds, independent of β . Instead of solving the primal problem $\max_x \min_u \varphi(x,u)$ we shall try to solve the dual problem $\min_u \max_x \varphi(x,u)$ which as a consequence of the duality will give the same solution. Let

$$\psi(u) = \max_x \varphi(x,u) = \varphi(x(u),u), \tag{6.5}$$

then

$$\frac{\partial \psi}{\partial u_i} = -f_i(x(u)), \tag{6.6}$$

so that the gradient vector of ψ is known and an unconstrained method can be applied to solve the dual problem

$$\min \psi(u). \tag{6.7}$$

It is not difficult to show that for β sufficiently large the function $\psi(u)$ is strictly convex.

Since each function evaluation entails the solution of the unconstrained problem $\max_x \varphi(x,u)$ it is essential to minimize work on the steplength calculation.

The simple way to avoid extensive linear searches which has been explained at the end of chapter 4 might be used here. Methods like this one which use a primal maximization to evaluate the dual function could be called primal-dual methods. They have been extensively studied by Buys (1972). For the special problem (6.3) it has been tried to solve the dual problem (6.7) by using a gradient method with predetermined steplength. Since from (6.4) it follows that

$$\nabla_x \varphi(x, \tilde{u}) = \nabla f(x) - \sum_i (\hat{u}_i + \beta f_i(x)) \nabla f_i(x)$$

and since the Lagrange multipliers must satisfy the relations:

$$\nabla f(x) - \sum_i u_i \nabla f_i(x) = 0$$

it is tempting to consider the following method:

1. Choose u^0 arbitrarily.
2. At step k :

   a. $x^k := \max_x \varphi(x, u^k)$,

   b. $u_i^{k+1} = u_i^k + \beta f_i(x^k)$.

This has been suggested by Powell (1969) and independently by Hestenes (1969). Powell also showed that under certain conditions the method would indeed converge. Since the dual problem will be solved by applying a steepest decent method with predetermined steplength convergence will never be very good. As we have seen more advanced unconstrained minimization techniques can be applied.

As an example (see Buys (1972), pp. 43-44) we shall consider the problem

$$\max \{x_1 x_2 \mid x_1 + x_2 = 1\}.$$

$$\varphi(x, u) = x_1 x_2 - u(x_1 + x_2 - 1) - \tfrac{1}{2}\beta(x_1 + x_2 - 1)^2,$$

$$\nabla_x \varphi(x, u) = 0 \implies x_1(u) = x_2(u) = \frac{\beta - u}{2\beta - 1} \qquad (\beta > \tfrac{1}{2} \text{ required})$$

$$\psi(u) = \frac{1}{2\beta - 1}(u^2 - u + \tfrac{1}{2}\beta), \quad \hat{u} = \tfrac{1}{2} \qquad \text{, hence } \hat{x}_1 = \hat{x}_2 = \tfrac{1}{2}.$$

Powell's method does not converge in this example for $\beta = 1$ but it does for $\beta = 2$.

We then obtain the sequence (starting with $u = 0$):

$\beta = 2$	u	x_1	x_2	$x_1 + x_2 - 1$
	0	$\frac{2}{3}$	$\frac{2}{3}$	$\frac{1}{3}$
	$\frac{2}{3}$	$\frac{4}{9}$	$\frac{4}{9}$	$-\frac{1}{9}$
	$\frac{4}{9}$	$\frac{14}{27}$	$\frac{14}{27}$	$\frac{1}{27}$

....

For an extension to inequality constraints we consider the problem:

$$\text{Max}\,\{f(x) \mid f_i(x) = 0,\ i \in I_1\ ;\ f_i(x) \leqq 0,\ i \in I_2\}. \tag{6.8}$$

Falk (1967) has studied the possibility of applying the same approach as described above with non-negativity requirements for the $u_i,\ i \in I_2$, so that the dual problem becomes

$$\text{Min}\,\{\psi(u) \mid u_i \geqq 0,\ i \in I_2\}, \tag{6.9}$$

a linearly constrained problem that can be solved by one of the methods of chapter 5. This will not further be explained in this article. Instead we shall follow a slightly different approach using another function $\varphi(x, u)$, originally introduced by Gould (1969):

$$\varphi(x, u) = f(x) - \sum_{i \in I} k(f_i(x), u_i) \tag{6.10}$$

with $I = I_1 + I_2$ and

$$k(f_i(x), u_i) = u_i f_i(x) + \tfrac{1}{2}\beta f_i(x)^2 \quad \text{if } i \in I_1 \text{ or}$$

$$\text{if } i \in I_2 \text{ and } u_i + \beta f_i(x) \geqq 0\,;$$

$$= -\frac{u_i^2}{2\beta} \quad \text{if } i \in I_2 \text{ and } u_i + \beta f_i(x) \leqq 0\,;$$

$$(\beta > 0). \tag{6.11}$$

For a discussion of this function the reader is referred to the papers of Gould and Buys.
Let $\hat{x}$ be a local maximum of the nonlinear programming problem, let

$I_2(\hat{x}) = \{i \in I_2 \mid f_i(\hat{x}) = 0\}$, let the vectors $\nabla f_i(\hat{x}),\ i \in I_1 + I_2(\hat{x})$ be linearly independent, let $\hat{u}_i$ be the associated dual variables (Lagrange multipliers), then

$$\hat{u}_i + \beta f_i(\hat{x}) \geqq 0 \qquad \text{if } i \in I_2(\hat{x}),$$

$$\hat{u}_i + \beta f_i(\hat{x}) < 0 \qquad \text{if } i \notin I_2(\hat{x}).$$

This follows from $f_i(\hat{x}) = 0$, $\hat{u}_i \geqq 0$ if $i \in I_2(\hat{x})$ and $f_i(\hat{x}) < 0$, $\hat{u}_i = 0$ if $i \notin I_2(\hat{x})$.

Hence

$$\varphi(\hat{x}, \hat{u}) = f(\hat{x}).$$

Moreover

$$\nabla_x \varphi(\hat{x}, \hat{u}) = 0.$$

In the neighbourhood of $\hat{u}$ we can again introduce the dual function $\psi(u) = \max_x \varphi(x, u)$ so that the dual problem $\min \psi(u)$ can be solved. For $i \in I_2$ there are no non-negativity requirements on the u_i. Again it can be proved that for $\beta > 0$ sufficiently large $\psi(u)$ is convex. Moreover

$$\frac{\partial \psi(u)}{\partial u_i} = -f_i(x(u)) \qquad \text{if } i \in I_1 \text{ or}$$

$$\text{if } i \in I_2 \text{ and } u_i + \beta f_i(x) \geqq 0;$$

$$= \frac{u_i}{\beta} \qquad \text{if } i \in I_2 \text{ and } u_i + \beta f_i(x) \leqq 0.$$

An unconstrained method can thus be applied to solve the dual problem. Again each function evaluation $\psi(\tilde{u})$ entails the solution of the unconstrained problem in x : $\max \varphi(x, \tilde{u})$.

Buys gives the following general algorithm:

1. Select u^0 , so that $\psi(u^0)$ is defined ; $k=0$.
2. At step k :
   a. compute $x^k := \max \varphi(x, u^k)$;
   b. if $|\nabla \psi(u^k)| \leq \varepsilon$ stop; else
   choose a direction s^k such that $\{\nabla \psi(u^k)\}^T s^k > 0$, a steplength λ_k and
   $u^{k+1} = u^k - \lambda_k s^k$;
   c. set $k = k+1$ and repeat.

Important questions are:
(1) Is it necessary to solve the subproblems $\max \varphi(x, \tilde{u})$ completely?
(2) How should the direction s^k be chosen?
(3) How should the steplength λ_k be chosen?

Buys also gives a second algorithm in which the subproblems $\max\limits_x \varphi(x, \tilde{u})$ are not completely solved. First a decreasing sequence $\varepsilon_k > 0$, converging to 0 is selected.
The method then proceeds along the following lines (we only mention step k):
1. Use an unconstrained maximization method to find a point x^k such that
   $|\nabla_x \varphi(x^k, u^k)| \leq \varepsilon_k$.
2. $u^{k+1} = u^k - \lambda_k s^k$.

For the determination of λ_k and s^k he makes several suggestions like
- a gradient step with predetermined steplength like in Powell's method for the case $I_2 = \emptyset$,
- the use of a rank one variable metric method, so that no linear search is required (for a discussion of this method, see Powell (1971)).
- by direct determination of u^{k+1} using a least squares approximation to the overdetermined system $\nabla f(x^k) - \sum u_i \nabla f_i(x^k) = 0$

Numerical results of these methods have been encouraging, at least for small problems. Further investigations and experiments with primal-dual methods seem to be justified. Related work has been reported by Fletcher (1970) and Miele et al.(1971). As a final remark we may state that linear constraints in the original problem (6.8) can be treated separately by considering linearly constrained problems of the type max $\{\varphi(x, \tilde{u}) \mid x \in L\}$. This might have computational advantages. Quite recently Robinson (1973) has reported another primal-dual method which looks quite promising.

REFERENCES

: J. Abadie and J. Carpentier 1969 Generalization of the Wolfe reduced gradient method to the case of nonlinear constraints, pp. 37-47 of R. Fletcher (ed.), Optimization, Academic Press.

: J. D. Buys 1972 Dual algorithms for constrained optimization problems, Thesis, University of Leiden.

: J. E. Falk 1967 Lagrange multipliers and nonlinear programming, J. Math.Anal. Appl. 19, 141-159.

: A.V. Fiacco and G.P.McCormick 1968 Nonlinear programming: sequential unconstrained minimization techniques, Wiley.

: R.Fletcher and M.J.D. Powell 1963 A rapidly converging descent method for minimization, Computer Journal 6, 163-168.

: R.Fletcher and C.M. Reeves 1964 Function Minimization by conjugate gradients, Computer Journal 7, 149-154.

: R. Fletcher 1970 A class of methods for nonlinear programming with termination and convergence properties, pp. 157-175 of J. Abadie (ed.), Integer and nonlinear programming, North-Holland.

: M. Frank and P. Wolfe 1956 An algorithm for quadratic programming, Naval Research Logistics Quarterly 3, 95-110.

: D. Goldfarb 1969 Extension of Davidon's variable metric method to maximization under linear inequality and equality constraints,

: D. Goldfarb 1969 SIAM J.Appl.Math. 17, 739-764.

: F. J. Gould 1969 Extensions of Lagrange multipliers in nonlinear programming, SIAM J. Appl.Math. 17, 1280-1297.

: M. R. Hestenes 1969 Multiplier and gradient methods, J. Optimization Theory and Appl. 4, 303-320.

: P. Huard 1967 Resolution of mathematical programming with nonlinear constraints by the method of centres, pp. 209-219 of J. Abadie (ed.), Nonlinear Programming, North-Holland.

: J. Kowalik and M.R.Osborne 1969 Methods for unconstrained optimization problems, Elsevier.

: A.Miele, E.E. Cragg, R.R. Iyer and A.V.Levy 1971 Use of the augmented penalty function in mathematical programming problems, J.Optimization Theory and Appl. 8, 115-153.

: W. Murray (ed.) 1972 Numerical methods for unconstrained optimization, Academic Press.

: O.Pironneau and E.Polak 1973 Rate of convergence of a class of methods of feasible directions, SIAM J. on Numerical Analysis,

: M.J.D. Powell 1969 A method for nonlinear constraints in minimization problems, pp. 283-298 of R. Fletcher (ed.), Optimization, Academic Press.

1971 Recent advances in unconstrained optimization, Math.Programming 1, 26-57.

1974 Unconstrained Minimization and Extensions for Constraints, this volume, pp.

: K. Ritter 1973 A superlinearly convergent method for minimization problems with linear inequality constraints, Math.Progr. 4, 44-71.

: S. M. Robinson 1972 A quadratically-convergent algorithm for general nonlinear programming problems, Math.Progr. 3, 145-156.

: J. D. Roode 1968 Generalized Lagrangean Functions, Thesis, University of Leiden.

: J. B. Rosen 1960 The gradient projection method for nonlinear programming, part I: linear constraints, SIAM J.Appl.Math. 8, 181-217.

: D. Topkis and A. Veinnott 1967 On the convergence of some feasible direction algorithms for nonlinear programming, J.Soc.Industr. and Appl.Math. Control 5, 268-279.

: P. Wolfe 1961 A duality theorem for nonlinear programming, Quart. of Appl.Math. 19, 239-244.

1963 Methods of nonlinear programming, pp. 67-86 of: R.L. Graves and P. Wolfe (eds,), Recent advances in mathematical programming, McGraw-Hill.

1972 On the convergence of gradient methods under constraint, IBM J. of Res. and Dev. 16, 407-411.

: W.I. Zangwill 1967 Nonlinear programming via penalty functions, Man.Science 13, 344-358.

1969 Nonlinear programming, Prentice-Hall.

: G. Zoutendijk 1960 Methods of Feasible Directions, Elsevier.

: G. Zoutendijk

1966 Nonlinear programming: a numerical survey, J. Soc. Industr. and Appl. Math. Control 4, 194-210.

1970a Nonlinear programming, computational methods, pp. 37-85 in J. Abadie (ed.), Nonlinear and integer programming, North-Holland.

1970b Some algorithms based on the principle of feasible directions, pp 93-122 of J.B. Rosen, O.L. Mangasarian and K. Ritter (eds.), Nonlinear Programming, Academic Press.

Mathematical Programming in Theory and Practice,
P.L. Hammer and G. Zoutendijk, (Eds.)

A simplicial method for nonlinear programming

by A. S. Gonçalves, University of Coimbra, Portugal

The problems of the real world fitted to be approached by a nonlinear programming model with linear constraints, used to be solved by using linear Programming techniques. However, as Faure and Huard say in [4], this is not a natural way of solving such problems and the appearance of nonlinear methods of solution, such as the reduced gradient method by Wolfe [11] and the gradient projection method by Rosen [9] has been very welcomed. They proved to be computationally more efficient than the linear techniques, since they work faster in providing good solutions, although these algorithms are not finite and even their theoretical convergence to a solution is only proved under further hypothesis, Faure and Huard [4].

In this paper, it is constructively demonstrated that under an assumption made about the objective function (see the existence assumption in section 2.2) the nonlinear problems of that kind have an optimal solution. (In [5] an attempt is made to generalize this algorithm to nonlinear programs with nonlinear constraints)

The algorithm established here is finite and in the case of a linear objective function it reduces to the simplex method.

When the nonlinear program is quadratic the algorithm leads to a version of Beale's method for Quadratic Programming [1] avoiding the "free variables". It is proved in [6] that this version is a profitable computational improvement of Beale's algorithm.

1. Preliminaries

Consider the nonlinear programming problem:

$$(P) \quad \begin{aligned} &\text{Maximize} \quad f(\mathbf{x}) \\ &\text{subject to} \quad \mathbf{Ax} = b \qquad (1) \\ &\qquad \mathbf{x} \geq 0 \end{aligned}$$

where the function $f : R^n \to R$ is concave with continuous derivatives, A is an $m \times n$ matrix and b is an m-vector.

Let $\bar{x}$ be a feasible solution in (P), i.e., let $\bar{x}$ satisfy all the constraints in (P).

Define the sets of indices, $N = \{1,2,\dots,n\}$ and

$$J = \{j \mid \bar{x}_j > 0\},$$

and let $I \subset N$ be such that the square submatrix A^I, which denotes the restriction of matrix A to its columns, the indices of which belong to I, is not singular. We say that A^I forms a basis, which we refer as the basis I.

We distinguish among the components of $\bar{x}$ those, the indices of which belong to I, forming the subvector $\bar{x}_I$ and the remaining ones forming the subvector $\bar{x}_{\bar{I}}$, where $\bar{I} = N - I$.

A variable x_i is called a basic variable if $i \in I$, or a nonbasic variable if $i \in \bar{I}$.

A feasible solution $\bar{x}$ may be of the following type:

(a) A nondegenerate basic solution if $I = J$

(b) A nondegenerate nonbasic solution if $I \subset J$

(c) A degenerate basic solution if $J \subset I$

(d) A degenerate nonbasic solution if $I - J \neq \emptyset$ and $J \not\subset I$.

According to the above notation, system (1) may be written in the form

$$(A^I, A^{\bar{I}}) . \begin{matrix} x_I \\ x_{\bar{I}} \end{matrix} = b$$

or, as A^I is nonsingular

$$x_I = t - T^{\bar{I}} . x_{\bar{I}}$$

where

$$\begin{aligned} T &= (A^I)^{-1} . A \\ t &= (A^I)^{-1} . b \end{aligned} \qquad (2)$$

Let us form for problem (P) the following Lagrangean expression:

$$L(u,x) = f(x) - u^T.(Ax - b) \tag{3}$$

where u is a column vector of m Lagrangean multipliers. This expression will always have the same value as the objective function, $f(x)$, if we have

$$u^T.(Ax - b) = 0.$$

This condition is satisfied if $Ax = b$.

Consider the effect on the Lagrangean of changing a variable x_j. This is done by differentiating $L(u,x)$ with respect to x_j:

$$(\partial L(u,x)/\partial x_j) = (\partial f/\partial x_j) - u^T.A^j = d_j, \quad \text{for } j\epsilon N \tag{4}$$

The last equality serves as definition of the column vector d of n elements which are the first derivatives of $L(u,x)$ with respect to the x variables.

The necessary and sufficient conditions for $\bar{x}$ to be an optimal solution of problem (P) are given by Kuhn and Tucker [8] as the following

$$\bar{x} \text{ is feasible in (P)}$$

$$\bar{d}_j = (\partial f/\partial \bar{x}_j) - u^T.A^j \leq 0 \quad \text{for all } j$$

$$\bar{x}^T.\bar{d} = 0$$

where u is the vector of Lagrangean multipliers and $(\partial f/\bar{x}_j)$ is the value of the partial derivative $(\partial f/\partial x_j)$ for $x = \bar{x}$.

2. Description of the method

2.1. Starting procedure

To start the procedure we must know a feasible solution $\bar{x}$ of (P) satisfying the following:

Complementarity Condition - there exists a column vector u, such that:

$$(\partial f/\partial \bar{x}_j) - u^T.A^j = 0 \quad \text{for all } j\epsilon K, \tag{5}$$

where $K = I \cup J$, with $J = \{j | \bar{x}_j > 0\}$ and I is a basic set.
(This condition is not required to start with in Wolfe's procedure of the reduced gradient method and in Rosen's gradient projection method.)
In particular, we have

$$(\partial f/\partial \bar{x}_j) - u^T.A^j = 0 \quad \text{for } j \in I.$$

and the solution of this linear system is

$$u^T = (df/d\bar{x}_I).(A^I)^{-1}, \tag{6}$$

where $(df/d\bar{x}_I)$ denotes the row vector the elements of which are

$$(\partial f/\partial \bar{x}_j), \quad \text{for } j \in I.$$

Substituting (6) for u in (5), and because of (2), we have for a given I and $\bar{x}$.

$$\bar{d}_j(I,\bar{x}) = (\partial f/\partial \bar{x}_j) - (df/d\bar{x}_I).\ T^j, \quad \text{for } j \in N \tag{7}$$

The vector d depends on the basis considered and is called the reduced gradient.

When a starting solution is not known, we can use some method, as the phase I of the simplex method (cf. [3] p. 100) to find a basic feasible solution of system (1), which is obviously a good one to start with.

A sequence of steps, in the algorithm below, leading to the construction of a new feasible solution satisfying the complementarity condition, will be called an iteration of the algorithm.

The basic idea of the procedure, which enables us to prove the finiteness of the algorithm, consists in constructing at each iteration always an improved feasible solution satisfying the complementarity condition.

2.2. Improvement of a feasible solution

If the feasible solution $\bar{x}$ satisfies the complementarity condition and $\bar{d}_j \leq 0$ for all j, then the Kuhn-Tucker conditions are satisfied. Admitting the concavity hypothesis for the objective function, we can say that $\bar{x}$ is an optimal solution for (P).
Otherwise, for some j, $\bar{d}_j > 0$, say $\bar{d}_s > 0$, and we can improve the solution $\bar{x}$. To do that a similar technique as in Wolfe's procedure is

used, but only one nonbasic variable, x_s, is allowed to vary independently at a time, keeping the quantities $d_j = 0$ for $j \in K$ and $j \neq s$.

Thus we shall look for a positive value of x_s which produces a better solution satisfying the following conditions:

$$
\begin{aligned}
&x_i \geq 0 && \text{for all } i \in N \\
&x_i = 0 && \text{for } i \neq s \text{ and } i \notin K \\
&d_j(I,x) = 0 && \text{for } j \in (K - I) \\
&x_I = t - T^{\bar{I}} . x_{\bar{I}} && \qquad (8)
\end{aligned}
$$

where t and T are defined by (2).

Now we assume the following

Existence Assumption - The objective function in problem (P) is such that the system of equations

$$d_j(I,x_1,\ldots,x_n) = 0 \quad \text{for } j \in K - I$$

defines with (8) a system of continuous functions

$$x_j = g_j(x_{\bar{K}}) \quad \text{for } j \in K \qquad (9)$$

for all feasible x and basis I (in [5] an attempt is made to weaken this assumption).

For example in the case of a quadratic objective, the existence assumption is satisfied.

According to that assumption, the components of the new solution vector can be written explicitly as functions of the variable x_s, where we make $x_s = \bar{x}_s + \theta$, with a non-negative parameter. We then have the vector $x(\theta)$ with the following components:

$$x_s = \bar{x}_s + \Theta$$

$$x_j = g_j(\Theta) \qquad \text{for } j\in K \tag{10}$$

$$x_i = 0 \qquad \text{for } i\notin K \text{ and } i \neq s$$

where expressions (10) are obtained from (9) putting $x_s = \bar{x}_s + \Theta$ and $x_i = 0$ for $i\notin K$ and $i \neq s$.

$x(\Theta)$ is a feasible solution in (P) for any value of Θ for which $x_i \geq 0$ for all $i\in N$, i.e., for $0 \leq \Theta \leq \Theta_M$ where

$$\Theta_M = \min_{j\in K} \{\Theta_j\} \tag{11}$$

and Θ_j is the smallest positive root of $g_j(\Theta)$. If,for some $j,g_j(\Theta)$ has no positive root, we put $\Theta_j = +\infty$.
Since,for all $j,g_j(\Theta)$ are continuous functions, it follows that $\Theta_M = 0$ if and only if,for some $j\in K$, $\bar{x}_j = 0$ and the function $g_j(\Theta)$ is decreasing at $\Theta = 0$. Thus, under the nondegeneracy assumption, Θ_M never vanishes.

Now it is essential to know how the objective function varies when $0 < \Theta \leq \Theta_M$.

<u>Theorem 1</u>: If at the feasible solution $\bar{x}$, the quantity d_s, as defined by (7), is positive, then, under the nondegeneracy assumption, there exist values $x_s > 0$ such that $f(x) > f(\bar{x})$.

(We refer to the appendix for the proofs of theorems).

Let us define the composite function

$$F(\Theta) = f[x(\Theta),$$

with $x(\Theta)$ as defined above. The value of x_s yielding the maximum increase of $f(x)$ is that corresponding to the value of maximizing $F(\Theta)$, i.e., the value of Θ_s such that

$$F(\Theta_s) = \max \{F(\Theta), \quad \text{for} \quad 0 < \Theta \leq \Theta_M\}, \tag{12}$$

which may be obtained by any classical procedure to compute the maximum of a function of one variable.

Theorem 2: If $d_s(I,\bar{x}) > 0$, and we improve the feasible solution by the above technique, then at $\Theta = \Theta_s$ we have $d(I,x(\Theta_s)) \geq 0$.

From the above theorems and under the nondegeneracy assumption it follows that if, for the feasible solution $\bar{x}$, the quantity $d_s(I,\bar{x}) > 0$, then $\bar{x}$ can always be improved, increasing x_s until either some x_j for $j\epsilon K$ decreases to zero, at $\Theta = \Theta_M$ or $F(\Theta)$ attains its absolute maximum, at $\Theta = \Theta_s$.
In any case we have $d_s(I,x(\Theta)) \geq 0$. However, since $F(\Theta)$ is not necessarily concave, this does not mean that while x_s is increased from $\bar{x}_s$ to $\bar{x}_s + \Theta_s$, the quantity $d_s(I,x(\Theta))$ is neither negative nor zero, for some intermediate feasible solution.

Actually this could not happen if instead of the existence assumption, the following strong condition used by Cottle [2] was imposed on the objective function: the Jacobian matrix of the mapping $d_{\bar{I}}$ of components $d_j(x_{\bar{I}})$ for $j\epsilon\bar{I}$ is positively bounded, where the expressions for $d_j(x_{\bar{I}})$ are obtained from the expressions for d_j given bij (7), substituting the basic variables x_I by $x_I = t - T^{\bar{I}}.x_{\bar{I}}$.

The definition of positive boundedness implies that for all feasible solutions the values of the partial derivative $(\partial d_s/\partial x_s)$ lies between two positive quantities, and therefore when x_s is increased from $\bar{x}_s$ to $\bar{x}_s + \Theta_s$, d_s is always increasing,in this case.

2.3. Renewal of the improvement procedure

Once Θ_s is computed the following cases may happen:

(a) $\Theta_s = \infty$, then problem (P) has no finite solution and the procedure terminates

(b) $\Theta_s < \infty$, in this case the following possibilities occur:

(b_1) $\Theta_s < \Theta_M$, then no variable x_j for $j \in K$ vanishes, and the improvement procedure can be restarted in the same way, taking as starting solution the vector $x(\Theta_s)$. The basic set is not changed.

(b_2) $\Theta_s = \Theta_M$, in this case some x_j with $j \in K$ vanishes in the solution $x(\Theta_s)$ and we must consider whether x_j is a basic or a nonbasic variable.

If the vanishing variable is nonbasic, it is excluded from the set K and the improvement procedure is restarted in the same way as in the previous case.

If the vanishing variable is basic, then the improvement procedure cannot be restarted in the same way, because this might lead to a vanishing Θ_s. To continue the procedure, the basis must be changed. This can be done in several ways according to the nature of the solution $\bar{x}$. In the following section a procedure will be described, defining a sequence of steps that leads, through one or more changes of basis, to a feasible solution satisfying the complementarity condition, which will be used to restart the improvement procedure at the next iteration. For that purpose, the intermediate solutions obtained will be such that either $d_j = 0$ for all $j \in (K - I)$ or only one, $d_s > 0$, for some $s \in (K - I)$.

2.4. Criteria to change the basis

Theorem 3: If in a solution $\bar{x}$ of problem (P) we interchange the basic variable x_r and the nonbasic x_s, the new reduced gradient vector, $d^*(I^*,\bar{x})$, relative to the new basis I^*, is expressed as a function of the old one, $d(I,\bar{x})$, by

$$d^*(I^*,\bar{x}) = \bar{d} - (\bar{d}_s / T_r^s).T_r ; \tag{13}$$

where $I^* = I - \{r\} + \{s\}$, I is the old basis, and T is the matrix defined by (2) when the basis is I.

Now assume that x_s is the nonbasic variable in $\bar{x}$ to be altered, and x_r is a basic variable vanishing in the solution $x(\Theta_s)$ computed by the above technique.

We shall consider the following cases:

(1) $\bar{x}$ is either a basic solution.

In this case make a pivot operation, as in the simplex method, using

as pivot the element T_r^s which is certainly positive, because otherwise the variable x_r would not have vanished.
Then use the vector $x(\Theta_s)$ as a feasible solution to restart the improving procedure at the next iteration. The set I is replaced by $I - \{r\} + \{s\}$.

(2) $\bar{x}$ is a nonbasic solution:
Compute the reduced gradient component of x_s at the point $x(\Theta_s)$:

$$d_s^{'}\left[I,x(\Theta_s)\right] = \left[\partial f/\partial x_s\right]_{x = x(\Theta_s)} - \left[df/dx_I\right]_{x = x(\Theta_s)} .T^s$$

a) If $d_s^{'} = 0$, then $x_s > 0$ and $d_j(I,x(\Theta_s)) = 0$ for $j\epsilon(K - I)$.

Define the set of indices $K_1 = (K - I) + \{s\}$.

The row vector T_r **cannot** be the null vector since x_r is forced to become zero at $x(\Theta_s)$; then choose an index $j\epsilon K_1$ such that $T_r^j \neq 0$ and make a pivot iteration using T_r^j as pivot.
As the variable x_r leaving the basis is vanishing, it is automatically eliminated from the set K.
Because the quantity d_j corresponding to the variable x_j entering the basis is zero, it follows from theorem 3, that the resulting quantities d_j at the point $x(\Theta_s)$ and for the new basis $I^{'} = I + \{j\} - \{r\}$ do not change, i.e., the feasible solution $x(\Theta_s)$ satisfies the complementarity condition.

b) If $d_s^{'} > 0$, compute the row vector $T_r^{K_1}$, with K_1 as defined above.

i) If $T_r^{(K - I)}$ is the null vector, then $T_r^s \neq 0$ and interchange x_s with x_r in the basis. d_s becomes zero since x_s is now basic, and it follows from theorem 3 that all d_i for $i\epsilon(K - I)$ remain zero.
Hence $x(\Theta_s)$ satisfies the complementarity condition for the new basis $I^{'} = I - \{r\} + \{s\}$.

ii) If $T_r^{(K-I)}$ is not the null vector, consider some $j\in(K-I)$ such that $T_r^j \neq 0$ and interchange x_j and x_r in the basis. As in case a), the quantity d_j corresponding to the variable x_j entering the basis is zero. From theorem 3 it follows that the resulting quantities d_j at the point $x(\Theta_s)$ and for the new basis $I' = I + \{j\} - \{r\}$ do not change.

However, this time the solution $x(\Theta_s)$ does not satisfy the complementarity condition, i.e., $d_i' = 0$ for $i\in K - \{s\}$, and $d_s = d_s' > 0$.
At this stage we must continue to improve the feasible solution, $x(\Theta_s)$ still choosing the variable x_s to be increased by the same technique, doing that as often as necessary to arrive at a solution satisfying the complementarity condition.

3. Finiteness under perturbation

There are two possible reasons for cycling in the above technique. The first is that we could fail to find an improved solution satisfying the complementarity condition, passing an infinite number of times through step b)-ii) in the same iteration. The following theorem excludes this possibility:

Theorem 4: The maximum number of times the procedure can pass through step b)-ii), in the same iteration, is equal to the number of positive nonbasic variables in the feasible solution.

The other possible reason for cycling is degeneracy. In fact, as in the simplex method, degeneracy in the above algorithm may occur when some basic variable vanishes.

If degeneracy does occur, then it is possible to have a sequence of iterations with no increase in the value of the objective function, therefore under such circumstances cycling may occur.

In the simplex method degeneracy is very common but cycling is extremely unusual. The same may be expected to occur in the algorithm presented here, for two reasons: First, the improvement procedure used in the algorithm uses the same technique as in Wolfe's method, and in this one cycling is also extremely unusual [4].

Secondly, the algorithm has many similarities with the simplex method, reducing to it in the case of a linear objective function.

However, when degeneracy occurs, cycling can be prevented - which is proved in theorem 5 - by using some perturbation technique (cf. [3], ch. 10) on the constraint set $Ax = b$. So we can state:

Theorem 5: The algorithm gives an optimal solution of problem (P) in a finite number of steps.

4. Linear and quadratic objective functions

When in problem (P) the objective function is linear, then we always have $\Theta_s = \Theta_M$ and, starting with a basic feasible solution, we always arrive at case 1) of the criterion to change the basis. Thus we obtain always basic feasible solutions at each iteration, making the same steps as in the simplex method.

If the criterion function is quadratic, then we may have nonbasic as well as basic solutions at each iteration, and we may arrive at any case considered above, to change the basis. This time, the resulting algorithm is either completely equivalent to the method by Beale, [1], for the case when, according to Beale's technique, no "free variable" has to be introduced into the basis, or it may be seen as a condensed way to accomplish Beale's method. In [6], this is proved and the resulting algorithm is considered in detail.

5. Numerical example

In order to illustrate the algorithm, we shall use a quadratic example devised by Beale [1], and used later also by Van de Panne and Whinston [10], for demonstrating their methods of quadratic programming.

$$\text{Maximize } C = 6x_1 - 2x_1^2 + 2x_1x_2 - 2x_2^2$$

subject to the constraints

$$x_1 \geq 0, \quad x_2 \geq 0$$

$$x_1 + x_2 \leq 2$$

Introducing the slack variable x_3 which will be considered as the starting basic variable, we can write

$$x_3 = 2 - x_1 - x_2$$

So we have

$$d_1 = 6 - 4x_1 + 2x_2$$

$$d_2 = \quad 2x_1 - 4x_2$$

starting with the solution $x_1 = x_2 = 0$, $x_3 = 2$, we have $d_1 = 6$ and $d_2 = 0$. Therefore the variable x_1 is increased until d_1 becomes zero at $x^*_1 = 3/2$. So we introduce the additional constraint

$$6 - 4x_1 + 2x_2 = 0$$

which with the previous ones gives

$$x_1 = 3/2 + 1/2x_2$$

$$x_3 = 1/2 - 3/2x_2$$

The new solution is $x_1 = 3/2$, $x_2 = 0$ and $x_3 = 1/2$. This time we increase x_2 being stopped by x_3 going to zero at $x^*_2 = 1/3$, and we have to remove x_3 from the basis.

Since

$$T^{1,2}_3 = (1,1)$$

and $d'_2 = 2 > 0$ at the new solution, we have arrived at case 2)-b)-ii), and we have to introduce x_1 into the basis.

From the initial constraint we obtain

$$x_1 = 2 - x_2 - x_3$$

and according to(7) we have

$$d_2 = 6 - 12x_2 - 6x_3$$

$$d_3 = 2 - 6x_2 - 4x_3$$

Now we continue to increase x_2 being stopped by d_2 going to zero at $x_2^* = 1/2$. As $d_3 = -1 < 0$, we have reached the final solution at this point:

$$x_1 = 3/2,\ x_2 = 1/2 \text{ and } x_3 = 0, \text{ with } C = 11/2.$$

APPENDIX

Proof of theorem 1:

Because $\bar{d}_s > 0$, the partial derivative of the Lagrangean $L(u,x)$, defined by (3), with respect to x_s, is positive. Hence, when x_s is increased the Lagrangean also increases.
Furthermore, the changes in the variables x_i, for $i \in K$, which result from the increasing of the variable x_s, do not affect the value of the Lagrangean, since the partial derivatives of the Lagrangean with respect to these variables are zero. The value of the Lagrangean is always equal to the value of the objective function if $Ax = b$ is satisfied, which is true for all feasible solutions.

Since under the nondegeneracy assumption Θ_M never vanishes, applying the above technique a more profitable solution in terms of the objective function can always be obtained when $d_s > 0$

q.e.d.

Proof of theorem 2:

Assume that $d_s(I,x(\Theta_s)) < 0$. Because $d_s(I,\bar{x}) > 0$ and $d_s(I,x(\Theta))$ is a continuous function of Θ, there exists some value $\Theta_o < \Theta_s$ such that $d_s(I,x(\Theta_o)) = 0$ and $d_s(I,x(\Theta)) < 0$ for all $\Theta_o < \Theta \leq \Theta_s$. As the objective function decreases when x_s is increased, as long as d_s is negative, it follows that $F(\Theta_o) > F(\Theta_s)$, which contradicts (12). Therefore $d_s(I,x(\Theta_s)) \geq 0$.

q.e.d.

Proof of theorem 3:

By definition

$$d_j^*(I^*,\bar{x}) = (\partial f/\partial \bar{x}_j) - \sum_{i\in I^*} (\partial f/\partial \bar{x}_i).T_i^{*j} \;;$$

for $j\in N$.

Where T^* is the matrix defined by (2) when the basis taken is I^*. The two matrices T and T^* are related by the following expression (cf. [3], pp. 197)

$$T^* = T + QT_r \;,$$

where Q is an m-vector whose components are:

$$Q_r = 1/T_r^s - 1 \quad \text{and} \quad Q_i = - T_i^s/T_r^s \quad \text{for } i \neq r.$$

Then we may write

$$d_j^* = (\partial f/\partial \bar{x}_j) - \sum_{i\in I^* i \neq s} (\partial f/\partial \bar{x}_i).\,[\, T_i^j - T_i^s \Big/ T_r^s \,.T_r^j \,] -$$

$$- (\partial f/\partial \bar{x}_s).T_r^j \Big/ T_r^s$$

or, if we add and subtract $(\partial f/\partial \bar{x}_r).T_r^j$:

$$d^*_j = (\partial f/\partial \bar{x}_j) - \sum_{i\in I^* i \neq s} [(\partial f/\partial \bar{x}_i).T_i^j - (\partial f/\partial \bar{x}_r).T_r^j] -$$

$$- [(\partial f/\partial \bar{x}_s) - \sum_{i\in I^* i \neq s} (\partial f/\partial \bar{x}_i).T_i^s - (\partial f/\partial \bar{x}_r).T_r^s].(T_r^j \Big/ T_r^s).$$

Finally, according to the definition of $d_j(I,\bar{x})$ we have

$$d_j^* = \bar{d}_j - \bar{d}_s.(T_r^j \;\; T_r^s)$$

q.e.d.

Proof of theorem 4:

Let q be the number of positive nonbasic variables in the feasible solution x. At each time we pass through step b), the number of positive nonbasic variables in the new solution decreases by one.

After q times there will be no positive nonbasic variable in the improved solution, besides x_s. Therefore, case 1) will necessarily occur and the resulting improved solution will satisfy the complementarity condition.

q.e.d.

Proof of theorem 5:

The algorithm terminates either when we arrive at a feasible solution with $d_j \leq 0$ for all j, or when we have $\Theta_s = \infty$.

Suppose it cycled indefinitely. Then, it would produce an infinite number of feasible solutions satisfying the complementarity condition, since it follows from theorem 4, that the number of improvement steps, needed to pass from a feasible complementarity solution to the next one, is finite.

Assuming nondegeneracy, then, every time a feasible solution is improved, the objective function increases by a positive amount.

Because the number of possible subsets K is finite, and everyone of these subsets defines a unique feasible complementary solution, we would necessarily arrive at the situation of having different optimum values corresponding to the same solution of the given problem, which is a contradiction.

q.e.d.

Acknowledgments:

This paper is based on chapter IV of the author's Ph.D. thesis, submitted to the University of Birmingham [7]. I am specially indebted to Professor S. Vajda for the valuable discussions and encouragements as my supervisor. The research was supported by NATO and the "Instituto de Alta Cultura, Portugal".

References

1. Beale, E.M.L.,
"On Quadratic Programming"
Naval Res. Logist. Quart., 6 (1959), 227-243.

2. Cottle, R.W.,
"Nonlinear Programs with Positive Bounded Jacobian"
J. SIAM Appl. Math., 14 (1966), 147-158.

3. Dantzig, G.B.,
"Linear Programming and Extensions"
Princeton University Press, 1963.

4. Faure, P. and P. Huard,
"Résolutions des Programmes Mathématiques à Fonction non Linéaire par la Méthode du Gradient Reduit"
Révue Française de Rech. Operat., 36 (1965), 167-206.

5. Gonçalves, A.S.,
"A Finite Reduced Gradient Method and its Application to the Nonlinear Complementarity Problem"
submitted for publication in Numerische Mathematik.

6. Gonçalves, A.S.,
"A Version of Beale's Method Avoiding the Free Variables"
presented at the A.C.M. National Conference, Chicago, 1971.

7. Gonçalves, A.S.,
"Primal-Dual and Parametric Methods in Mathematical Programming"
Ph.D. thesis, University of Birmingham (1969), 218-230.

8. Kuhn, H.W. and Tucker
"Nonlinear Programming"
Second Berkeley Symp. in Math. Stat. and Probability (1951), 481-492.

9. Rosen, J.B.,
"The Gradient Projection Method for Nonlinear Programming: Part I, Linear Constraints"
J. Soc. Indust. Appl. Math., 8 (1960), 181-217.

10. Van de Panne, C. and A. Whinston,
"The Simplex and the Dual Method for Quadratic Programming"
Op. Res.Q., 15 (1964), 355-388.

11. Wolfe, P.,
"Methods of Nonlinear Programming"
In R.L. Graves and P. Wolfe (eds),
Recent Advances in Mathematical Programming, McGraw-Hill, New York, 1963.

Mathematical Programming in Theory and Practice,
P.L. Hammer and G. Zoutendijk, (Eds.)

SOME APPLICATIONS OF NONLINEAR OPTIMIZATION

Lothar Collatz

In this survey some special applications of optimization problems and their theory to different fields of mathematics, physics, engineering, economy, traffic and other topics are collected.

I. Nonlinear Optimization

1. General formulation in the real field for optimization problems with finitely many unknowns $x = \{x_1, \ldots, x_n\}$ and finitely many restrictions

(1.1) $$f_j(x) \leq 0 \qquad (j=1,\ldots,m)$$

Function to be minimized

(1.2) $$Q(x) = \text{Min.}$$

Often one assumes

(1.3) $$x \geq 0 \text{ , that means } x_k \geq 0 \text{ for } k=1,\ldots,n.$$

2. Special types

The problem (1.1) (1.2) is called convex, pseudoconvex, ... , if Q and all f_j are convex, pseudoconvex, ...

A function $\varphi(x)$, defined in a convex domain B, is called (Mangasarian [69], Stoer-Witzgall [7o], ..)
(In the following x^T means the transposed vector to x; for some of the types mentioned in fig.1 short definitions are given. Sometimes the restrictions are supposed to be linear. For all details see for instance Collatz-Wetterling [71])

pseudoconvex , if $\varphi(x) \geq \varphi(y)$ follows from $(x-y)^T \operatorname{grad} \varphi(y) \geq 0$

pseudoconcave, if $\varphi(x) \leq \varphi(y)$ follows from $(x-y)^T \operatorname{grad} \varphi(y) \leq 0$

quasiconvex , if $\varphi(\alpha x+(1-\alpha)y) \leq \varphi(y)$ follows from $\varphi(x) \leq \varphi(y)$ for all $\alpha \in [0,1]$

strict quasiconvex , if $\varphi(\alpha x+(1-\alpha)y) < \varphi(y)$ follows from $\varphi(x) < \varphi(y)$ for all $\alpha \in [0,1]$

quasiconcave , if $\varphi(\alpha x+(1-\alpha)y) \geq \varphi(y)$ follows from $\varphi(x) \geq \varphi(y)$ for all $\alpha \in [0,1]$

convex , if $\varphi(\alpha x+(1-\alpha)y) \leq \alpha\varphi(x)+(1-\alpha)\varphi(y)$ for all $\alpha \in [0,1]$

concave , if $\varphi(\alpha x+(1-\alpha)y) \geq \alpha\varphi(x)+(1-\alpha)\varphi(y)$ for all $\alpha \in [0,1]$

and for all $x,y \in B$

The optimization is called separable, if Q and f_j are sums of functions of only one independent variable or if this can be reached by a linear nonsingular transformation of the independent variables; for instance $Q = Q_1(x_1) + Q_2(x_2) + Q_n(x_n)$ and correspondingly

$$f_j = \sum_{k=1}^{n} f_{jk}(x_k) .$$

Fig. 1 holds for differentiable function and every arrow goes from a class of optimization problems to a more general class.

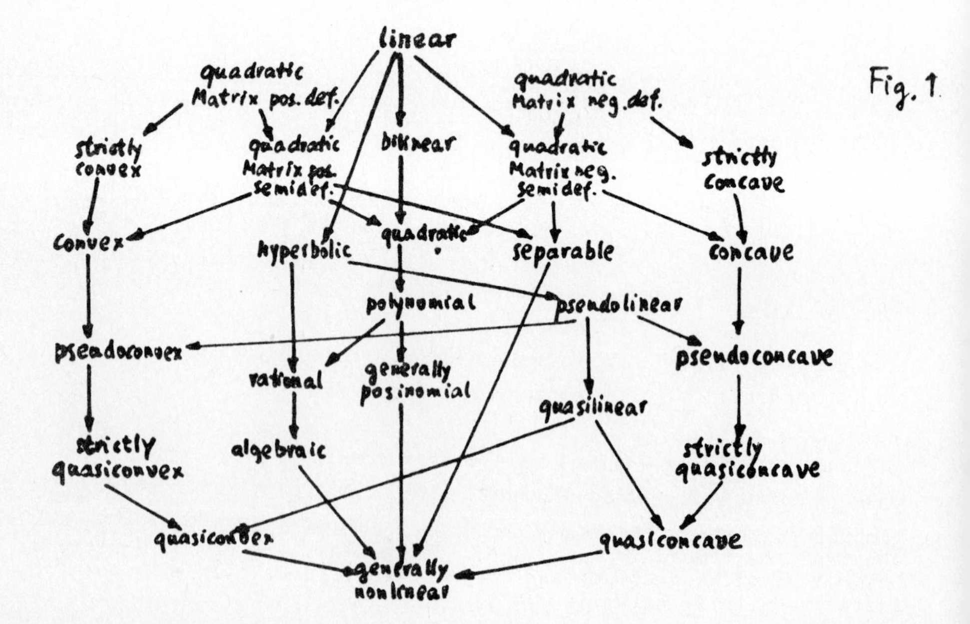

3. About the variants of convexity

a) Variants of convexity occur very often; for instance $f(x) = x^2$ is convex and $g(x) = -x^2$ is concave, and $\exp(x^2)$ is convex but $\exp(-x^2)$ is pseudoconcave and not concave, fig. 2 .

Lemma: Let be $\varphi(x)$ real concave and differentiable on the convex domain B and let be W the range of $\varphi(x)$ if x runs through B . Let be $h(z): W \to \mathbb{R}^1$ with $h'(z) > 0$. Then $\phi(x) = h(\varphi(x))$ is pseudoconcave on B (see for instance Collatz-Wetterling [71], p.93).

Lemma: (Mangasarian [69]): Let be $Z(x)$, $N(x)$ differentiable on the convex domain B and $Z(x)$ convex, $N(x) > 0$ on B . Furthermore at least one of the two following assumptions may be true

(i) N is affin linear,

(ii) N is convex and $Z(x) \leq 0$ in B ,

then $\varphi(x) = \frac{Z(x)}{N(x)}$ is pseudoconvex

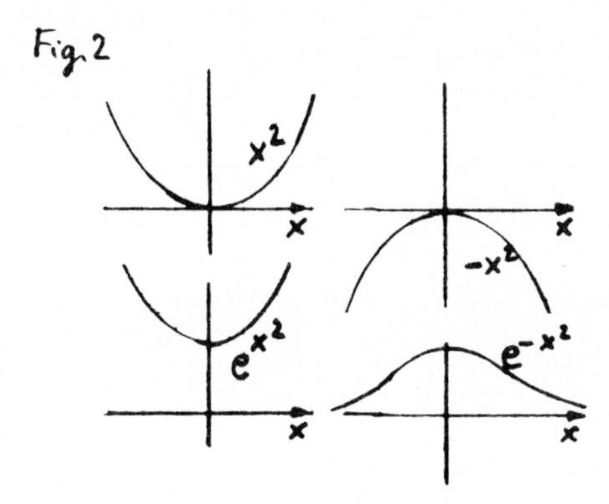

A consequence of this lemma is, that every hyperbolic optimization is pseudoconvex and also pseudoconcave and quasilinear.

b) Many theorems for convex and concave functions can be generalized for variants of these functions (Stoer-Witzgall [70]) for instance:

For every quasiconvex function $\varphi(x)$ defined on a convex domain B holds: the set M of minimal points is convex and every local strict minimum is a global strict minimum.

For every strictly quasiconvex function $\varphi(x)$ defined in a convex domain B holds:

Every local minimum is a global minimum.

4. Occurence in the applications

a) Economy, rentability $Q = \dfrac{\text{win}}{\text{costs for investment}} = \dfrac{a_o + \sum_j a_j x_j}{b_o + \sum_j b_j x_j}$

usually a quotient of two linear functions; together with certain linear restrictions this is a hyperbolic optimization problem.

Many other nonlinear optimization problems in the economy are well known.

b) Geometrical and traffic problems

α) Division of a celler in n parts of equal area. The sum of the lengths of all walls one has to build for the division should be as small as possible, fig.3, for a cellar of quadratic form and $n=3$.

Fig.3 $n=3$

If all parts are asked to be convex, for $n = 4$ one sees a solution in fig.4.

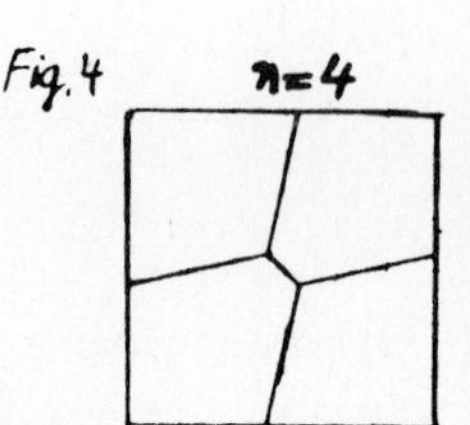

For great values of n, for instance $n=20$ or $n=50$, the solution on a computer is difficult, because the topological structure of the solution is unknown

ß) Given villages P_j $(j=1,\ldots,n)$, fig. 5, and a lake L or, fig. 6, two lakes L_1, L_2 .

Where should be the situation of a factory F , so that the sum

$\sum_{j=1}^{n} q_j d(P_j,F)$ is a minimum, where q_j are given weights and $d(P_j,F)$ is the length of a road from P_j to F , avoiding the lakes.

In fig. 5 one has two distinct solutions, therefore the optimization cannot be convex.

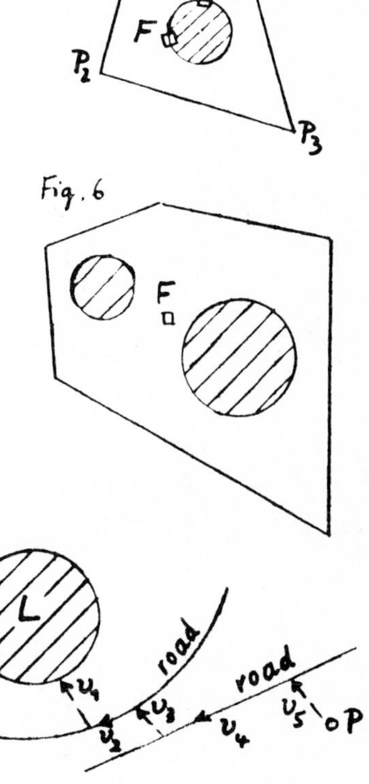

γ) Way from starting point P to a lake L , fig. 7. Velocities v_1,v_3,v_5 on the acres, v_2,v_4 on the roads.

Time-Minimum-Problem, usually separable optimization.

δ) One wishes to build a road of smallest lenth from a village P_1 to another village P_2 with the conditions: (Kubik [71])
The road avoids the area L_1 (in fig. 8 a circle) and meets the area L_2 (in fig. 8 a circle). The curvature ρ is a continuous piecewise linear function of the arc-length s ; there is a given bound for the modulus of ρ . If one admittes m pieces, one has $(3m-1)$ parameters.

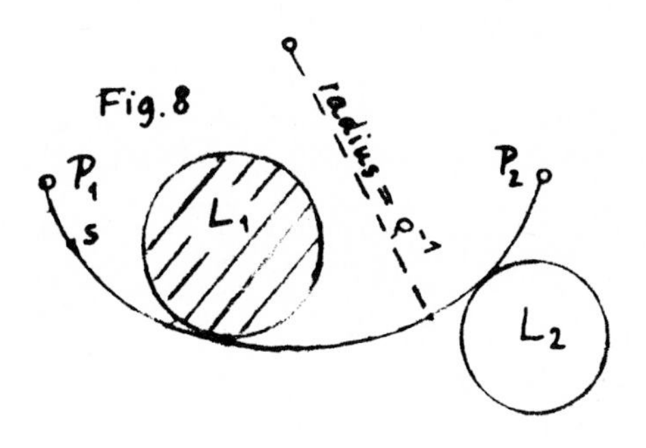

c) Optimal control problems

$x(t) = \{x_1(t),\dots,x_n(t)\} \in R^n$

$u(t) = \{u_1(t),\dots,u_m(t)\} \in R^m$, $\dot{x} = G(t,x(t),u(t))$,

initial point $x(t_1) = a$, end point $x(t_2) = b$, furthermore restrictions for x and u , possible t_2 unknown,

$$F = \int_{t_1}^{t_2} \varphi(t,x(t),u(t))dt = \text{Min.}$$

Discretising one gets a classical optimization problem.

d) Physics, Mechanical systems, fig.9, minimum of the potential energy for equilibrium, principle of shortest light-time (f.i. Collatz-Wetterling [71],p.97, a.o.)

Fig.9.

e) Approximation. Example: Approximate $f(x) = e^{-x}$ in the interval $J = [0,1]$ by a rational function $w(x,a) = \frac{1}{a+x}$, so that $\phi(a) = \| w-f \| = \text{Min.}$

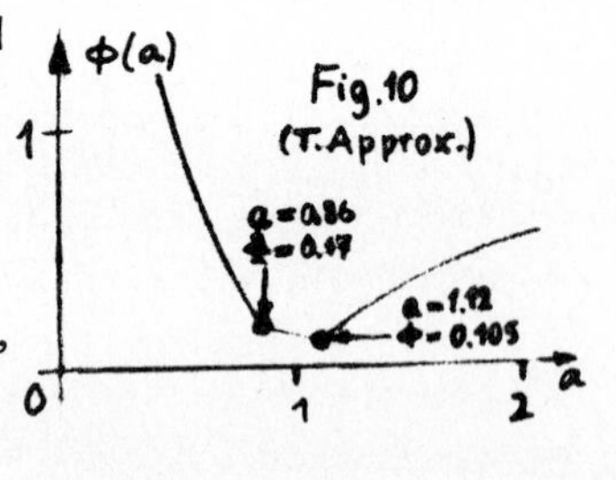

Fig.10 (T.Approx.)

For the norm $\| g \| = \underset{x\in J}{\text{Max}} \ |g(x)|$ (Tschebyscheff Approximation), fig.10, we have in this example a quasiconvex optimization.

For the norm $\|w-f\| = \left[\int_J (w-f)^2 dx\right]^{1/2}$ (L_2-Approximation), fig.11, we have a pseudoconvex optimization.

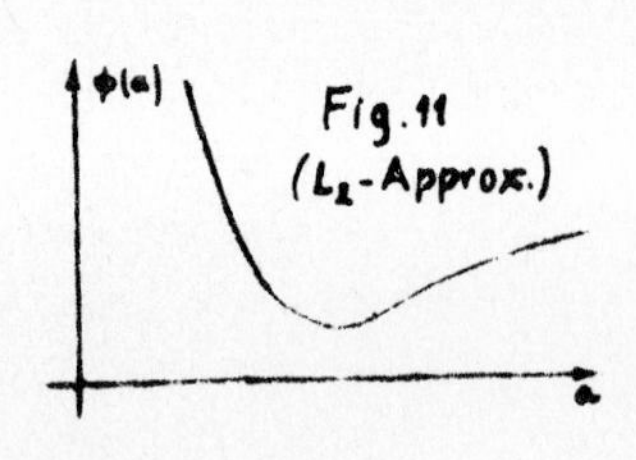

Fig.11 (L_2-Approx.)

5. <u>Integer Optimization</u>

a) Some of the independent variables are restricted to be integer.
b) The function Q to be minimized should be integer.

The problems b) are usually more complicated than the problems a) and often they are difficult even to formulate in a form suitable for computers. Example:

fig.12

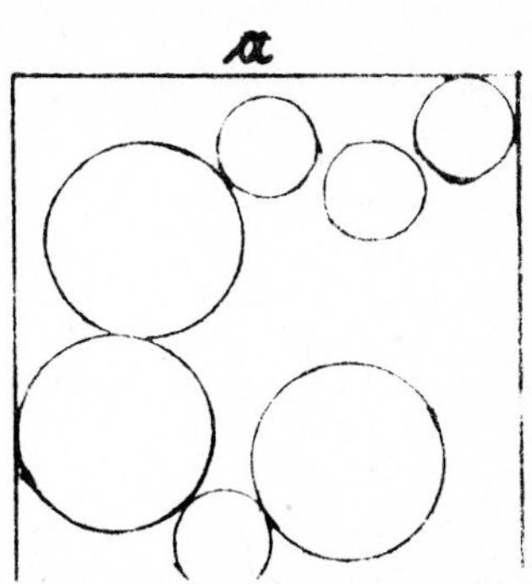

There are m circles of radius 1 and n circles of radius 2 in a square of side-length a; no two of the m+n circles have a common inner point, fig. 12.

First problem: m, n given, a = Minimum, problem easy to formulate.

Second problem: a given, $Q = 4n+m$ = Maximum (area covered by the circles should be as great as possible); difficult to handle on a computer.

The corresponding problems with spheres in the three-dimensional space are even more complicated but more interesting.

II. Duality in optimization problems with infinitely many restrictions

(Krabs [68], [71], Lempio [71], Collatz [71]).

1. Simplest problems of Tschebyscheff-Approximation

B given closed domain in the real n-dimensional space R^n of points $x = \{x_1,\dots,x_n\}$; C(B) = linear space of continuous functions g(x) defined on B; C(B) : Banach space by using the maximum-norm

$$\|g\| = \operatorname*{Max}_{x\in B} |g(x)| \tag{2.1}$$

$a = \{a_1,\dots,a_p\}$ parametervector $\in A$;
A = given subset of the real p-dimensional pointspace R^p.
$W = \{w(x,a),\ a\in A\}$ given subset of C(B).

Tschebyscheff-Approximation (shortly T.A.) for a given function $f(x) \in C(B)$

$$\rho_0 = \inf_{w\in W} \|w-f\| = \text{minimal distance} \tag{2.2}$$

$\hat{w} \in W$ with $\|\hat{w}-f\| = \varrho_0$ is a minimal solution .

2. Formulation as optimization problems

Infinitely many restrictions:

$$(2.3) \qquad \begin{cases} -\delta_1 \le w(x,a)-f(x) \le \delta_2 & \text{for all } x \in B \\ \delta_1 \ge 0\ , \ \delta_2 \ge 0 \end{cases}$$

Infimum-problem

$$(2.4) \qquad Q = \delta_1 = \delta_2 = \text{infimum}$$

Special cases:

a) $\delta_1 = \delta_2$ classical T.A.
b) $\delta_1 = 0$ onesided T.A. from above } (shortly T_1A.)
c) $\delta_2 = 0$ onesided T.A. from below }
d) no restrictions about δ_1, δ_2 , (only $\delta_1 \ge 0$, $\delta_2 \ge 0$)
unsymmetric T.A. (shortly T_uA.)

3. Example for unsymmetric Tschebyscheff-Approximation

In this example the coefficients of the polynomial $f(x)$ are selected in such a way that one has big differences between the special types of approximation, see fig.13, in the interval $1 \le x \le 5$.

Let be

$$f(x) = \tfrac{1}{2}(x^3-5x^2+6), \qquad w(x,a) = a \cdot x \ .$$

Then the best unsymmetric T.A. is $w = \frac{1}{2}x$; fig. 13 shows also the best T.A. and the best T_1.A. from above and from below.

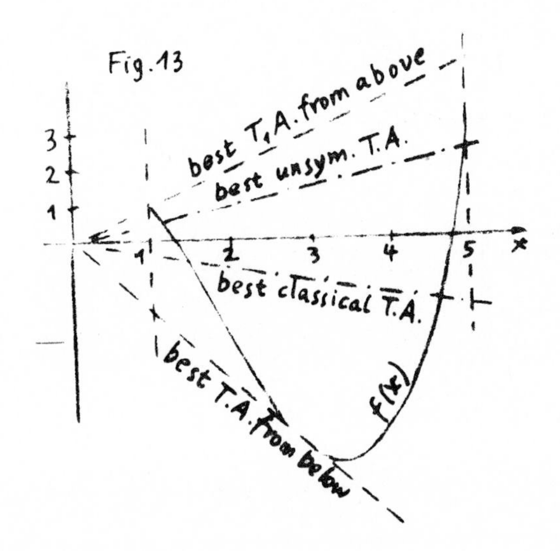

4. Weak duality for linear problems

Primary problem: for the linear onesided Tschebyscheff Approximation with fixed functions $w_\nu(x) \in B$:

Restrictions:

$$0 \leq \sum_{\nu=1}^{p} a_\nu w_\nu(x) - f(x) \leq \delta \quad \text{for all} \quad x \in B \tag{2.5}$$

Infimum problem

$$\delta = \text{Infimum} \tag{2.6}$$

Let be $R^\times$ the dual space of linear bounded functionals $\ell(h)$ for $h \in C(B)$. We introduce the cone K of nonnegative functions $h(x)$, for which $h(x) \geq 0$ for $x \in B$; furthermore we consider the cone $K^\times$ of nonnegative functionals $\ell(h)$, that means the functionals $\ell(h)$ with $\ell(h) \geq 0$ for $h \in K$; for these functionals we write $\ell(h) > \theta$.

Especially if we select certain points $P_j \in B$ $(j=1,\ldots,r)$ then

(2.7) $\quad \ell(h) " \sum_{j=1}^{r} q_j h(P_j)$ with $q_j > 0$, P_j fixed $\in B$

is a nonnegative functional.

A pair $\ell_j > \Theta$ $(j=1,2)$ is called admissable, if

(2.8) $\quad \ell_1(w_v) = \ell_2(w_v) \quad (v=1,\ldots,p)$, $\ell_2(1) = 1$

Dual problem: Restrictions (2.8),

(2.9) $\quad \lambda = \ell_1(f) - \ell_2(f) = \text{supremum}$

Weak duality: Let be $a_1,\ldots,a_p,\delta$ admissable as in (2.5) and the nonnegative ℓ_1, ℓ_2 admissable as in (2.8), then

(2.1o) $\quad \ell_1(f) - \ell_2(f) \leq \delta$

Generally it is not necessary to consider general nonnegative functionals but it is sufficient to look on point functionals and their combinations. The reason is the

Theorem: (compare Lempio [72]) Let be $q_1(x), \ldots, q_n(x)$ linearly independent functions of $C(B)$ and S the linear subspace generated by the $q_y(x)$. Then every real linear functional $\ell(h)$ on S can be represented in the form

$$\ell(h) = \sum_{j=1}^{n} \lambda_j h(x_j) \quad \text{for all} \quad w \in S$$

where λ_j are real constants and x_j points of B .

5. Weak duality for nonlinear unsymmetric Tschebyscheff-Approximation

Primary problem as stated in (2.3) (2.4). Dual space R^{x} and functionals $\ell(h)$ as above.

A pair $\ell_j > \Theta$ $(j=1,2)$ is called admissable, if

(2.11) $$A = \ell_1(w)-\ell_2(w) \leq 0 \quad \text{for all} \quad w \in W$$

$$\text{and if} \quad \ell_1(1) = \ell_2(1) = 1 .$$

Dual problem for admissable pairs ℓ_1, ℓ_2 :

(2.12) $$Q^{x} = \ell_1(f)-\ell_2(f) = \text{supremum}$$

Weak duality: Let be $a_1,\dots,a_p,\delta_1,\delta_2$ admissable as in (2.3) and the nonnegative ℓ_1, ℓ_2 admissable as in (2.11), then

(2.13) $$\ell_1(f)-\ell_2(f) \leq \delta_1+\delta_2 .$$

6. Example for the duality theorem:

B as interval $[-d,d]$; $f(x) \in C(B)$.
Class $W = \{w(x,a) = (a_1+a_2x)^2\}$.

Let be α,β real numbers with $0<\alpha<\beta \leq d$.
We take as functionals

$$\ell_1(h) = \sigma h(-\beta)+(1-\sigma)h(\alpha)$$

$$\ell_2(h) = \sigma h(\beta)+(1-\sigma)h(-\alpha) \quad \text{with} \quad 0<\sigma<1 .$$

Then one calculates immediately

$$\ell_1(w) = \ell_2(w) \quad \text{for all} \quad w \in W$$

if

$$\beta\sigma = (1-\sigma)\alpha$$

For given α,β one can determine σ .

Here is $A \equiv 0$ and one can change ℓ_1,ℓ_2, so that (2.13) gives the result

$$|\ell_1 f-\ell_2 f| \leq \delta_1+\delta_2 .$$

III. Optimization in Differential- and Integral Equations

1. Example for onesided Tschebyscheff Approximation

Nonlinear hyperbolic equation for a function $u(x,y)$ in the domain B : $\{(x,y)$ with $x>0,\ y>0\}$: $Tu = \frac{\partial^2 u}{\partial x \partial y} - (1+u^2) = 0$

with initial conditions $u(x,0) = u(0,y) = 0$ on ∂B .

Approximate solution in the form $w(x,y,a_1,a_2) = xy(a_1+a_2(x+y))$

For the triangle D : $(x>0,\ y>0,\ x+y<1)$, fig.14, one has $w \leq u$ from $Tw \leq 0$ in D, $D \subset B$.

One wishes to determine a_1, a_2 so that $-\delta \leq Tw \leq 0$ in D, δ = Min

For $a_1 = 0.9411$, $a_2 = 0.0294$ one has $-0.0589 \leq Tw \leq 0$;

similarly $a_1 = 1$, $a_2 = 15-\sqrt{224} \approx 0.0334$ gives $0 \leq Tw \leq 0.0668$

especially for $y = 1-x$ one has

$$0.9705\ x(1-x) \leq u(x,1-x) \leq 1.0334\ x(1-x).$$

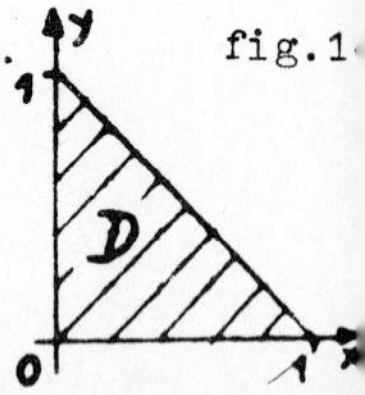

fig.14

Many other examples in Collatz [70].

2. Other types of Approximation problem

Syn-Approximation: One has to approximate $T_1 f, T_2 f, \dots, T_s f$ together, for instance f, $\frac{\partial f}{\partial x}$ and $\frac{\partial f}{\partial y}$;

Simultan-Approximation (Bredendiek [69]: One has different domains, for instance B and the boundary ∂B .

Combi-Approximation: One uses different classes $W_1, W_2, \dots, W_s$, for instance one uses in the domain B another class as on the boundary ∂B . (More details in Collatz-Krabs [72])

3. General Field Approximation

Many of the types mentioned above can be written as field approximation:

$$0 \le \phi(x,a) \le \delta \; ; \quad \delta = \text{Min.}$$

with the restrictions

$$0 \le G_j(x,a) \qquad (j=1,\dots,m)$$

for all $x \in B$, $a \in A$ with given functions ϕ, G_j .

Special cases are the approximations with restricted ranges, Taylor [69].

4. Monotonically decomposible operators (M.D.O.)

Let be R a partially ordered Banach space with an ordering relation " $<$ " , and T_j $(j=1,2)$ compact operators, defined on a convex domain D. Let be T_1 syntone (from $v < w$ follows $T_1 v < T_1 w$) and T_2 antitone (from $v < w$ follows $T_1 v > T_2 w$) .

For solving the equation

$$u = Tu + r \quad \text{with} \quad T = T_1 + T_2$$

we start with two elements $v_o, w_o \in D$ and calculate

$$v_1 = T_1 v_o + T_2 w_o + r$$

$$w_1 = T_1 w_o + T_2 v_o + r$$

If then holds

$$v_o < v_1 < w_1 < w_o$$

it follows as consequence of the Schauder fix-point theorem, that there exists at least one solution u with the error estimation

$$v_1 < u < w_1 \quad .$$

One wishes to get good error bounds; this is the optimization (as field approximation)

$$0 \le w_1 - v_1 \le \delta ; \quad \delta = \text{Min.}$$

with the restrictions

$$0 \le v_1 - v_o , \qquad 0 \le w_o - w_1 \qquad \forall \quad x \in B \ .$$

Here we have used as ordering $f < g$ the classical ordering $f(x) \le g(x)$ for all $x \in B$.

Example: Let be given the Urysohn-Equation

$$u(x) = Tu = \int_o^1 \frac{dt}{3-x-t+u(t)}$$

We take as domain D the class of functions $f(x)$, which are continuous and nonnegative in $0 \le x \le 1$.

Starting with $v_o = 0$, $w_o = 1$, we get

$$w_1 = Tv_o = \int_o^1 \frac{dt}{3-x-t} = \ell n \frac{3-x}{2-x}$$

$$v_1 = Tw_o = \int_o^1 \frac{dt}{3-x-t+1} = \ell n \frac{4-x}{3-x}$$

Then $v_o \le v_1 \le w_1 \le w_o$ is satisfied, compare fig. 15, and therefore exists at least one solution $u(x)$ of the considered equation in the strip

$$v_1 \le u(x) \le w_1$$

For getting better bounds one can introduce parameters in v_o, w_o and carry out the described field-optimization.
Other numerical examples in Collatz [71a] .

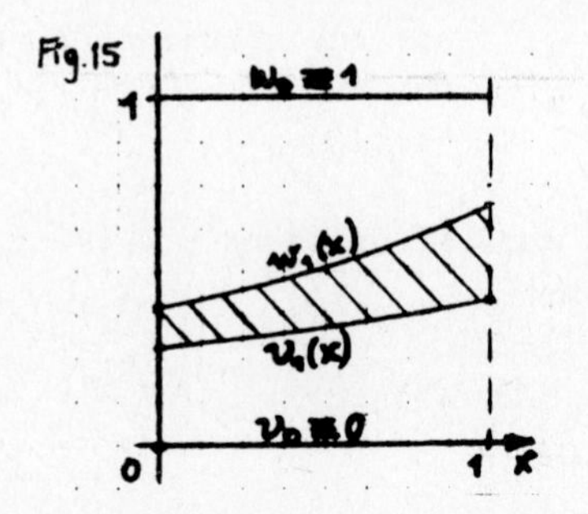

IV. Approximation by functions of fewer variables

1. General formulation

One problem of David Hilbert asks for the representation of continuous functions of several variables by aggregates of functions of only two independent variables; but here the problem is the approximation of continuous functions $f(x) = f(x_1,\dots,x_n)$ by functions of the form

$$(4.1) \qquad \phi(u_1(x_{11},x_{12},\dots,x_{1m_1}),\dots,\ u_q(x_{q1},x_{q2},\dots,x_{qm_q}))$$

where every x_{jk} is identical with a certain x_ν, and $m_\mu < n$ for $\mu = 1,\dots,q$.

2. Special cases, which occur frequently

a) Sumtype-Approximation

$u_\mu = u_\mu(x_\mu)$ depends only on x_μ (or more general on some x_ν)

$$(4.2) \quad \phi = \sum_{\mu=1}^{n} u_\mu \quad \text{or} \quad \phi = \sum_{\nu=1}^{q} u_\nu(x_{\nu 1},x_{\nu 2},\dots,x_{\nu m_\nu})$$

b) Productsumtype-Approximation

$$(4.3) \quad \phi = \sum_{\nu=1}^{q} u_\nu(x_1,\dots,x_m)\ v_\nu(x_{m+1},\dots,x_n) \quad \text{with} \quad m < n$$

c) Parametrictype-Approximation

$$(4.4) \quad \phi = \sum_{\nu=1}^{n} u_\nu(x_{\nu 1},x_{\nu 2},\dots,x_{\nu m_\nu})\cdot w_\nu(x_{\nu,m_{\nu+1}},\dots,x_{\nu,n},a_{\nu 1},\dots,a_{\nu p_\nu})$$

with w_ν as given functions

Of course, there are further types.

3. Occurence of these approximation problems (some selected examples)

a) Computer: For the input of a given function $f(x)$ into a computer one substitutes $f(x)$ by a function of sumtype or productsumtype or another expression the computer can calculate easily.

b) Hammerstein's nonlinear integral equation

$$u(x) = \int_B K(x,t)\, \varphi(u(t))\, dt \tag{4.5}$$

where K, φ are given.

One approximates K by a degenerated kernel K^*

$$K(x,t) \approx K^*(x,t) = \sum_{\nu=1}^{q} u_\nu(x)\, v_\nu(t) \quad \text{with} \quad x = \{x_1,\ldots,x_n\},\; t = \{t_1,\ldots,t_n\} \tag{4.6}$$

Then one can calculate an approximate solution u^* of $u^* = \int K^* \varphi(u^*(t))\, dt$ by solving a system of nonlinear equations for the c_ν in the expression $u^* = \sum c_\nu u_\nu(x)$.

In the linear case $\varphi(z) = \gamma_1 + \gamma_2 z$ with γ_1, γ_2 as constants one has the error estimation of Kantorowitsch Kryloff [56]

$$|\varepsilon(x)| = |u^*(x) - u(x)| \leq \max_{\substack{x\in B\\ t\in B}} |K(x,t) - K^*(x,t)| \cdot C \tag{4.7}$$

(Tschebyscheff-Approximation),

where C depends on the data of the integral equation.

c) Dirichlet-Problem

$$u = \sum_{j=1}^{n} \frac{\partial^2 u}{\partial x_j^2} = f(x,\ldots,x_n) \quad \text{for} \quad x \in B \tag{4.8}$$

$u = \psi(x)$ for $x \in \partial B$ (boundary of B). One approximates

$$f \approx \varphi(x) = \sum_{v=1}^{n} u_v(x_v) ;$$

and chooses $h_v(x_v)$ with $\frac{d^2h_v}{dx_v^{\,2}} = u_v$, and z_σ with

$\Delta z_\sigma = 0 \quad (\sigma=1,\ldots,s)$

then $z(x) = \sum_{\sigma=1}^{s} \alpha_\sigma z_\sigma(x) + \sum_{v=1}^{n} h_v$ satisfies $\Delta z = \varphi$ and on ∂B one has a Tschebyscheff-Approximation for the α_σ .

d) variable coefficients

(4.9) $$Lu = K(y)\, u_{xx}+u_{yy} = f(x,y)$$

with K,f as given functions. One has the productsumtype-Approximation

(4.1o) $$f(x,y) \approx \hat{f}(x,y) = (a_1+a_2y)K(y)u_1(x)+(a_3+a_4x)u_2(y)$$

then $$Z = (a_1+a_2y)h_1(x) + (a_3+a_4x)h_2(y)$$

with $h_1''=u_1$, $h_2''=u_2$

solves the equation

$$LZ = \hat{f} ;$$

for $K(y) = 1,-1,y,k(y)$ one has the Laplace-, Wave-, Tricomi-, Tschaplygin-equation; similar $Lu = u_t-\Delta u$, $Lu = u_{tt}+u_{xxxx}$,

$Lu = \sum_{j=1}^{n} L_j u$, with L_j as differential operators with respect to x_j with variable coefficients can be treated.

e) Inverse problems

Observed temperature $f(x,\ldots,x_n)$ for $n=2$ or 3 ;

the temperature $u(x_1,\ldots,x_n)$ may satisfy an equation $-\Delta u = q = \text{const}$, where q is to be determinated. One has the parametric type-Approximation

$$(4.11) \qquad f(x) \approx -\frac{q}{2n} \sum_{j=1}^{n} x_j^2 + w(x_v)$$

where $\Delta w = 0$, for instance $w = \text{Re}\,\varphi(x_1+ix_2)$ for $n=2$; one wishes to determine q (φ as holomorphic function)

4. Procedure

a) Discretization:

$$(4.12) \qquad -\delta_1 \leq f-\phi \leq \delta_2, \quad \delta_1 \geq 0, \quad \delta_2 \geq 0, \quad \delta_1+\delta_2 = \text{Min.}$$

This is a nonlinear optimization problem

b) Approximating the free functions, for instance $u_v(x_v)$ by parametrizing

$$(4.13) \qquad u_v(x_v) \approx \sum_u \beta_{v\mu}\, b_{v\mu}\,(x_v), \quad \mu=1,\ldots,M$$

then Field-Approximation (written as optimization problem with infinitely many restrictions)

$$(4.14) \quad \begin{cases} -\delta_1 \leq G(x,a,b) \leq \delta_2 \quad \delta_1 \geq 0, \ \delta_2 \geq 0, \ \delta_1+\delta_2 = \text{Min.} \\ 0 \leq G_j(x,a,b) \quad j=1,\ldots,J \quad \text{for all } x \in B \end{cases}$$

a,b as parameter vectors

c) Sumtype-Approximation (Sprecher [68] a.o.) Example:

Approximate $f(x,y,z) = \dfrac{1}{4-x-y^2z}$ by $w = u(x,y) + v(z)$

in the domain $B = 0 \leq \{x,y,z\} \leq 1$.

One best Tschebyscheff-Approximation is

$$w = \frac{1}{2}\left[\frac{1}{4-x} + \frac{1}{4-x-y^2}\right] - \frac{1}{24} + \frac{z}{12} \quad \text{with the minimal distance}$$

$$\rho_o = \underset{B}{\text{Max}} \; |w-f| = \frac{1}{24}$$

d) Productsum-Type and Parametric Type

$$(4.15) \quad f(x,t) \approx w(x,t) = \sum_{v=1}^{q} u_v(x)v_v(t) \quad \text{with} \quad x=\{x_1,\ldots,x_n\}, \; t=\{t_1,\ldots,t_m\}.$$

L_2-Approximation for $n=m$ gives an eigenvalue problem, Golomb [62] ; the Tschebyscheff-Approximation is more complicated, even in the case $n=m=1$ with $w(x,t) = u_1(x)\,v_1(t) + u_2(x)\,v_2(t)$; but let us take v_1 and u_2 as known functions, as $a(t)$ and $b(x)$, this means

$$(4.16) \qquad w = u(x)\,a(t) + b(x)\,v(t)$$

Let be $a(t) > 0$, $b(x) > 0$ for $x \in B$, $t \in B$.

One can apply the Theory of H-sets.
By considering a rectangle R with corners P_j $(j=1,2,3,4$ as in figure 16) and $f_j = f(P_j)$ one can calculate the term

$$(4.17) \qquad \delta = - \frac{a_1b_of_1+a_ob_of_2-a_ob_1f_3+a_1b_1f_4}{(a_o+a_1)\,(b_o+b_1)}$$

and $|\delta|$ is then a lower bound for the minimal distance ρ_o of the Tschebyscheff-Approximation for f in the class (4.16), one has $|\delta| \leq \rho_o$.

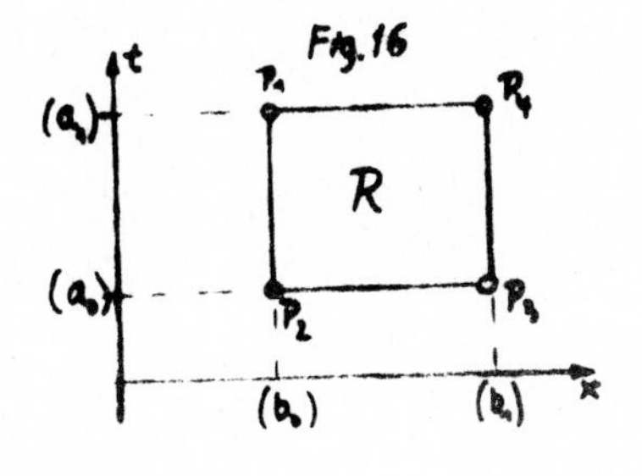

Fig.16

(a_ν,b_ν are the values of the functions a,b at the points P_ν)

References

J. Abadie [7o]: Integer and nonlinear programming, North-Holland Publishing Company 197o, 544 S.

G. Aumann [68]: Approximation von Funktionen, in R. Sauer, I. Szabó Mathematische Hilfsmittel des Ingenieurs, Teil III, Springer 1968, 32o-351.

E. Bredendiek [69]: Simultan-Approximation, Arch. Rat. Mech. Anal. 33 (1969), 3o7-33o.

J. Bracken - G.P. McCormick [68]: Selected applications of nonlinear programming, New York, John Wiley & Sons 1968, 11o p.

L. Collatz [7o]: Applications of nonlinear optimization to approximation problems, in Abadie [7o]: Integer and nonlinear programming, North-Holland Publishing Company 197o, p. 285-3o8.

L. Collatz [71]: Anwendungen der Dualität der Optimierungstheorie auf nichtlineare Approximationsaufgaben, Proc. Symposium Optimierungsaufgaben Oberwolfach Nov. 1971, Birkhäuser Verlag, Internat. Ser. Num. Math. 17 (1973), 21-27.

L. Collatz [71a]: Some applications of Functional Analysis to Analysis particularly to Nonlinear integral equations, Proc. Symp. Nonlin. Funct. Anal., edited by Rall, Academic Press 1971, 1-43.

L. Collatz - W. Wetterling [71]: Optimierungsaufgaben, Springer, 2nd ed., 1971, 222 p.

L. Collatz - W. Krabs [73]: Approximationstheorie, Tschebyscheffsch Approximation mit Anwendungen, Teubner, Stuttgart 1973, 2o8 p.

U. Golomb [59]: Approximations by functions of fewer variables in Proc. Symp. On numerical approximation, edited by R.E. Langer, Univ. of Wisconsin, Madison 1959, p. 275-327.

L.W. Kantorowitsch - W.I. Kryloff [56]: Näherungsmethoden der höheren Analysis, Berlin 1956, 611 p.

W. Krabs [68]: Lineare Optimierung in halbgeordneten Vektorräumen, Num. Math. 11 (1968), 22o-231.

K. Kubik [71]: Optimale Linienführung innerhalb eines Korridors, ein nichtlineares Optimierungsproblem, Birkhäuser Verlag, Internat. Ser. Num. Math. 17 (1973), 91-1oo.

F. Lempio [71]: Lineare Optimierung in unendlichen dimensionalen Vektorräumen, Computing 8 (1971), 284-29o.

F. Lempio [72] : Dualitätssätze und einseitige Tschebyscheff Approximation, to appear.

O.L. Mangasarian [69] : Nonlinear programming, New York, McGraw Hill 1969, 22o p.

D.A. Sprecher 68 : On best approximations of functions of two variables, Duke Mathem. Journal 35 (1968), 391-397.

J. Stoer - C. Witzgall [7o] : Convexity and optimization in finite dimensions I., Springer 197o, 293 p.

Lothar Collatz
Universität Hamburg
Institut für Angewandte Mathematik
2ooo Hamburg 13
Rothenbaumchaussee 41

Mathematical Programming in Theory and Practice,
P.L. Hammer and G. Zoutendijk, (Eds.)

NONCONVEX QUADRATIC PROGRAMS, LINEAR COMPLEMENTARITY PROBLEMS, AND INTEGER LINEAR PROGRAMS(*)

F. Giannessi(†) – E. Tomasin(††)

1. Introduction.

This paper is concerned with the general quadratic programming problem, i.e. the problem of finding the minimum of a quadratic function, under linear constraints. As it is well known, such a problem occurs in many fields of mathematics, mechanics, economics, and so on.

When the function to be minimized is convex, the problem is well understood both theorically and computationally [3, 7, 9, 10]. Instead, when it is not convex, the problem is open at least as it concerns its resolution. In such a case the known methods for convex problems obtain a stationary point or, at most, a local minimum. Up to date there are several proposals to solve the general case [4, 8, 15, 17]. They belong essentially to two kinds of approaches, enumerative methods [4, 15] and cutting plane methods [8, 17]. The former is attractive and the most widely pursued, but, to date, the efficiency of the existing methods does not seem to be satisfactory. The latter has not produced up to date a general finite and rigorous method.

Here a finite method is proposed to solve the general quadratic programming problem, by solving the equivalent linear complementarity problem. Such a method, which is

---

(*) Research supported by the National Group of Functional Analysis and its Applications of Mathematics Committee of C.N.R.

(†) Dept. of Operations Research and Statistical Sciences - Univ. of PISA - Via S. Giuseppe, 22 - PISA (ITALY).

(††) Institute of Mathematics - Univ. of Venice - Ca' Foscari - Venice (ITALY)

a cutting plane method, enables one to find, among the vertices of a convex polyhedron, those which minimize a linear function and satisfy a given condition, for instance a complementarity condition or a zero-one integer condition. Thus, the method can be used to solve the general linear complementarity problem, which may be met indipendently from the quadratic problem, and to solve directly zero-one programs, without formulating them as concave quadratic programs. Some properties are stated to justify the method and to investigate the structure of the complementarity problem corresponding to the quadratic one. Such properties are used to define a branch and bound procedure to make pratically efficient the above mentioned method [18].

Moreover, it is shown how such a procedure can be accelerated, by taking advantage of a known optimality condition [19], and of other known devices. At last, a numerical example is given to clarify the exposition.

2. The general quadratic programming problem.

The problem with which we are concerned can be stated without loss of generality as[1]

$$P : \min \phi(x) = c^T x + \frac{1}{2} x^T Dx, \qquad x \in \{x : Ax \geqslant b;\ x \geqslant 0\} \ ,$$

where A is a matrix of order $(m \times n)$ and $D = D^T$. The following well known theorems hold.

Theorem 2.1. (Kuhn and Tucker [1, 14]). *If x is a local minimum for P, there exist vectors y, u, v, such that*

$$(2.1a) \qquad c + Dx - A^T y - u = 0$$

$$(2.1b) \qquad Ax - v = b$$

$$(2.1c) \qquad x^T u = y^T v = 0$$

$$(2.1d) \qquad x, y, u, v \geqslant 0 \ .$$

---

[1] The "T" as superscript and "min" will denote transposition and global minimum, respectively.

Theorem 2.2. ([7], p. 146). *If (x, y, u, v) is a solution of (2.1), then the equality*

$$\phi(x) = \frac{1}{2}(c^T x + b^T y) \tag{2.2}$$

holds.

Define

$$\hat{A} = \begin{pmatrix} D & -A^T \\ A & 0 \end{pmatrix} ; \quad \hat{b} = \begin{pmatrix} -c \\ b \end{pmatrix} ; \quad \hat{c} = \frac{1}{2}\begin{pmatrix} c \\ b \end{pmatrix} ; \quad z_I = \begin{pmatrix} x \\ y \end{pmatrix} ; \quad z_{II} = \begin{pmatrix} u \\ v \end{pmatrix} ; \quad z = \begin{pmatrix} z_I \\ z_{II} \end{pmatrix}$$

Theorem 2.3. *Let P have optimal solutions. The linear complementarity problem*

$$\min \hat{c}^T z_I, \quad z_I \in \{z : \hat{A} z_I - z_{II} = \hat{b} ; \ z_I^T z_{II} = 0; \ z \geqslant 0\} \tag{2.3.}$$

is equivalent[2] *to P.*

Theorem 2.3., which is a straightforward consequence of theorems 2.1., 2.2., shows that what we need to solve P is to solve a linear complementarity problem [11, 13]. For this reason the linear complementarity problem will concern us in the following sections.

3. The linear complementarity problem.

For sake of simplicity and without fear of confusion with the preceding notation, define now $x_I^T = (x_i)$, $x_{II}^T = (x_{m+i})$, $A = (a_{ij})$, $a^T = (a_{io})$, $c^T = (c_j)$, $i = 1,....,m$; $j = 1,...$ $.....$, $n = 2m$,

$$x^T = (x_I^T, x_{II}^T) ;$$

[2] In the sense that the first n elements of an optimal solution of the latter one are an optimal solution of the former one.

and consider the problem

$$Q : \min c^T x, \quad x \in R \triangleq \{x : Ax = a; \; x_I^T x_{II} = 0, \; x \geqslant 0\}.$$

We will now be concerned with Q, instead of (2.3), which is a particular case of Q. To this aim define $X \triangleq \{x : Ax = a; \; x \geqslant 0\}$, and let X^* be the convex closure of the points of X, which verify $x_I^T x_{II} = 0$; the following evident theorems hold.

Theorem 3.1. *The set of optimal solutions of Q is set of faces (in particular vertices) of X.*

PROOF. A vector $x \in R$ must have at least $n - m = m$ zero elements. Then, either an optimal solution (shortly, o.s.) of Q is a vertex of X, or it belongs to a face of X, whose points are o.s. of Q. The result follows.

As a conseguence of theorem 3.1. we find the following

Theorem 3.2. *Q is equivalent*[3] *to the linear problem*

$$\min c^T x, \quad x \in X^* \quad . \tag{3.1}$$

Of course, if we knew X^*, the problem (3.1) could be solved, instead of Q. Unfortunately, X^* is unknown in the general case; fortunately, it is sufficient to know only a subset of X^* containing an o.s. of Q. Then, it is natural the idea to determine such a subset to reduce Q to a linear programming problem.

4. A cutting plane method to solve Q. The first cut.

Consider the problem

$$Q_0 : \min c^T x, \quad x \in X.$$

---

[3] In the sense that they have the same o.s. and the same minimum.

Without loss of generality, Q_0 *can be assumed to have finite optimal solutions.* In fact, if $X = \emptyset$, then $R = \emptyset$, and then, by theorem 2.1, P has not finite o.s. If Q_0 has no finite o.s., Q and P may have finite o.s.. In such a case, let $\hat{Q}$ and $\hat{Q}_0$ denote respectively Q and Q_0, both under the additional constraints:

$$x_1 + \ldots\ldots + x_n + x_{n+1} = a_{00}, \qquad x_{n+1} \geqslant 0 \ . \tag{4.1}$$

Then, by the following self-evident

Theorem 4.1. *Q has finite optimal solutions, iff a real* a_{00} *exists, such that* $\hat{Q}$ *has finite optimal solutions satisfying the inequality* $x_{n+1} > 0$;

$\hat{Q}_0$ can be considered instead of Q_0, when Q_0 has no finite o.s., if a_{00} has been selected large enough. Theorems 3.1. and 3.2. hold for $\hat{Q}$ too. In fact, the o.s. of $\hat{Q}$, as well as Q, must have $m = n + 1 - (m + 1)$ zero elements. Theorem 3.1. follows in the present case too.

Let V_0 be a vertex of X, whose coordinates are a finite o.s. of Q_0, and let the reduced form

$$x_i + \alpha_{i,m+1}\, x_{m+1} + \ldots\ldots + \alpha_{in}\, x_n = \alpha_{io}, \qquad i = 1,\ldots\ldots, m \ , \tag{4.2}$$

of system $Ax = a$ give, at $x_{m+1} = \ldots\ldots = x_n = 0$, the coordinates of V_0. If the nonnegative n-vector

$$(\alpha_{10}, \ldots\ldots, \alpha_{mo}, 0, \ldots\ldots, 0) \ , \tag{4.3}$$

satisfies $x_I^T x_{II} = 0$, then it is an o.s. of Q too.

Otherwise, it is possible to determine a convex polyhedron, say it $\overline{X}$, which has only the vertices of X, but V_0.

To this aim define for a linear programming problem, like Q_0, a *valid cut* as an inequality (halfspace) which is satisfied by all the extreme points of the feasible region, but one, i.e., the present o.s. A *minimal valid cut* is a valid cut which is not implied by any other valid cut, i.e. it is a valid halfspace not contained in any other valid halfspace.

An *extreme valid cut* is a valid cut whose boundary cannot be expressed as a proper convex combination of boundaries of valid cuts.

Assume nondegeneracy for V_0, i.e.[4]

$$\alpha_{i0} > 0\,, \qquad i = 1,\ldots\ldots, m\,; \tag{4.4}$$

and define

$$\bar{x}_j \triangleq \sup\,\{x_j : x_i = \alpha_{i0} - \alpha_{ij}\, x_j \geqslant 0, \quad i = 1,\ldots\ldots, m\}\,, \qquad j = m+1,\ldots\ldots, n\,;$$

$$\alpha_{m+1,j} \triangleq \begin{cases} 0, & \text{if } \bar{x}_j = +\infty \\ \dfrac{1}{\bar{x}_j} & \text{if } \bar{x}_j < +\infty \end{cases} \quad ; \quad j = m+1,\ldots\ldots, n\,;$$

$$\alpha^T \triangleq (\alpha_{m+1,m+1}, \ldots\ldots, \alpha_{m+1,n})\,.$$

The inequality

$$\alpha^T x_{II} \geqslant 1 \tag{4.5}$$

will be said an *adjacent cut* for X (as it is weakly verified by all the vertices of X adjacent to V_0).

Consider the set $S \triangleq \{x_{II} : \alpha^T x_{II} < 1;\ x_{II} \geqslant 0\}$, which is a (not closed) simplex in a subspace of $\mathbb{R}^m$ (without fear of confusion, S will denote the above set of $\mathbb{R}^m$, and the set $\{x : \alpha^T x_{II} < 1;\ x \geqslant 0\}$ of $\mathbb{R}^n$).

Theorem 4.2. (i) *The adjacent cut (4.5) is a valid cut, and it is an extreme valid cut, if* $\alpha > 0$. (ii) $X - S = \bar{X}$.

PROOF. (i) Because of (4.4) V_0 does not verify (4.5), and the inequality $\alpha \geqslant 0$ holds, so that S does not contain any vertex of X, but V_0; thus (4.5) is a valid cut. Moreover, it is an extreme

[4] The case of degeneracy will be considered in section 6.

cut too, as it is weakly verified by all the vertices of X adjacent to V_0. (ii) is a straightforward consequence of (i).

Remark that, if $\alpha = 0$, then S is the nonnegativity orthant, and $X - S = \overline{X} = \emptyset$; in fact, now V_0 is of course the only vertex of X. In such a case $R = \emptyset$, as V_0 does not satisfy the complementarity condition.

Denote by Q_1 the problem obtained by adding to Q_0 the constraints

$$(4.6) \qquad \alpha^T x_{II} - x_{n+1} = 1 \ ; \qquad x_{n+1} \geqslant 0 \ .$$

As the vector (4.3) does not verify $x_I^T x_{II} = 0$, by theorem 4.1., the addition of (4.6) to Q does not change its feasible region. Then, Q_1 is now associated to Q, in place of Q_0. Obviously, Q_1 has finite o.s. too. Indeed, it is easy to show that the o.s. of Q_1 weakly verify (4.5); so that, by theorem 4.2, they are elements of the set, $A_X(V_0)$, of the vertices of X, which are adjacent to V_0. Assume that the vertex $V_1 \in A_X(V_0)$ is an o.s. of Q_1, and assume that x_h, $(h > m)$, is the nonbasic variable (shortly, n.v.) that must be made positive in (4.2) to go from V_0 to V_1 (this implies $\overline{x}_h > 0$). Then, (4.2) can be put in the following form[5]

$$(4.7) \quad \begin{cases} x_i + \alpha'_{i,m+1} x_{m+1} + \dots + \alpha'_{i,h-1} x_{h-1} + \alpha'_{i,h+1} x_{h+1} + \dots + \alpha'_{i,n+1} x_{n+1} = \alpha'_{i0}, & i = 1,\dots, m; \\ x_h + \alpha'_{m+1,m+1} x_{m+1} + \dots + \alpha'_{m+1,h-1} x_{h-1} + \alpha'_{m+1,h+1} x_{h+1} + \dots + \alpha'_{m+1,n+1} x_{n+1} = \alpha'_{m+1,0} . \end{cases}$$

If V_1, given by (4.7) at $x_j = 0$, $j = m+1,\dots, n+1$, $j \neq h$, verifies $x_I^T x_{II} = 0$, then V_1 is an o.s. of Q too. Otherwise, it is natural to think of iterating the preceding

---

[5] We have $\alpha'_{m+1,n+1} = -1/\alpha_{m+1,h}$; moreover, for every $i = 1,\dots, m$ and $j = 0, m+1,\dots, n$, the equalities

$$\alpha'_{m+1,j} = \frac{\alpha_{m+1,j}}{\alpha_{m+1,h}} \ ; \quad \alpha'_{ij} = \alpha_{ij} - \frac{\alpha_{m+1,j}}{\alpha_{m+1,h}} \alpha_{ih} ; \quad \alpha'_{i,n+1} = \frac{\alpha_{ih}}{\alpha_{m+1,h}}$$

hold.

method, i.e. of building an inequality like (4.5) to exclude V_1 but not the other vertices of the feasible region of Q_1. This will be realized in section 6.

5. Some properties of the adjacent cut.

In the system (4.2) let x_p be a basic variable (shortly, b.v.), which equals zero at $x_h = \bar{x}_h$ (this implies $\alpha'_{po} = 0$); and let x_k, $(k > m;\ k \neq h)$, be a generic n.v. such that at $x_k = \bar{x}_k$ the basic solution of (4.2) gives us a vertex, say V_k, adjacent to V_0 in X; moreover, let $x_q = 0$, $(q \leqslant m)$, in (4.2) at $x_k = \bar{x}_k$.

Theorem 5.1. *V_1 and V_k are adjacent in $\bar{X}$.*

PROOF. The definitions of (4.6) and V_k imply

$$(5.1) \qquad \frac{1}{\alpha_{m+1,k}} = \frac{\alpha_{qo}}{\alpha_{qk}} = \min \left\{ \frac{\alpha_{io}}{\alpha_{ik}} : \alpha_{ik} > 0;\quad i = 1,\dots\dots, m \right\}.$$

Moreover, for every $i = 1,\dots\dots, m$, such that $\alpha'_{ik} > 0$[6], the inequality

$$(5.2) \qquad \frac{\alpha'_{io}}{\alpha'_{ik}} \geqslant \frac{\alpha'_{m+1,0}}{\alpha'_{m+1,k}}$$

holds. In fact, because of the very definition of the α'_{ij}, (5.2) is equivalent to $\alpha_{io} \geqslant \alpha_{ik}/\alpha_{m+1,k}$. Because of (4.4) and $\alpha_{m+1,k} > 0$, such an inequality is trivial, if $\alpha_{ik} \leqslant 0$; if $\alpha_{ik} > 0$, it is equivalent to the inequality $\alpha_{io}/\alpha_{ik} \geqslant 1/\alpha_{m+1,k}$, which holds, because of (5.1) (remark that the latest inequality is weakly verified at $i = q$). (5.2) implies

$$\frac{\alpha'_{m+1,0}}{\alpha'_{m+1,k}} = \frac{\alpha'_{qo}}{\alpha'_{qk}} = \min \left\{ \frac{\alpha'_{io}}{\alpha'_{ik}} : \alpha'_{ik} > 0;\ i = 1,\dots\dots, m \right\}.$$

These equalities show that in (4.7) x_h and x_q are the b.v. which first become zero, when the n.v.

---

[6] The inequality $\alpha'_{ik} > 0$ holds for at least one $i = 1,\dots\dots, m$, otherwise x_k should not be bounded in (4.7), as it has been assumed.

x_k increases. This completes the proof.

Let $A_{\overline{X}}(V_1)$ denote the set of vertices, which are adjacent to V_1 in $\overline{X}$. Then, theorem 5.1 can be restated as

Theorem 5.1. $A_X(V_0) - \{V_1\} \subseteq A_{\overline{X}}(V_1)$.

Let us now state the following

Theorem 5.2. *In the system (4.7) the inequalities (i)* $\alpha'_{p,n+1} > 0$; *(ii)* $\alpha'_{pj} \leqslant 0$, $j = m + 1, \ldots\ldots, n$, $j \neq h$ *hold*[7].

PROOF. (i) The definitions of (4.6) and V_1 imply $\alpha_{m+1,h} > 0$ and $\alpha_{ph} > 0$; thus, $\alpha'_{p,n+1} = \alpha_{ph}/\alpha_{m+1,h} > 0$. (ii) By the definition of α'_{pj}, the current of (ii) is equivalent to

$$\alpha_{pj} - \frac{\alpha_{m+1,j}}{\alpha_{m+1,h}}\,\alpha_{ph} \leqslant 0\,. \tag{5.3}$$

By definition of $\alpha_{m+1,j}$, $\alpha_{m+1,j} = 0$ implies $\alpha_{ij} \leqslant 0$, $i = 1, \ldots\ldots, m$; thus, if $\alpha_{m+1,j} = 0$, (5.3) becomes $\alpha_{pj} \leqslant 0$, and it holds.
If $\alpha_{m+1,j} > 0$, and $\alpha_{pj} \leqslant 0$, (5.3) is trivial, as $\alpha_{m+1,h}$, α_{ph} are positive. If $\alpha_{m+1,j} > 0$, and $\alpha_{pj} > 0$, then the definitions of $\alpha_{m+1,j}$ and V_1 imply $1/\alpha_{m+1,j} = \bar{x}_j \leqslant \alpha_{po}/\alpha_{pj}$, $1/\alpha_{m+1,h} = \alpha_{po}/\alpha_{ph}$, and thus (5.3), as $\alpha_{ph} > 0$. This completes the proof.

Theorem 5.1. and 5.2. are useful to show that, after a second cut has been made (if necessary), (4.6) become superfluous.

6. The case of degeneracy.

Consider the system (4.7), and assume that V_1 does not verify $x_I^T x_{II} = 0$. Then,

---

[7] As a consequence of theorem 5.2., it is easy to show that either $\alpha'_{p,m+1} + \ldots\ldots + \alpha'_{p,h-1} + \alpha'_{p,h+1} + \ldots\ldots$ $\ldots\ldots + \alpha'_{pn} < 0$, or the elements of $A_X(V_0)$ and V_0 are the only vertices of X, so that Q is now a trivial problem.

it is natural to try to determine a cut, like in section 4, to exclude V_1 too. To this aim the definition of adjacent cut has now to be enlarged, as $A_{\bar{X}}(V_1)$ may contain more than $n + 1 - (m + 1) = m$ elements (as there are more then m distinct edges incident to V_1) so that it could be impossible to determine an inequality like (4.5), weakly verified by all the elements of $A_{\bar{X}}(V_1)$. Remember that in (4.7) $\alpha'_{po} = 0$, so that to define an adjacent cut in the present case is equivalent to define an adjacent cut when degeneracy occurs, i.e. when (4.4) are not verified.

Then, assume to have the following linear programming problem

$$\text{(6.1a)} \qquad \min (c_1 x_1 + \ldots\ldots + c_N x_N)$$

subject to the constraints

$$\text{(6.1b)} \qquad x_i + \alpha_{i,M+1} x_{M+1} + \ldots\ldots + \alpha_{iN} x_N = \alpha_{io}, \qquad i = 1,\ldots\ldots, M,$$

$$\text{(6.1c)} \qquad x_j \geqslant 0 \qquad , \qquad j = 1,\ldots\ldots, N;$$

and that the vector

$$\text{(6.2)} \qquad (x_i = \alpha_{io}, \; i = 1,\ldots\ldots, M; \quad x_i = 0, \; i = M + 1,\ldots\ldots, N)$$

is an o.s. of (6.1) and does not verify $x_I^T x_{II} = 0$. When (4.4) are not verified we put $M = m$, $N = n$, and we identify (6.1b) with (4.2). After the first cut (4.6) has been determined, by suitable positions and by putting $M = m + 1$, $N = m + 1$, (6.1b) is identified with (4.7). Without loss of generality, we assume[8]

$$\text{(6.3)} \qquad \alpha_{io} > 0, \; i = 1,\ldots\ldots, \bar{M}; \; \alpha_{io} = 0, \; i = \bar{M} + 1,\ldots\ldots, M, \quad (0 < \bar{M} < M).$$

We will now show how to define and how to build a cut, which excludes, among the vertices of (6.1b, c), only the vertex (6.2). To this aim we need an efficient way to determine a convex polyhedral cone.

---

[8] In fact, if $\alpha_{io} = 0$, $i = 1,\ldots\ldots, M$, then (6.1b, c) has only one vertex, which does not verify $x_I^T x_{II} = 0$, and then Q has not o.s.; if $\alpha_{io} > 0$, $i = 1,\ldots\ldots, M$, we are in the case of section 4.

7. Determination of the edges of a convex polyhedron.

Consider the latest $M - \overline{M}$ of (6.1b), which may be equivalently rewritten as

$$\alpha_{i,M+1} x_{M+1} + \ldots\ldots + \alpha_{iN} x_N \leqslant 0, \qquad i = \overline{M} + 1, \ldots\ldots, M. \tag{7.1}$$

Define $C_0 \triangleq \{(x_{M+1}, \ldots\ldots, x_N) : x_{M+1} \geqslant 0; \ldots\ldots; x_N \geqslant 0\}$; and C_r as the intersection among C_0 and the first r halfspaces (7.1), $(r = 1, \ldots\ldots, M - \overline{M})$. The first step to determine a cut in the present case consists in determining the convex polyhedral cone $C = C_{M-\overline{M}}$. This will be realized in a gradual way (See also [6]). Denote by

$$x_j = \beta^r_{ij} t, \; j = M + 1, \ldots\ldots, N; \; t \geqslant 0; \; i = 1, \ldots\ldots, k_r; \; (\beta^r_{ij} \geqslant 0), \tag{7.2}$$

the parametric equations of the edges, say respectively

$$H^r_1, \ldots\ldots, H^r_{k_r}, \quad \text{of} \quad C_r, \; r = 0, \; 1, \ldots\ldots, M - \overline{M}. \tag{7.3}$$

Remark that, at $r = 0$, (7.2) are trivially known, and their order is such that H^0_i and H^0_{i+1}, $i = 1, \ldots\ldots, k_0$, $(H^0_{k_0} + 1 \equiv H^0)$, are *adjacent*, in the sense that they weakly verify $N - M - 1$ linearly independent inequalities of those which define C_0, $N - M - 2$ of them being the same. Now we assume to know C_r, i.e. to know (7.2), and that all the inequalities which define C_r are non-redundant[9]. It will be shown how to determine the convex polyhedral cone C_{r+1}. To this aim consider the $(r + 1)$ - th of (6.4), i.e.

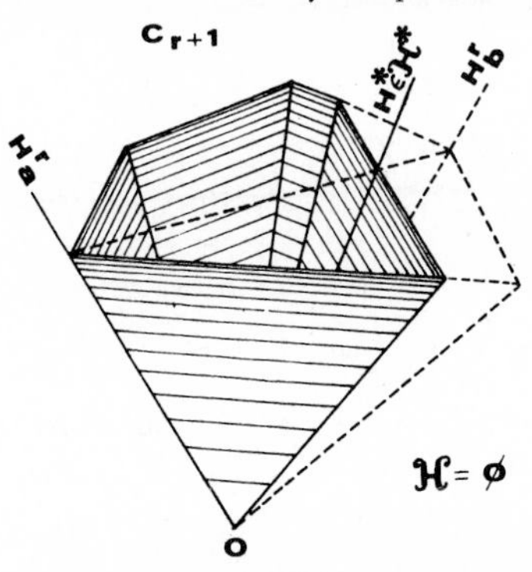

$$\alpha_{\overline{M}+r+1,M+1} x_{M+1} + \ldots\ldots + \alpha_{\overline{M}+r+1,N} x_N \leqslant 0. \tag{7.4}$$

Let H^r_a, H^r_b denote two generic edges among

[9] An inequality, and the corresponding hyperplane and halfspace defining C_r, is said non-redundant (redundant) for C_r, if C_r, i.e. its edges, is (not) modified, when it is deleted.

(7.3), such that H^r_a verifies (7.4) and H^r_b does not verify it; to this 2-uple the ray

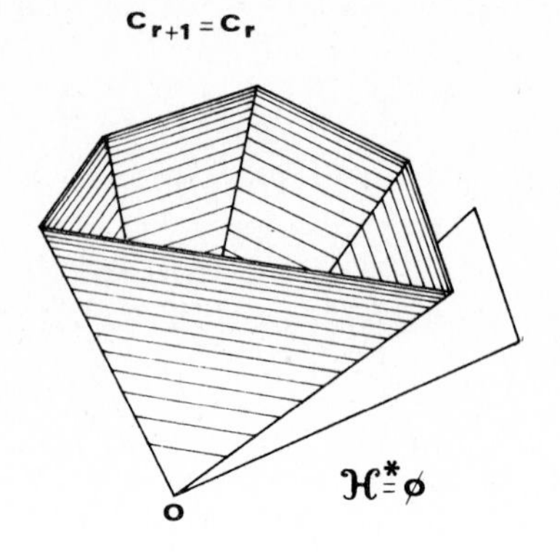

$$H^* = \mu^* H^r_a + (1 - \mu^*)\, H^r_b \quad ,$$

with

$$\mu^* = \frac{\alpha_{\overline{M}+r+1,M+1}\ \beta^r_{b,M+1} + \ldots\ldots + \alpha_{\overline{M}+r+1,N}\ \beta^r_{bN}}{\alpha_{\overline{M}+r+1,M+1}\ (\beta^r_{b,M+1} - \beta^r_{a,M+1}) + \ldots\ldots + \alpha_{\overline{M}+r+1,N}\ (\beta^r_{bN} - \beta^r_{aN})}$$

is associated, which weakly verifies (7.4). Repeat this for all the possible 2-uple of the kind H^r_a, H^r_b, and denote by $\mathbf{H}^*$ the set of rays $\mu^* H^r_a + (1 - \mu^*)\, H^r_b$ which are obtained this way. Moreover, let $\mathbf{H}^+$, $\mathbf{H}^0$ denote the sets of edges of C_{r+1}, which respectively strongly and weakly verify (7.4). Remark that the elements of $\mathbf{H}^+$ are obviously the edges (7.3), which strongly verify (7.4). About $\mathbf{H}^0$ the following theorems hold.

Theorem 7.1. *If* $\mathbf{H}^* \neq \emptyset$, *then* $\mathbf{H}^0 \subseteq \mathbf{H}^*$.

PROOF. Let $H^{r+1}_p \in \mathbf{H}^0$. As $H^{r+1}_p \in C_r$, H^{r+1}_p can be expressed as convex combination (shortly, c.c.) of (7.3):

$$H^{r+1}_p = \mu_1 H^r_1 + \ldots\ldots + \mu_{k_r} H^r_{k_r} \quad , \tag{7.5}$$

with

(7.6a) $\mu_1 + ... + \mu_{k_r} = 1; \quad \mu_i \geq 0, \quad i = 1,....., k_r$.

Moreover, $H_p^{r+1} \in \mathbf{H}^0$ implies

$$(7.6b) \qquad \left(\sum_{i=1}^{k_r} \sum_{j=M+1}^{N} \alpha_{\overline{M}+r+1,j} \; \beta_{ij}^r \right) \mu_i = 0 \; .$$

Remark that (7.6) defines a bounded convex polyhedron, so that its feasible solutions can be expressed as c.c. of its extreme points, which have at most two positive coordinates. It will be shown that (i) if $\mathbf{H}^* \neq \emptyset$, an element of $\mathbf{H}^*$ corresponds to an extreme point of (7.6). In fact, a c.c., say H, of two of (7.3), say H_i^r, H_j^r, corresponds to an extreme point of (7.6). As H weakly verifies (7.4), either both H_i^r, H_j^r weakly verify (7.4), or they are a 2-uple of the kind H_a^r, H_b^r, which exists if $\mathbf{H}^* \neq \emptyset$. In the former case, H cannot be a proper c.c. (i.e. a c.c. with positive coefficients), otherwise the assumption that it corresponds to an extreme point of (7.6) is contradicted. Thus, as everyone of (7.3), which weakly verifies (7.4), can be expressed as c.c. of 2-uple H_a^r, H_b^r, $H \in \mathbf{H}^*$. In the latter one, $H \in \mathbf{H}^*$ is trivial. (i) follows. From (7.5) and (i) it follows that H_p^{r+1} can be expressed as c.c. of the elements of $\mathbf{H}^*$. Now remark that the convex hull of $\mathbf{H}^*$ belongs to C_{r+1}. Then, H_p^{r+1} can be expressed as proper c.c. of at least two elements of $\mathbf{H}^*$, if it does not coincide with any element of $\mathbf{H}^*$; and the same happens in C_{r+1}. Thus, the assumption that H_p^{r+1} is an edge of C_{r+1} is contradicted. This completes the proof.

Theorem 7.2. *Let $C_{r+1} \neq \emptyset$. (i) The inequality (7.4) is redundant for C_{r+1}, iff $\mathbf{H}^* = \emptyset$. (ii) The s-th inequality of (7.1), $s = 1,....., r$, is redundant for C_{r+1}, iff everyone of (7.3), which weakly verifies the s-th of (7.1), does not strongly verify (7.4).*

PROOF. (i) is a straightforward conseguence of theorem 6.1. (ii) *Sufficiency.* Assume that the s-th of (7.1), i.e.

$$(7.7) \qquad \alpha_{s,M+1} \; x_{M+1} + ... + \alpha_{sN} \; x_N \leq 0 \; ,$$

is non-redundant for C_{r+1}; so that an edge of C_{r+1}, say H, exists, which weakly verifies (7.7) and is deleted, when (7.7) is. Either H strongly, or it weakly verifies (7.4). In the former case H is also an edge of C_r, and the assumption of (ii) is contradicted. In the latter one H cannot be an edge of C_r^-, otherwise it should not be deleted, when (7.7) is; then H must be a proper c.c. of at least two edges of C_r weakly verifying (7.7) and (7.4) and the assumption that H is an edge of

C_{r+1} is contradicted.

Necessity. Let the necessity of (ii) be false. Then, as all the inequalities which define C_r are assumed to be non-redundant for C_r itself, at least one of (7.3), say H, weakly verifies (7.7) and strongly verifies (7.4). Then, H is an edge of C_{r+1} and it looses such a property, when (7.7) is deleted. The assumption that (7.7) is redundant is contradicted. This completes the proof.

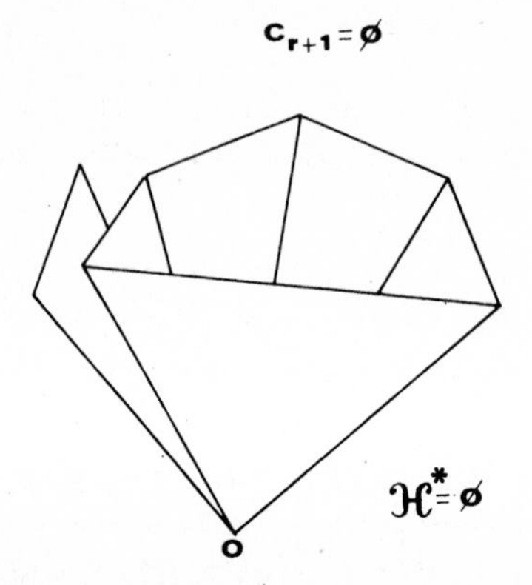

Theorem 7.1. and 7.2. enable one to determine easily C_{r+1}, and the set of non-redundant inequalities defining it. In fact, if $H^* = \emptyset$, either $C_{r+1} = \emptyset$, or $C_{r+1} \equiv C_r$; these two ways are easily checked. If $H^* \neq \emptyset$, the edges of C_{r+1} are quickly given by the union of H^* and the subset of H^*, whose elements weakly verify $N - M - 1$ linearly independent inequalities of those which define C_{r+1}. Then, by theorem 7.2, it is easy to eliminate the redundant inequalities for C_{r+1}. At $r = M - \overline{M}$, (7.2), which now are denoted by

$$x_j = \beta_{ij} t, \quad j = M + 1, \ldots\ldots, N; \quad t \geqslant 0; \quad i = 1, \ldots\ldots, k; \quad (\beta_{ij} \geqslant 0), \tag{7.8}$$

are the parametric equations of the edges of C, which now are denoted by

$$H_1, \ldots\ldots, H_k, \tag{7.9}$$

respectively.

If necessary, it is possible to obtain the edges of C_{r+1} satisfying the adjacent property above mentioned for C_0.

8. A generalized adjacent cut.

We are now ready to define an adjacent cut in the case of degeneracy. To this aim we will determine the set, say $A_X(V_0)$, of vertices of the convex polyhedron X, defined by (6.1b, c), which are adjacent to the vertex V_0, given by (6.2)[10].

To this aim put

$$\text{(8.1)} \qquad \bar{t}_s \triangleq \sup \{t : \left(\sum_{j=M+1}^{N} \alpha_{ij} \beta_{sj} \right) t \leqslant \alpha_{i0}, \quad i = 1,\ldots\ldots, \bar{M}\}, \qquad s = 1,\ldots\ldots, k .$$

Without loss of generality, (8.1) *can be assumed to by finite*[11]. Then, the elements of $A_X(V_0)$ are the points[12]

$$\text{(8.2)} \qquad V_s \triangleq (\bar{x}_{sj} = \beta_{sj}\, \bar{t}_s, \; j = M+1,\ldots\ldots, N), \qquad s = 1,\ldots\ldots, k .$$

If $k \leqslant N - M$, a suitable use of the method of section 4 leads us to a cutting hyperplane like (4.6), which is said an *adjacent cut* here too. If $k > N - M$, an inequality like (4.5), weakly verified by (8.2), may not exist. To overcome this difficulty we accept the possibility to determine more than one inequality like (4.5), if $k > N - M$. In such a case define $I = \{1,\ldots\ldots, k\}$; $\alpha^T = \{\alpha_{M+1},\ldots\ldots, \alpha_N\}$; and consider the set

$$A \triangleq \{\alpha : V_s^T \alpha \geqslant 1; \; s \in I\} .$$

Theorem 8.1. *(i)* $A \neq \emptyset$ *(ii)* $\alpha^T x_{II} \geqslant 1$ *is an extreme valid cut for* X *iff* α *is a vertex of* A.

PROOF. Because of the very definition of (8.2), the relations $V_s \geqslant 0$; $V_s \neq 0$ hold for any $s \in I$; thus there exists a $(N - M)$ -uple $\epsilon > 0$, large enough, such that $\epsilon \in A$, so that $A \neq \emptyset$.
(ii) Define

$$S \triangleq \{x_{II} : \alpha^T x_{II} < 1; \; x_{II} \geqslant 0\} ;$$

---

[10] Without fear of confusion, the notation X, V_0, $A_X(V_0)$ of sections 3, 4, is used again here.

[11] In fact, (6.1b) can be enlarged (if necessary) with the constraints $x_{M+1} + \ldots\ldots + x_N + x_{N+1} = \alpha_{00}$; $x_{N+1} \geqslant 0$, α_{00} large enough, like (4.1).

[12] Without fear of confusion V_s denotes both the point, and the $(N - M)$ - uple.

Sufficiency. Every vertex, say V^*, of X, but V_0, belongs to $X - S$. In fact, for any $s \in I$, $V_s \in X - S$, $V_0 \notin X - S$, and thus there exists a point $\tilde{V}_s$ of the segment $[V_0, V_s]$, such that $\tilde{V}_s^T \alpha = 1$. The assumption that $V^* \notin X - S$ implies that V^* can be expressed as proper c.c. of points of the convex closure of S, and it contraddicts the fact that V^* is an extreme point of X. It follows that $\alpha^T x_{II} \geqslant 1$ is a valid cut.
It is also extreme as α cannot be expressed as a proper c.c. of elements of **A**. *Necessity.* α cannot be expressed as proper c.c. of any other $(N - M)$ -uples, corresponding to valid inequalities, and thus, a fortiori, belonging to **A**. It follows that α is a vertex of **A**. This completes the proof.

Let $\bar{X}$ denote the union of the difference between X and the convex hull of its vertices, and the convex hull of the vertices of X, but V_0.

By using theorem 8.1. it is possible to construct a cut also in the present case to obtain the polyhedron $\bar{X}$, like in section 4. To this aim let $\{\alpha^1, \ldots\ldots, \alpha^p\}$ be the set of all the extreme points of **A**. Everyone of the inequalities

$$(8.3) \qquad \alpha^{i^T} x_{II} \geqslant 1 , \qquad i = 1, \ldots\ldots, p ,$$

is, by theorem 8.1., an extreme valid cut for X; the system (8.3) will be said a *generalized adjacent cut* for X. To justify such a definition it will be shown that the intersection among X and the halfspaces defined by (8.3) is the very polyhedron $\bar{X}$, like in section 4.
To this aim consider the sets

$$S^i \triangleq \{x_{II} : \alpha^{i^T} x_{II} < 1;\ x_{II} \geqslant 0\}, \qquad i = 1, \ldots\ldots, p .$$

Theorem 8.2. $X - \bigcup_{i=1}^{p} S^i = \bar{X}$.

PROOF. Put $S \triangleq \bigcup_{i=1}^{p} S^i$. Theorem 8.1. implies $\bar{X} \subseteq X - S$, as everyone of (8.3) is a valid cut. Assume $(X - S) - \bar{X} \neq \emptyset$, and consider $V \in (X - S) - \bar{X}$. As V_0, $V \notin \bar{X}$, there exists an hyperplane, say $\hat{\alpha}^T x_{II} = 1$, which strictly separates the segment $[V_0, V]$ and the convex set $\bar{X}$ ($\bar{X}$ is contained in the halfspace $\hat{\alpha}^T x_{II} \geqslant 1$). As $\hat{\alpha}^T x_{II} \geqslant 1$ is a valid cut for X, $\hat{\alpha} \in$ **A**; thus $\hat{\alpha}$ can be expressed as a c.c. of the extreme points of **A**. It follows that $\hat{\alpha}^T x_{II} \geqslant 1$ is implied by system (8.3); this contraddicts the fact that V does not satisfy $\hat{\alpha}^T x_{II} \geqslant 1$. Then, $(X - S) - \overline{X} = \emptyset$. This completes the proof.

The method exposed in the previous sections can be summarized by the following steps.

Step 0. The problem P is given. Determine the problem Q_0 ($\hat{Q}_0$, if necessary). Put $r = 0$, and go to step 1.

Step 1. Q_r is solved (if $r > 0$, an o.s. of Q_r is easily obtained, starting with an o.s. of Q_{r-1}). If an o.s. of Q_r verifies the complementarity condition, an o.s. of P is obtained as a subvector of it; terminate the algorithm. Otherwise, go to step 2.

Step 2. An adjacent cut is determined (by the method either of section 4 or of the present one). The constraints, which form the present cut (the inequality (4.5) or the system (8.3)), are in addition to Q_r to define a new Q_r. Go to step 1.

The preceding algorithm is *finite*. In fact, at step 2 the number of inequalities to be determined is finite, because of theorem 8.1. Thus, every iteration of steps 1 and 2 requires a finite number of elementary operations.

Of course the crucial point is the number of inequalities (8.3), i.e. the number of extreme points of A, which may be huge in pratical problems. The following sections are devoted essentially to show how the present method can be used for practical purposes.

9. Relaxation properties of the complementarity problem Q.

Computational experience has shown that the most part of the cuts like (8.3), required to solve P by the method of the preceding sections, is necessary to eliminate the vertices due to the introduction of the capacity constraint (4.1). These vertices generally are not vertices of X, so that they have to be cut off. Then, it is useful to know something more about the problem Q_0, i.e. about the linear program obtained by relaxing the complementarity condition in the problem Q. To this aim let us come back to the notation of section 2.

Theorem 9.1. *If the feasible region of P is nonempty, bounded and does not reduce to the origin of* $\mathbb{R}^n$, *then the equality*

$$(9.1) \qquad \inf \left\{ \frac{1}{2} (c^T x + b^T y) : (x, y) \in R \right\} = -\infty$$

holds for Q_0, *i.e. for problem (2.3) relaxed by deleting the complementarity condition.*

PROOF. Remark that Q_0 and its dual problem can be put respectively in the following forms[13]:

$$(9.2a) \quad \begin{cases} \min \, (c^T x + b^T y) \\ Dx - A^T y \geqslant -c \\ Ax \geqslant b \\ x, y \geqslant 0 \end{cases} \quad ; \qquad (9.2b) \quad \begin{cases} \max \, (-c^T w + b^T z) \\ -Dw - A^T z \geqslant -c \\ Aw \geqslant -b \\ w, z \geqslant 0 \end{cases}$$

To show (9.1) it is enough to show that the feasible region of (9.2b) is empty. The assumptions made for the feasible region of P, i.e. $\{x : Ax \geqslant b;\ x \geqslant 0\}$, imply $\{x : Ax \geqslant -b;\ x \geqslant 0\} = \emptyset$ [18]. It follows that the feasible region of (9.2b) is empty. This completes the proof.

Of course, the set of problems for which (9.1) holds is completely characterized by means of problems (9.2): the feasible regions of (9.2) to be respectively nonempty and empty is necessary and sufficient condition for (9.1) to hold.

Unfortunately the assumptions of theorem 9.1 are usually satisfied in practical problems; if P is strictly concave, then its feasible region must be bounded to have finite extremum. It follows that usually the constraint (4.1) must be introduced and the efficiency of the method is drastically reduced. In section 12 it will be shown how to

---

[13] The objective function of (9.2a) is twice the one of Q_0. This makes no difference to show (9.1)

overcome such a difficulty by using the preceding method as a subalgorithm in a branch and bound procedure. To this aim we need a deeper knowledge about the relaxed problem Q_0.

As Q_0 usually has infinite extremum, and as a branch may be obtained by putting equal to zero in Q_0 either a variable or its complement, the following question is natural.

Given a subspace of $\mathbb{R}^{2(n+m)}$, has the restriction of Q_0 to it finite extremum?

When the assumptions of theorem 9.1 are satisfied the only unbounded variables are the elements of y and u; some of these elements have to be put equal to zero. Let the subsets $\bar{I} \subseteq I \triangleq \{1,\ldots\ldots, m\}$; $\bar{J} \subseteq J = \{1,\ldots\ldots, n\}$ be given. The following theorem gives an answer to the preceding question.

Theorem 9.2. *Let the feasible region of P be nonempty, bounded. The problem*

$$(9.3) \qquad Q_0, \quad \text{where} \quad y_i = 0, \ i \in \bar{I}; \quad u_j = 0, \ j \in \bar{J},$$

has finite extremum iff the systems

$$(9.4a) \quad \begin{cases} Dx - A^T y - u = -c \\ Ax - v = b \\ x, y, u, v \geqslant 0 \\ y_i = 0, \ i \in \bar{I} \\ u_j = 0, \ j \in \bar{J} \end{cases} \qquad (9.4b) \quad \begin{cases} \sum_{j=1}^{n} a_{ij} w_j \geqslant -\frac{1}{2} b_i, \ i \in I - \bar{I} \\ w_j \geqslant 0, \quad j \in J - \bar{J} \end{cases}$$

posses solutions.

PROOF. Consider the dual of problem (9.3) (where the u and v are regarded as slack variables and used to transform the constraints of (9.3) into inequalities):

(9.5)
$$\begin{cases} \max\ (-c^T w + b^T z) \\ Dw + A^T z \leqslant \frac{1}{2} c \\ \sum_{j=1}^{n} a_{ij} w_j \geqslant -\frac{1}{2} b_i, & i \in I - \bar{I} \\ z \geqslant 0; \quad w_j \geqslant 0, & j \in J - \bar{J}\ . \end{cases}$$

By the fundamental duality theorem, (9.3) has finite extremum iff the feasible regions of (9.3) and (9.5) are nonempty. Such a condition can be simplified. To this aim remark that (9.5) can be written in the following form

(9.6)
$$-c^T w + \max\ b^T z$$
$$\begin{cases} A^T z \leqslant \frac{1}{2} c - Dw \\ z \geqslant 0 \end{cases}$$

where w is regarded as fixed and satisfying (9.4b).

The dual of (9.6) is the problem

(9.7)
$$-c^T w + \min\ (\frac{1}{2} c - Dw)^T x$$
$$\begin{cases} Ax \geqslant b \\ x \geqslant 0 \end{cases}$$

which, by the two assumptions of theorem 9.2., has finite extremum. Thus, the same happens for (9.6), for any w. It follows that (9.3) has finite extremum iff its feasible region, i.e. (9.4a), is nonempty, and a w exists, satisfying (9.4b). This completes the proof.

As it will be shown in section 12, the theorem 9.2 suggests a rule to choose the variables on which the branch has to be done in order to reduce the above mentioned consequences due to the introduction of the constraint (4.1).

10. Connections between nonconvex and concave quadratic problems.

When the nonconvex problem P is not strictly concave, it is interesting to decompose P in subproblems either convex or strictly concave. If this is possible, then the first ones can be solved using one of the well-known methods, while for the second ones the algorithm previously described can be used. Exploiting this way the structure of a nonconvex quadratic problem one can use the above cutting plane method to solve efficiently also large-scale problems, provided that the "strictly concave part" of the problem is small enough.

Such a decomposition can be realized by a tree like procedure [5], based on the following properties.

Theorem 10.1. *If $\phi(x)$ is not strictly convex on the feasible region of P, i.e. on*

$$\{x : Ax \geqslant b; \quad x \geqslant 0\}, \tag{10.1}$$

then it attains its global minimum on the relative boundary of the feasible region of P.

Theorem 10.2. *The relative boundary of an n-dimensional polyhedron is the union of its $(n-1)$-dimensional closed faces.*

A facial decomposition of (10.1), and then of P, is naturally suggested by the preceding theorems. First the convexity of $\phi(x)$ is tested on (10.1); if $\phi(x)$ is convex, P is a convex problem and so it is easily solved. Otherwise, an o.s. of P lies on at least one of the $(n-1)$–dimensional faces of (10.1). Then P can be replaced by the subproblems, which have as feasible regions the very $(n-1)$–dimensional faces of (10.1), respectively, and as objective functions the restrictions of $\phi(x)$ on them. Such a procedure is iteratively used for each of the current subproblems. A tree, whose nodes correspond to the several subproblems, is then constructed; the arcs correspond to a problem obtained by putting equal to zero a certain variable. The optimal solutions of the subproblems, which are either convex or strictly concave, so that they are not further decomposed, but are solved, must be recorded. The values of $\phi(x)$ corresponding to them are at last compared to find the global minimum of P.

It remains to be shown how to obtain the $(n-1)$–dimensional faces of a given polyhedron. To this aim denote by B_i the i-th row of the matrix $\binom{A}{I}$; by β_i the i-th element of the vector $\binom{b}{0}$; then (10.1) can be rewritten as

$$\{x : B_i\, x \geqslant \beta_i\,, \qquad i = 1,\ldots\ldots, m + n\}\ . \tag{10.2}$$

Let I^0 be a subset of $\{1,\ldots\ldots, m+n\}$, and consider the relaxation of (10.2)

$$\{x : B_i\, x \geqslant \beta_i\,, \qquad i \in I^0\}\,. \tag{10.3}$$

I^0 is said *minimal*, if (10.3) is equal to (10.2), and if I^0 has the lowest cardinality.

Now it is easy to see that (10.1) has $|\,I^0\,|$ $(n-1)$–dimensional faces, given by

$$\{x : B_h\, x = \beta_h;\ B_i\, x \geqslant \beta_i,\quad i \in I^0 - \{h\}\}\,, \qquad h \in I^0\ . \tag{10.4}$$

From a computational viewpoint the set I^0 has to be found. This can be obtained by solving the following linear programs

$$\begin{cases} \min\ B_r\, x \\ B_i\, x \geqslant \beta_i, \quad i = 1,\ldots\ldots, m+n; \quad i \neq r \end{cases} \tag{10.5}$$

$$r = 1,\ldots\ldots, m+n\ .$$

$r \in I^0$, iff the minimum in (10.5) is less than β_r.

The resolution of problems (10.5) may take advantage by the use of dual simplex algorithm, or of a sequential one [12]. In such a way the constraints can be considered in an increasing number in order to avoid to work on superfluous inequalities.

11. A sufficient condition of optimality for the concave quadratic program.

In the preceding section it has been shown how to reduce the resolution of a non-convex quadratic program to strictly concave quadratic problems, which appear to be the crucial part.

Then, it is important to have more insight about the strictly concave case.

Numerical experience has shown that after a certain number of cuts like (8.3) have been generated, i.e. a certain problem Q_j has been written, often the (x, v) coordinates of its o.s., corresponding to the variables of P, give us an o.s. of P, even if such an o.s. of Q_j does not yet satisfy the complementarity condition. Thus, a lot of additional cuts have to be produced with a considerable amount of computational effort. This computations can be eliminated, when $\phi(x)$ is concave. In such a case, as soon as the coordinates of a Q_j, corresponding to the variables (x, v) of P, give us a vertex of the feasible region of P, a known condition can be used to state whether this vertex is an o.s. of P or not [19, 16].

Assume $\phi(x)$ to be concave. Let (x^0, v^0, y^0, u^0) be an o.s. of the current r-th problem Q_r of section 8, and let (x^0, v^0) be a vertex of the feasible region of P, rewritten as

$$\{(x, v) : Ax - v = b; \ x, v \geqslant 0\} \ . \tag{11.1}$$

Assuming nondegeneracy, there are exactly n edges of (11.1) incident to (x^0, v^0); without loss of generality assume (11.1) to be bounded, so that there are exactly n vertices of (11.1) adjacent to (x^0, v^0), say them (x^i, v^i), $i = 1, \ldots, n$.

Denote by ξ the n-vector of the n.b. corresponding to (x^0, v^0); by $\mathbb{R}^n_+$ the nonnegativity orthant in the ξ-space; and by $\tilde{\phi}(\xi)$ the restriction of $\phi(x)$ to the ξ-space. Let ξ^i be the point of $\mathbb{R}^n_+$ corresponding to (x^i, v^i), $i = 0, 1, \ldots, n$; of course $\xi^0 = 0$.

Remark that there is no loss of generality in assuming that the inequalities

$$\tilde{\phi}(\xi^i) \geqslant \tilde{\phi}(\xi^0), \qquad i = 1, \ldots, n \quad , \tag{11.2}$$

hold[14]. Remark that (11.2) imply that (x^0, v^0) is a global minimum over the simplex defined by ξ^i, $i = 0, 1, \ldots, n$. Now it is natural to look for a simplex containing the preceding one, over which (x^0, v^0) is still a global minimum. Define the points $\hat{\xi}^i$, $i = 1, \ldots, n$, such that

$$\tilde{\phi}(\hat{\xi}^i) = \tilde{\phi}(\xi^0), \qquad i = 1, \ldots, n .$$

It is easy to remark that the $\hat{\xi}^i$ alway exist, are linearly independent, belong to $\mathbb{R}^n_+$; moreover $\hat{\xi}^i$ has only the i-th component different from zero[15].

Denote by $\hat{S}$ the simplex defined by the points ξ^0, $\hat{\xi}^i$, $i = 1, \ldots, n$. Remark that ξ^0 is still a global minimum over $\hat{S}$, and that no other simplex properly containing $\hat{S}$ exists, over which (x^0, v^0) is a global minimum.

A straightforward conseguence is that, if (11.1) is contained in $\hat{S}$, then (x^0, v^0) is a global minimum for P and the algorithm of section 8 can be terminated, even if a vertex of X, satisfying the complementarity condition, has not yet reached. In the contrary case, it is true at least that $\hat{S} - \{\hat{\xi}^i\}$ does not contain any global minimum of P. Then, the inequality

$$\sum_{i=1}^{n} \frac{1}{\hat{\xi}^i} \xi_i \geq 1 \tag{11.3}$$

can be added as new constraint to P. At this point either the method of section 8 is applied to the new problem P obtained this way, or (11.3) is in addition to the current Q_r of section 8.

The case of degeneracy for (x^0, v^0) can be handled with the same device used in integer programming in [2]. It consists essentially in eliminating a suitable set of the inequalities $Ax \geq b$, $x \geq 0$ in order to have exactly n vertices of (11.1) adjacent to (x^0, v^0).

---

[14] If $\tilde{\phi}(\xi^h) < \tilde{\phi}(\xi^0)$, by pivoting (x^0, v^0) can be replaced by (x^h, v^h) and so on, untill (11.2) hold; this procedure is justified by the obvious fact that (x^0, v^0) is not global minimum.

[15] The i-th element of $\hat{\xi}^i$ is quickly obtained as the positive root of the second degree equation

$$\tilde{\phi}(0, \ldots, 0, \xi_i, 0, \ldots, 0) = \tilde{\phi}(\xi^0) .$$

12. A branch and bound procedure to solve P.

Let us now come back to the general nonconvex problem P, to which the problem Q is associated as indicated in sections 3-8. Now it will be shown how to define a branch and bound procedure to solve Q, on which several devices (some of them have been shown in sections 10-11) can be applied.

Using the notation of section 2, z_i indicates an element of z_I, and z_{m+i} its complementary variable (element of z_{II}); the complementarity condition requires $z_i z_{m+i} = 0$, $i = 1,\ldots, m$. $[*]_{z_h}$ denotes the problem obtained from * by putting $z_h = 0$ in it.

By a *branch* on problem * containing z_i, z_{m+i} we mean the replacement of * by the subproblems $[*]_{z_i}$ and $[*]_{z_{m+i}}$. Starting from Q a finite tree can be associated to this branching procedure; the nodes correspond to the subproblems; the arcs to give zero value to a variable.

Assume to be at a certain node and to have found an optimal solution, with the method of section 8, of the corresponding problem; thus a *feasible* solution $\bar{z}$ of Q is available. Then $\bar{q} \triangleq \hat{c}^T\bar{z}_I$, i.e. the value of the objective function of Q corresponding to $\bar{z}$, can be used as a *bound* and $\bar{z}$ as the incumbent solution of Q. In fact, assume to be at another node of the tree, and to begin to solve the corresponding problem by the methos of section 8. The first step consists in relaxing it, by deleting the complementarity condition, and in solving the outcoming linear program; denote by q^* its minimum. If $q^* \geqslant \bar{q}$, then the subtree originated by the current node can be deleted. In such a way a branch and bound procedure is defined.

The facial decomposition outlined in section 10 can be connected with the present procedure. At a certain node of the tree assume that in the corresponding problem there are elements of (x, v) at zero value and belonging to the minimal set I^0 of section 10. Consider all of them and the corresponding face on which the convexity of $\phi(x)$ is tested. If $\phi(x)$ is convex, then the tree originated by the current node can be deleted. If not all of the above variables are considered then a greater dimensional face is met and a greater part of the tree may be deleted.

The implementation of the branch and bound procedure outlined above requires a lot of further rules for the choice of the variables on which to branch. Of course, large numerical experience is necessary to have much more insight in this crucial point.

13. Applications to integer linear problems.

Let $\tilde{A}, \tilde{b}, \tilde{c}$ be assigned integer matrices of orders $(m \times n)$, $(m \times 1)$, $(n \times 1)$, respectively. Consider the integer problem[16]

$$(13.1) \qquad \min \tilde{c}^T x, \quad x \in \{x : \tilde{A}x \geqslant \tilde{b}; \quad x \in B^n\}$$

where $B = \{0, 1\}$; $B^n = B \times B \times \dots \times B$ (n times).

The results of the preceding sections can be used to obtain two different approaches to solve (13.1), both of the cutting plane type.

Theorem 13.1. (Raghavachari [16]). *Without loss of generality, assume* $c \geqslant 0$. *Then the integer problem (13.1) is equivalent*[17] *to the continuous problem*

$$(13.2) \qquad \min\, [\tilde{c}^T x + \mu x^T (e - x)]\,, \qquad x \in \{x : \tilde{A}x \geqslant b; \quad 0 \leqslant x \leqslant e\},$$

where $e^T = (1, \dots, 1)$, *and* μ *is a positive large enough number.*

With the position (I_n is the identity matrix):

$$(13.3) \qquad A = \begin{pmatrix} \tilde{A} \\ -I_n \end{pmatrix}; \quad b = \begin{pmatrix} \tilde{b} \\ -e \end{pmatrix}; \quad c = \tilde{c} + \mu e\,; \qquad D = -2\mu I_n\,;$$

the problem P of section 2 coincides with (13.2).

Thus, by theorem 13.1, (13.1) is equivalent to the problem P, under (13.3), whose objective function is concave.

A first approach to solve (13.1) consists in solving the equivalent problem (13.2)

---

[16] Every integer linear problem can be obviously formulated in terms of 0-1 variables.

[17] See footnote 3.

by means of the method previously exposed. In such a way a cutting plane method to solve (13.1) is obtained; in fact, the inequality (4.5), or more generally the system (8.3), is a *valid cut* for the feasible region of (13.1), i.e. it does not exclude any integer feasible solution of (13.1). The method can be simplified by exploiting the particular structure of the quadratic problem (13.2). Moreover, when (13.1) is the formulation of a particular class of integer problems, the method can be further specialized[18].

A second approach consists in a straightforward application to (13.1) of the algorithm of sections 4-8. More precisely, without loss of generality, assume that an o.s., say x^0, of the problem

$$\text{(13.4)} \qquad \min \tilde{c}^T x,\ x \in \tilde{X} = \{x : \tilde{A}x \geqslant b;\ 0 \leqslant x \leqslant e\}\ ,$$

is known. If x^0 has integer co-ordinates, then it is an o.s. of (13.1) too. Otherwise, as an o.s. of (13.1) is a vertex of X, the polyhedron X of section 4 is identified with $\tilde{X}$, and (4.5) is addition to (13.4). As now it is not restrictive to assume that the coefficients of (4.5) be integers (with $\alpha_{m+1,0}$ as right-hand side, and not necessarily 1), the slack variable of (4.5) must be integer too. Thus, an o.s. of (13.1) can be obtained by comparing the integer elements of $A_X(V_0)$ (if any exist) with the o.s. of (13.1), having $\alpha^T x_{II} \geqslant 1 + \alpha_{m+1,0}$ (the elements of $A_X(V_0)$ have been enumerated directly and can be excluded; thus the slack variable of (4.5) can be constrained to be $\geqslant 1$) as additional constraint. It follows that, after $\alpha^T x_{II} \geqslant 1 + \alpha_{m+1,0}$ has been added to (13.4), the inequalities (4.4) are again satisfied and degeneracy is avoided. Let x^1 be an o.s. of (13.4) enlarged this way. If x^1 has integer co-ordinates, it is compared with the incumbent one, and the algorithm terminates. Otherwise, the method of section 4 is applied again.

If degeneracy is avoid by the preceding device instead of using sections 6-8, then convergence is not guaranteed.

14. Numerical examples.

Example 1. Assume $m = 1$, $n = 2$, $x^T = (x_1, x_2)$, and consider the concave problem

$$P : \min \phi(x) = \frac{1}{2}(x_1 + x_2 - x_1^2 - x_2^2),\quad x \in \{x : -2x_1 - x_2 \geqslant -6;\ x \geqslant 0\}\ ,$$

whose feasible region (the shaded area) and some levels of $\phi(x)$, corresponding to the stationary points of $\phi(x)$ are shown in Fig. 1.

System (2.1) becomes

[18] This is under study for the truck dispatching problem.

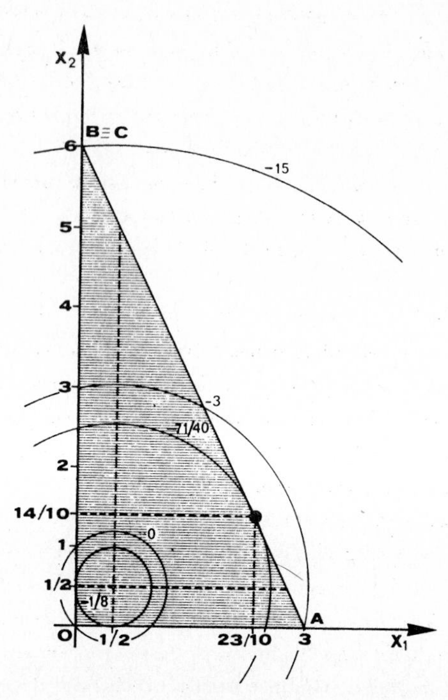

Fig. 1

(14.1a)
$$\begin{cases} \frac{1}{2} - x_1 \quad + 2y_1 - u_1 = 0 \\ \frac{1}{2} \quad - x_2 + y_1 \quad - u_2 = 0 \end{cases}$$

(14.1b)
$$-2x_1 - x_2 - v_1 = -6$$

(14.1c)
$$x_1 u_1 = x_2 u_2 = y_1 v_1 = 0$$

(14.1d)
$$x_1, x_2, u_1, u_2, y_1, v_1 \geqslant 0;$$

so that, by theorem 2.3., P is equivalent to the problem, which consists in finding the global minimum of $\left(\frac{1}{4}x_1 + \frac{1}{4}x_2 - 3y_1\right)$, under the constraints (14.1). By defining $x_3 = y_1$, $x_4 = u_1$, $x_5 = u_2$, $x_6 = v_1$, such a problem, which is the problem Q of section 3, becomes

(14.2a) $$\min \left(\frac{1}{4}x_1 + \frac{1}{4}x_2 - 3x_3\right)$$

under the constraints

(14.2b) $$\begin{cases} -x_1 + 2x_3 - x_4 = -\frac{1}{2} \\ -x_2 + x_3 - x_5 = -\frac{1}{2} \\ 2x_1 + x_2 + x_6 = 6 \end{cases}$$

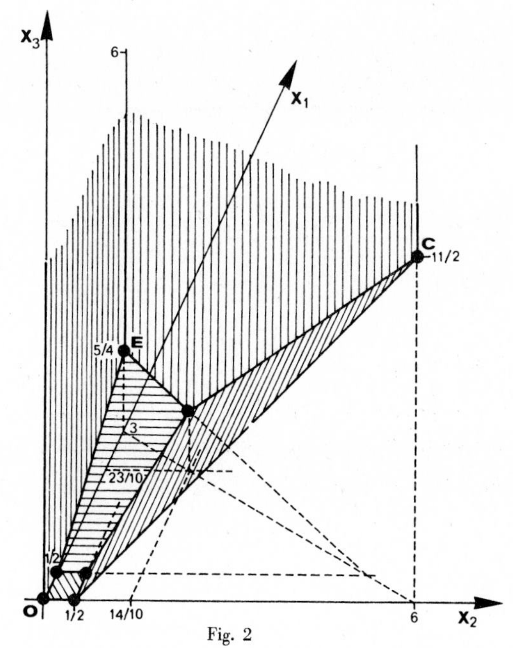

Fig. 2

(14.2c) $$x_j \geqslant 0, \qquad j = 1,\dots, 6 \ ;$$

(14.2d) $$x_j x_{j+3} = 0, \qquad j = 1, 2, 3 \ .$$

Fig. 2 shows a geometric interpretation of the feasible region of Q_0, i.e. the problem (14.2a, b, c) in the (x_1, x_2, x_3)–space; the shaded area is a part of the frontier; the full (empty) points are the vertices which (do not) verify the complementarity condition (14.2d). It is easily seen that Q_0 has no finite o.s., so that the additional constraints (4.1) have to be introduced:

(14.3) $$x_1 + \dots + x_6 + x_7 = 25 \ ; \qquad x_7 \geqslant 0 \ .$$

$\hat{Q}$ and $\hat{Q}_0$ are now problems (14.2), (14.3) and (14.2a, b, c), (14.3), respectively. Remark that, when (14.2b) hold, (14.3) are equivalent to $-2x_1 - x_2 + 4x_3 \leqslant 18$. Fig. 3 shows a geometric interpretation of the feasible region of $\hat{Q}_0$.

The system (4.2) is now

(14.4) $$\begin{cases} x_1 + \frac{1}{2}x_2 + \frac{1}{2}x_6 = 3 \\ x_3 + \frac{1}{4}x_6 + \frac{1}{4}x_7 = 6 \\ -\frac{1}{2}x_2 + x_4 + \frac{1}{2}x_7 = \frac{19}{2} \\ x_2 + x_5 + \frac{1}{4}x_6 + \frac{1}{4}x_7 = \frac{13}{2} \end{cases}$$

The b.s. given by (14.4), i.e: ($x_1 = 3$, $x_2 = 0$, $x_3 = 6$, $x_4 = \frac{19}{2}$, $x_5 = \frac{13}{2}$, $x_6 = x_7 = 0$), which corresponds to the point A both of Fig. 1 and Fig. 3, is an o.s. of Q_0 and it does not verify (14.2d); then (4.5) has to be introduced.
Here we have $\bar{x}_2 = \bar{x}_6 = 6$; $\bar{x}_7 = 19$; so that (4.5) becomes

(14.5) $$\frac{1}{6}x_2 + \frac{1}{6}x_6 + \frac{1}{19}x_7 \geqslant 1$$

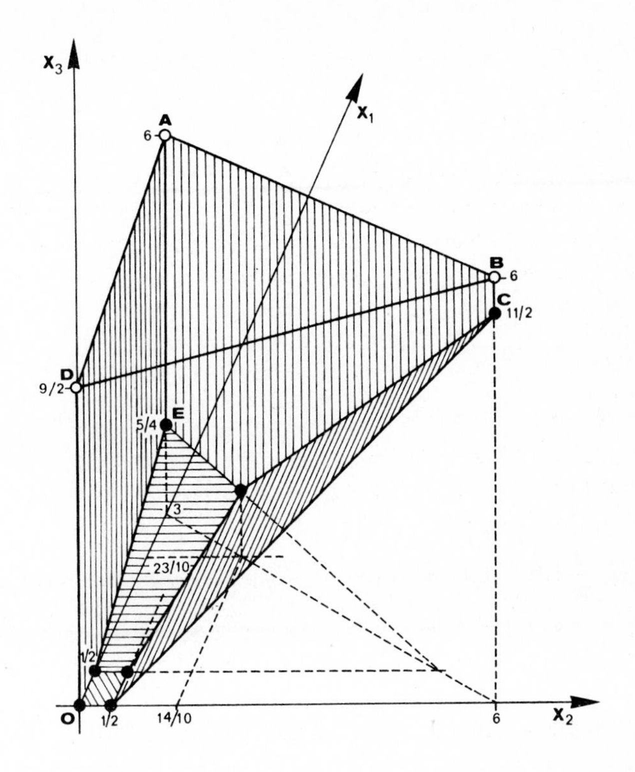

Fig. 3

and it is an addition to (14.4). Remark that, when (14.4) hold , (14.5) is equivalent to $\frac{13}{3}x_1 - x_2 + 4x_3 \leq 18$. Fig. 4 shows a geometric interpretation of system (14.4), (14.5) and the nonnegativity inequalities. Now remark that the system (14.4), (14.5) can be put in the following form

$$(14.6)\quad \begin{cases} x_1 - \frac{3}{19}x_7 + 3x_8 = 0 \\ x_2 + x_6 + \frac{6}{19}x_7 - 6x_8 = 6 \\ x_3 + \frac{1}{4}x_6 + \frac{1}{4}x_7 = 6 \\ x_4 + \frac{1}{2}x_6 + \frac{25}{38}x_7 - 6x_8 = \frac{25}{2} \\ x_5 - \frac{3}{4}x_6 - \frac{5}{76}x_7 + 6x_8 = \frac{1}{2} \end{cases}$$

The b.s. $(x_1 = 0,\ x_2 = x_3 = 6,\ x_4 = \frac{25}{2},\ x_5 = \frac{1}{2},\ x_6 = x_7 = x_8 = 0)$, given by (14.6), which corresponds to the point B both of Fig. 1 and Fig. 4, is an o.s. of the problem $\hat{Q}_1$, i.e. $\hat{Q}_0$ with the additional constraint (14.5), and it does not verify (14.2d). Then, the method of sections 6-8 must be applied. Here $M-\overline{M} = 1$, and (7.1) become $-\frac{3}{19}x_7 + 3x_8 \leqslant 0$; so that C is now the cone spanned by the rays $(x_6 = t,\ x_7 = x_8 = 0)$, $(x_6 = 0,\ x_7 = t,\ x_8 = 0)$, $(x_6 = 0,\ x_7 = 19t,\ x_8 = t)$, $t \geqslant 0$, which we identify with (7.8) at $k = 3$; (8.2) become respectively

$$(\overline{x}_{16} = 6, \overline{x}_{17} = \overline{x}_{18} = 0),\quad (\overline{x}_{26} = 0, \overline{x}_{27} = 19, \overline{x}_{28} = 0),\quad (\overline{x}_{36} = 0, \overline{x}_{37} = 2, \overline{x}_{38} = \frac{2}{19}),$$

and they correspond to the points D, E, C of Fig. 4, respectively. Remark that B and E are adjacent in the polyhedron of Fig. 4, while they were not in the polyhedron of Fig. 3. The same happens for D, E. As here $N - M = 8 - 5 = 3 = k$, we can avoid to find all the b.s. of Q, by applying again the method of section 4. Thus, (4.5) becomes now

$$(14.7)\qquad \frac{1}{6}x_6 + \frac{1}{19}x_7 + \frac{17}{2}x_8 \geqslant 1\ .$$

Remark that, when (14.6) hold, (14.7) is equivalent to $13x_1 - 2x_2 + 12x_3 \leqslant 54$. Fig. 5

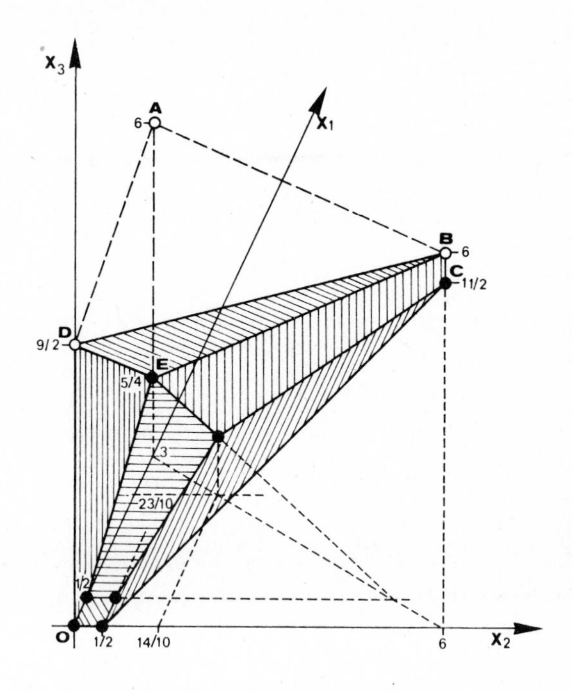

Fig. 4

shows a geometric interpretation of (14.6), (14.7) and the nonnegativity inequalities; the points C, E and D, E are adjacent in Fig. 5, while they are not in Fig. 4. The following form of (14.6), (14.7)

$$\begin{cases} 2x_1 + x_2 & & & + x_6 & & & & = 6 \\ \frac{17}{12}x_1 & + x_3 & & + \frac{1}{2}x_6 & & & + \frac{1}{2}x_9 & = \frac{11}{2} \\ \frac{23}{6}x_1 & & + x_4 & + \frac{1}{3}x_6 & & & + x_9 & = \frac{23}{2} \\ -\frac{7}{12}x_1 & & & x_5 + \frac{5}{6}x_6 & & & -\frac{1}{2}x_9 & = 0 \\ -\frac{17}{3}x_1 & & & + \frac{1}{3}x_6 + x_7 & & & + 2x_9 & = 2 \\ \frac{2}{57}x_1 & & & -\frac{1}{57}x_6 & & + x_8 & -\frac{2}{19}x_9 & = \frac{2}{19} \end{cases}$$

gives us the b.s. $(x_1 = 0,\ x_2 = 6,\ x_3 = 5,\ x_4 = \frac{23}{2},\ x_5 = x_6 = 0,\ x_7 = 2,$ $x_8 = \frac{2}{19},\ x_9 = 0)$ which is an o.s. of $\hat{Q}_2$, i.e. of the problem $\hat{Q}_1$ with the additional constraint (14.7). As such an o.s., which corresponds to the point C both of Fig. 1 and Fig. 5, verifies (14.2d), then it is an o.s. of $\hat{Q}$ and, because of the inequality $x_7 = 2 > 0$, of Q too. Thus, the point $B \equiv (x_1 = 0,\ x_2 = 6)$ of Fig. 1 is a global solution of the problem P.

Example 2. The use of the decomposition procedure of section 10 is useless if the problem is strictly concave. In such a case it leads to enumerate all the vertices of the feasible region of P (as it is easily seen in the preceding example). Instead, if P is neither convex nor strictly concave, then such a procedure can considerably simplify the resolution of P. For instance, assume $m = 1,\ n = 3,\ x^T = (x_1, x_2, x_3)$ and consider the problem.

$$(14.8)\qquad \begin{cases} \min \phi(x) = \frac{1}{2}(x_1 + x_2 + x_3 - x_1^2 - x_2^2 + x_3^2) \\ -2x_1 - x_2 - x_3 \geqslant -6 \\ x_j \geqslant 0, \qquad j = 1, 2, 3\ . \end{cases}$$

Remark that the minimal set I^0 of section 10 is $I^0 = \{1, 2, 3, 4\}$; so that there are

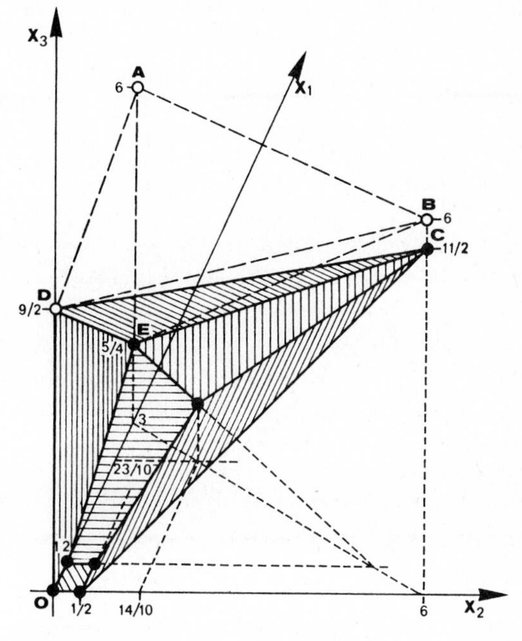

Fig. 5

$|I^0| = 4$ $(n-1)$-dimensional faces, and (10.4) become

$$F_1 = \{x : -2x_1 - x_2 - x_3 = -6; \quad x_j \geqslant 0, \ j = 1, 2, 3\};$$

$$F_{i+1} = \{x : x_i = 0; \ -2x_1 - x_2 - x_3 \geqslant -6; \ x_j \geqslant 0, \ j = 1, 2, 3, \ j \neq i\}; \ i = 1, 2, 3.$$

For every $j = 1, 2, 3, 4$, denote by P_j the problem, which has F_j as feasible region and the restriction of $\phi(x)$ on it as objective function. Because of theorems 10.1 and 10.2, P can be solved by finding, among the o.s. of P_j, the best ones.

Consider P_1, the objective function of which is given by $\frac{1}{2}(42 - 25x_1 - 12x_2 +$

$+ 3x_1^2 + 4x_1 x_2)$; as this form is nonconvex the facial decomposition must be carried on. By repeating the considerations of section 10, F_1 is replaced by its $(n - 2)$-dimensional faces

$$F_{11} = \{x : - x_2 - x_3 = -6;\ x_2, x_3 \geqslant 0;\ x_1 = 0\};$$

$$F_{12} = \{x : - 2x_1 - x_3 = -6;\ x_1, x_3 \geqslant 0;\ x_2 = 0\};$$

$$F_{13} = \{x : - 2x_1 - x_2 = -6;\ x_1, x_2 \geqslant 0;\ x_3 = 0\}.$$

These three faces define three subproblems. Anyway, it is easy to remark that every one of such $(n - 2)$-dimensional faces is contained in at least one face among F_2, F_3, F_4, so that the corresponding problems can be disregarded.

Let us now consider P_2, whose objective function is $\frac{1}{2}(x_2 + x_3 - x_2^2 + x_3^2)$; as this form is nonconvex, the decomposition of section 10 can be applied again. The feasible region of P_2 has three $(n-2)$-dimensional faces, which define in the usual way three subproblems of P_2, having the following o.s.

$$(x_1 = x_2 = x_3 = 0);\quad (x_1 = 0;\ x_2 = 6;\ x_3 = 0);\quad (x_1 = 0;\ x_2 = 6;\ x_3 = 0),$$

and, respectively, the minima $0, -15, -15$.

Let us now consider P_3, whose objective function is $\frac{1}{2}(x_1 + x_3 - x_1^2 + x_3^2)$; as this form too is nonconvex, the decomposition of section 10 leads us to three subproblems, having the following o.s.

$$(x_1 = x_2 = x_3 = 0);\quad (x_1 = 3;\ x_2 = x_3 = 0);\quad (x_1 = 3,\ x_2 = x_3 = 0),$$

and, respectively, the minima $0, -3, -3$.

Let us now consider at last P_4. Remark that this problem is that one of example 1, so that its o.s. is $(x_1 = 0,\ x_2 = 6,\ x_3 = 0)$, and the corresponding minimum is -15. By comparing the o.s. of the several subproblems, it turns out that $(x_1 = 0, x_2 = 6, x_3 = 0)$ is the o.s. of (14.8).

Remark 1. The b.s. given by (14.6) corresponds to point B of Fig. 1. At this point (11.2) are satisfied; the points of Fig. 1 corresponding to the $\hat{\xi}^i$ are now

$$(14.8) \qquad (x_1 = 0,\ x_2 = -5);\quad (x_1 = \frac{23}{5},\ x_2 = -\frac{16}{5}).$$

Then, the inequality, which corresponds to (11.3) in fig. 1, is now

$$\frac{9}{115}\,x_1 - \frac{1}{5}\,x_2 \geqslant 1\,.$$

As the addition of this inequality as a new constraint to P makes its feasible region empty, it can be concluded that $(x_1 = 0,\ x_2 = 6)$ is the o.s. of P, without introducing a new cut, and saving a lot of computations.

Remark 2. If we make a branch on problem (14.2) according to section 12, the introduction of the capacity constraint (14.3) can be avoided. For instance, the subproblem obtained by putting $x_4 = 0$ has finite minimum (see fig. 2). On the subproblem obtained by putting $x_1 = 0$ a new branch can be made; $x_5 = 0$ produces a new subproblem with finite minimum; $x_2 = 0$ produces a trivial problem.

REFERENCES

[1] ABADIE J., *On the Kuhn-Tucker Theorem.* In "Nonlinear Programming", *J.* Abadie (ed.), North-Holland Publ. Co., 1967, pp. 19-36.

[2] BALAS E., *Intersection cuts - A new type of cutting planes for integer programming.* Operations Research, vol. 19, 1971, pp. 19-39.

[3] BEALE E.M.L., *Numerical Methods.* In "Nonlinear Programming", J. Abadie (ed.), North-Holland Publ. Co., 1967, pp. 133-205.

[4] BURDET C.A., *General Quadratic Programming.* Carniege-Mellon University, Paper W.P. -41-71-2, Nov. 1971.

[5] BURDET C.A., *The facial decomposition method.* Graduate School of Industrial Administration, Carniege-Mellon Univ., Pittsburg, Penn., May 1972.

[6] CHERNIKOVA N.V., *Algorithm for finding a general formula for the nonnegative solution of a system of linear inequalities.* USSR Computational Mathematics and Mathematical Physics, V. I, pp. 228-233 (1965).

[7] COTTLE R.W., *The principal pivoting method of quadratic programming.* In "Mathematics of the decision sciences, Part. I, eds. G.B. Dantzig and A.F. Veinott Jr., American Mathematical Society, Providence, 1968, pp. 144-162.

[8] COTTLE R.W. and W.C. MYLANDER, *Ritter's cutting plane method for nonconvex quadratic programming.* In "Integer and nonlinear programming", J. Abadie (ed.), North-Holland Publ. Co., 1970, pp. 257-283.

[9] DANTZIG G.B., *Linear Programming and Extensions.* Princeton Univ. Press, 1963.

[10] DANTZIG G.B., A.F. VEINOTT, *Mathematics of the Decision Sciences.* American

Mathematical Society, Providence, 1968.

[11] EAVES B.C., *On the basic theorem of complementarity.* "Mathematical Programming", Vol. 1, 1971, n. 1, pp. 68-75.

[12] GIANNESSI F., *A sequential method for large-scale structured linear programs.* In "Studi e modelli di ricerca operativa" M. Volpato (ed.), UTET, Torino (Italy), 1972. An abstract in Proceedings of the 5-th IFORS Conference, J. Lawrence (ed.), Tavistock 1969.

[13] KARAMARDIAN S., *The complementarity problem.* "Mathematical Programming", Vol. 2, 1972, n. 1, pp. 107-123.

[14] KUHN H.W. and A.W. TUCKER, *Nonlinear Programming.* In: Second Berkeley Symp. Mathematical Statistics and Probability, ed. J. Neyman, Univ. of California Press, Berkeley, 1951, pp. 148-492.

[15] LEMKE C.E., *Bimatrix Equilibrium Point and Mathematical Programming.* "Management Science", Vol. 11, 1965, pp. 681-689.

[16] RAGHAVACHARI M., *On connections between zero-one integer programming and concave programming under linear constraints.* Operations Research, vol. 17, n.4, 1969.

[17] RITTER K., *A method for solving maximum problem with a nonconcave quadratic objective function.* Z. Wharscheinlichkeitstheorie, Vern. Geb. 4, 1966, pp. 340-351.

[18] TOMASIN E., *Global optimization in nonconvex quadratic programming and related fields.* Dept. of Operations Research and Statistical Sciences, Univ. of PISA. Paper N. 15. 1974.

[19] TUI HOANG, *Concave programming under linear constraints.* Soviet Math., 1964, pp. 1437-1440.

[20] ZWART P.B., *Nonlinear Programming: Counterexamples to global optimization algorithms proposed by Ritter and Tui.* Washington Univ., Dept. of Applied Math. and Computer Science. Research Report No. COO-1493-32.

Mathematical Programming in Theory and Practice,
P.L. Hammer and G. Zoutendijk, (Eds.)

MIN-MAX PROBLEMS

B. Lemaire*

The problem dealt with is that of minimizing the supremum of a compact set of functionals over a subset of a linear space. The main result connects the min-max problem with a convenient saddle-point problem. Other conditions for optimality are derived as also a constructive iterative procedure. Applications are mentioned to approximatinn theory, selection of a Pareto minimum and to optimal control with perturbations.

1. Definitions and statement of the problem.

Let E be a real linear space and $\mathfrak{R}$ and K two subsets of E such that K is contained in $\mathfrak{R}$.

Denote $\mathcal{F}(\mathfrak{R})$ the space of all functionals on $\mathfrak{R}$. supplied with the pointwise convergence topology. Define

$$\mathfrak{J} = \{j_\alpha \in \mathcal{F}(\mathfrak{R}) \mid \alpha \in A\}$$

where A is a topological compact space. Assume

$$\alpha \to j_\alpha(e) \qquad \text{continuous} \quad \forall\, e \in \mathfrak{R}$$

Define $\Psi \in \mathcal{F}(\mathfrak{R})$ by

$$\Psi(e) = \max_{\alpha \in A} j_\alpha(e) \,, \quad \forall\, e \in \mathfrak{R},$$
$$A(e) = \{\alpha \in A \mid j_\alpha(e) = \Psi(e)\}$$

Obviously $A(e)$ is a compact subset of A.

What we call the min-max problem is the following :

Find $\bar{e} \in K$ such that

$$\Psi(\bar{e}) = \inf_{e \in K} \Psi(e) \overset{\Delta}{=} \delta$$

Such an $\bar{e}$ will be called a min-max solution.

---

* Université MONTPELLIER II, Mathématiques,
Place E. Bataillon
34060 MONTPELLIER CEDEX, FRANCE.

2 - Conditions of optimality.

Let $\mathcal{C}(A)$ be the Banach space of all continuous real functions on A and $\mathcal{M}(A)$ the topological dual of $\mathcal{C}(A)$ that is to say the space of Radon measures on A, supplied with the weak topology $\sigma(\mathcal{M}(A), \mathcal{C}(A))$.

Denote by $\mathcal{M}^1_+(A)$ the set of Radon probabilities on A which, as one says, is a convex compact subset of $\mathcal{M}(A)$. Define

$$J : K \to \mathcal{C}(A) \quad \text{by} \quad J(e)(\alpha) = j_\alpha(e).$$

Let $\bar{e} \in K$ and $\bar{\lambda} \in \mathcal{M}^1_+(A)$ and consider the two following conditions

(1) $\quad < \lambda, J(\bar{e}) > \leq < \bar{\lambda}, J(\bar{e}) > \leq < \bar{\lambda}, J(e) >, \quad \forall e \in K, \forall \lambda \in \mathcal{M}^1_+(A)$

where $<.,.>$ = duality between $\mathcal{M}(A)$ and $\mathcal{C}(A)$,

(2) $\quad < \bar{\lambda}, J(\bar{e}) > = \inf\limits_{e \in K} < \bar{\lambda}, J(e) >$ and $\bar{\lambda} \in \mathcal{M}^1_+(A(\bar{e}))$.

Note 1 - Condition (1) means that $(\bar{e}, \bar{\lambda})$ is a saddle point of the functional $\mathcal{H}$ on $K \times \mathcal{M}^1_+(A)$ where

$$\mathcal{H}(e,\lambda) = < \lambda, J(e) > = \int_A j_\alpha(e)\, d\lambda(\alpha).$$

Condition (2) means that $\bar{e}$ minimizes (in a generalized sense which reduces to the usual one if A is a finite set) a convex combination of the j_α, $\alpha \in A$, where the "coefficients" vanish outside $A(\bar{e})$.

The main result is then the following

Theorem 1

1° Conditions (1) and (2) are equivalent.

2° Each of (1) or (2) implies $\bar{e}$ is a min-max solution and $< \bar{\lambda}, J(\bar{e}) > = \delta$.

3° The converse of 2° is true with additional convexity assumptions namely : if $\mathcal{R}$ and K are convex sets, if j_α is convex for every α in A and if $\bar{e}$ is a min-max solution then there exists $\bar{\lambda}$ in $\mathcal{M}_+^1(A)$ such that condition (1) holds.

proof : We first note that

$$(3) \qquad \max_{\alpha \in A} j_\alpha(e) = \max_{\lambda \in \mathcal{M}_+^1(A)} < \lambda, J(e) > \quad , \quad \forall e \in \mathcal{R}.$$

1° We derive from measure theory that the first inequality in (1) is equivalent to state that $\bar{\lambda}$ is concentrated on $A(\bar{e})$.

2° Quite obvious with property (3).

3° Obviously, there exists $\bar{\lambda}$ in $\mathcal{M}_+^1(A)$ such that

$$(4) \qquad \inf_{e \in K} < \bar{\lambda}, J(e) > = \max_{\lambda \in \mathcal{M}_+^1(A)} \inf_{e \in K} < \lambda, J(e) >$$

and from (3) we get

$$(5) \qquad \max_{\lambda \in \mathcal{M}_+^1(A)} < \lambda, J(\bar{e}) > = \inf_{e \in K} \max_{\lambda \in \mathcal{M}_+^1(A)} < \lambda, J(e) >$$

By a theorem of MOREAU ([8]), the right members of (4) and (5) are equal. Then it is classical ([5]) that $(\bar{e}, \bar{\lambda})$ is a saddle-point of $\mathcal{H}$.

Note 2 If we consider α as a random parameter, the min-max problem is that of minimizing the maximum risk and, in the convex case, theorem 1 says that it is equivalent to minimizing the expectation of $J(e)$ for a convenient probability on A.

Let us give now conditions of optimality involving differentiability or subdifferentiability. Assume the convexity properties as in th. 1, 3°. Suppose E is a topological real linear space and $\mathcal{R}$ is open. Denote $j'_\alpha(e\ ;\ h)$ the differential of j_α at e in the direction h (which exists for every e and h in $\mathcal{R}$ because j_α is convex and $\mathcal{R}$ is open), and $\partial\, j_\alpha(e)$ the subdifferential of j_α at e.

Theorem 2.

1° $\bar{e}$ is a min-max solution iff

$$(6)\qquad \inf_{e \in K}\ \max_{\alpha \in A(\bar{e})}\ j'_\alpha(\bar{e}\ ;\ e - \bar{e}) \geq 0.$$

2° If j_α is continuous or has a Gateaux derivative at $\bar{e}$ (so $\partial_{j_\alpha}(\bar{e}) \neq \emptyset$) for every α in $A(\bar{e})$, then $\bar{e}$ is a min-max solution iff

$$(7)\qquad \begin{cases} \exists\ \chi \in \overline{co}\{\partial j_\alpha(\bar{e})\ |\alpha \in A(\bar{e})\} \text{ such that} \\ \displaystyle\inf_{e \in K}\ < \chi,\ e - \bar{e} >_{E'E} \geq 0 \end{cases}$$

where $\overline{co}$ denotes the closed convex hull in E' topological dual of E supplied with the weak topology $\sigma(E',\ E)$.

proof :

1° $\psi(e) - \psi(\bar{e}) \geq j_\alpha(e) - j_\alpha(\bar{e}) \geq j'_\alpha(\bar{e}\ ;\ e - \bar{e})\ ,\ \forall\ \alpha \in A(\bar{e}).$

So (6) implies $\bar{e}$ is a min-max solution. For the converse, apply theorem 1. So (2) holds. Then, using the directional differential of $< \bar{\lambda},\ J(.) >$ at $\bar{e}$ (obtained by derivation under the sign $\int$, which has sense by the dominated convergence theorem of Lebesgue), (2) is classicaly equivalent to

$$\int_{A(\bar{e})} j'_\alpha(\bar{e}\ ;\ e - \bar{e})\quad d\lambda(\alpha) \geq 0\ ;\quad \forall\ e \in K$$

which obviously implies (6).

2° (6) and (7) are equivalent. For that, use the fact that

$j'_\alpha(e\,;\,h) = \max_{\chi \in \partial j_\alpha(e)} <\chi, h>$ ([9]) and apply the theorem of MOREAU above-mentioned.

Note that because $A(\bar{e})$ is compact and because the Krein-Milman theorem, the $\overline{co}$ mentioned in (7) is compact.

Note 3. We can also derive th.2 from a result due to ROCKAFELLAR ([11]) which describes the directional differential and the subdifferential of Ψ, namely

$$\Psi'(e\,;\,h) = \max_{\alpha \in A(e)} j'_\alpha(e\,;\,h)$$

$$\partial\Psi(e) = \overline{co}\,\{\partial j_\alpha(e) \,|\, \alpha \in A(\bar{e})\}$$

3. A Lagrangian principle.

Now assume K defined by a cone of constraints :

(8) $K = \{e \in \mathcal{R} \,|\, f(e) \le 0, \ \forall f \in \mathcal{L}\}$

where $\mathcal{L}$ is a closed pointed convex cone in $\mathcal{F}(\mathcal{R})$. There is no restriction to consider a cone of constraints. For instance if

(9) $K = \{e \in \mathcal{R} \,|\, f_t(e) \le 0 \ , \ \forall t \in T\}$

for some set T, then K can be defined as (8) taking for $\mathcal{L}$ the closed (in $\mathcal{F}(\mathcal{R})$) convex cone spanned by the set $\{f_t \ t \in T\}$.

Now define the Lagrangian

$$L : \mathcal{R} \times (\mathcal{M}^1_+(A) \times \mathcal{L}) \to \mathbb{R} \quad \text{by}$$

$$L(e\,;\,\lambda, f) = <\lambda, J(e)> + f(e).$$

Let $\mathcal{P}$ be the closed convex pointed cone of non negative functions of $\mathcal{F}(\mathcal{R})$ and consider the following condition

(10) $L(\bar{e} ; \lambda, f) \le L(\bar{e} ; \bar{\lambda}, \bar{f}) \le L(e ; \bar{\lambda}, \bar{f}), \forall e \in \mathcal{R}, \forall \lambda \in \mathcal{M}^1_+(A), \forall f \in \mathcal{L}$,

that is to say $(\bar{e} ; \bar{\lambda}, \bar{f})$ is a saddle-point of L.

Theorem 3.

1° Condition (10) implies $\bar{e}$ is a min-max solution and

$$L(\bar{e} ; \bar{\lambda}, \bar{f}) = \Psi(\bar{e}) = \delta .$$

2° The converse is true under additional convexity assumptions and a "regularity" condition on $\mathcal{L}$, namely : if $\bar{e}$ is a min-max solution, if $\mathcal{R}$ is convex, if j_α is convex $\forall \alpha \in A$, if f is convex $\forall f \in \mathcal{L}$, and if $\mathcal{P}-\mathcal{L}$ is closed in $\mathcal{F}(\mathcal{R})$ then there exist $\bar{\lambda} \in \mathcal{M}^1_+(A)$ and $\bar{f} \in \mathcal{L}$ such that (10) holds.

proof :

1° (10) implies (1). Then apply th.1, 2°.

2° By th.1, 3°, there exists $\bar{\lambda} \in \mathcal{M}^1_+(A)$ such that (1) holds. Then apply to the function $< \bar{\lambda}, J(.) >$ the Lagrangian principle for minimizing a real convex function subject to convex constraints ([2]).

Note 4. Condition "$\mathcal{P}-\mathcal{L}$" is closed is quite general. For instance it is proved in [2] that it contains the slater's regularity condition when K is defined as in (9) with T finite or T compact and $t \to f_t(e)$ continuous $\forall e \in \mathcal{R}$:

$$\exists\, e_0 \in \mathcal{R} \text{ such that } f_t(e_0) < 0, \forall t \in T.$$

4. Algorithms.

By th.1, in the convex case, the min-max problem is equivalent to a saddle-point problem. So one can use algorithms for such saddle-point problems. Let us give an example :

Assume E be a Hilbert space and K is closed. Let P_K the projector onto K. Suppose the Gateaux derivative $j'_\alpha(e)$ exist $\forall\, e \in K$, $\forall\, \alpha \in A$. The second algorithm of Arrow-Hurwicz([1]) suggests the following iterative procedure to approximate the min-max solution. give $e^o \in K$ and construct the sequence $(e^n, \lambda^n)_{n\in \mathbb{N}}$

a) $e^n \to \lambda^n$ by

$$< \lambda^n, J(e^n) > = \max_{\lambda \in \mathcal{M}_+^1(A)} < \lambda, J(e^n) >, \quad \lambda^n \in \mathcal{M}_+^1(A)$$

that is to say choose λ^n in $\mathcal{M}_+^1(A(e^n))$ (see the proof of th. 1, 1°).

b) $(e^n, \lambda^n) \to e^{n+1}$ by

$$e^{n+1} = P_K\{e^n - \rho < \lambda^n, J'(e^n) >\}, \quad \rho > 0$$

where $< \lambda, J'(e) >$ = G-derivative of $< \lambda, J(.) >$ by derivation under the sign $\int$ which has sense as above-mentioned.

In fact, one recovers the algorithm of subgradient submitted in [3] and adapted in [7], applied to Ψ, because (ROCKAFELLAR above-mentioned)

$$\partial\, \Psi(e) = \overline{co}\, \{j'_\alpha(e) \,|\, \alpha \in A(e)\} \quad \text{that is to say}$$

$$\partial\, \Psi(e) = \{ < \lambda, J'(e) > = \int_{A(e)} j'_\alpha(e) d\lambda(\alpha) \,|\, \lambda \in \mathcal{M}_+^1(A)\}$$

This algorithm of subgradient is the following :

(i) $e^n \to \overset{n}{\chi} \in \partial\Psi(e^n)$

(ii) $(e^n, \overset{n}{\chi}) \to e^{n+1} = P_k\{e^n - \rho\, \overset{n}{\chi}\}, \quad \rho > 0.$

Practically, in a) one can choose for λ^n the Dirac measure at $\alpha_n \in A(e^n)$. So the algorithm becomes

a') $e^n \to \alpha_n$ by

$$j_{\alpha_n}(e^n) = \max_{\alpha \in A} j_\alpha(e^n), \quad \alpha_n \in A$$

b') $(e^n, \alpha_n) \to e^{n+1} = P_K\{e^n - \rho\, j'_{\alpha_n}(e^n)\}$, $\rho > 0$.

Under this form, it is proved in [7] that, for a given ball B centred at $\bar{e}$, one can choose $\rho > 0$ such that e^n belong to B for all n sufficiently large.

5. Applications.

5.1. Tchebycheff approximation.

Let $\{\varphi_i\}_{i=1,n}$ a subset of linearly independant functions of $\mathcal{C}(A)$ and E the finite dimensional subspace of $\mathcal{C}(A)$ spanned by the φ_i. Give $g \in \mathcal{C}(A)$ and take $K = E$. The Tchebycheff approximation problem is then the min-max problem with

$$j_\alpha(e) = |g(\alpha) - e(\alpha)|^2$$

Conditions (6) and (7) are written here

$$(6') \quad \inf_{e \in E} \max_{A(\bar{e})} (g(\alpha) - \bar{e}(\alpha))e(\alpha) \geq 0.$$

$$(7') \quad 0 \in \text{co}\left\{\{(g(\alpha) - \bar{e}(\alpha))\varphi_i(\alpha)\}_{i=1,n} \,|\, \alpha \in A(\bar{e})\right\}$$

and th.2 reduces to the well-known Kolmogorov's theorem ([4]).

5.2. Selection of a Pareto minimum.

Definition. $\bar{e} \in K$ is a Pareto minimum iff there exist no $e \in K$ such that $j_\alpha(e) < j_\alpha(\bar{e}) \quad \forall\, \alpha \in A$.

When A is a finite set, the Pareto optimal problem appears in the Kuhn-Tucker's paper [6] where it is called the vector optimization problem. Conditions of optimality given there were extended by AUBIN ([2]).

What is the connection with the min-max problem ? Simply the following holds :

if $\bar{e}$ is a min-max solution then $\bar{e}$ is a Pareto minimum

(proof by contradiction).

Even the min-max solution is unique it may exist more than one Pareto minimum. So, in this case the min-max problem furnishes a method of selection of a Pareto minimum.

A Lagrangian principle holds for the Pareto minimum problem ([6], [2]) which looks like our th.3. But here we get more information about the so-called Pareto multiplier $\bar{\lambda}$.

5.3. Optimal control with perturbations.

Formally let us state the problem as the following : a system is governed by the state equation

$$\Lambda y = Be + F(\alpha) \tag{11}$$

$e \in K$ set of admissible controls,

$\alpha \in A$ set of allowed perturbations.

Suppose equation (11) define a map $y(e,\alpha)$ from $K \times A$ to the state space. Let g a real function defined on the state space and consider the criteria

$$j_\alpha(e) = g(y(e,\alpha)).$$

Then the min-max problem consists of finding an admissible control optimal for the maximum risk due to the allowed perturbations. This problem has been extensively studied in [7] where (11) is a linear partial differential equation with boundary conditions and the criteria the distance in an Hilbert space between $y(e,\alpha)$ and a desired target. In particular, condition (6) leads to a characterization of the optimal control by means of an adjoint state $p(\bar{e},\alpha)$, $\alpha \in A(\bar{e})$. See also [10].

BIBLIOGRAPHY

[1] ARROW (K.J.), HURWICZ (L.), UZAWA (H.). Studies in linear and non linear programming. Stanford U.P., 1958.

[2] AUBIN (J.P.). A Pareto minimum principle, in Differential games and related topics, ed. Kuhn and Szegö, North-Holland, 1971, pp. 147-175.

[3] AUSLENDER (A.). Methodes numériques pour la résolution des problèmes d'optimisation avec contraintes. Thèse, Grenoble, 1969.

[4] CHENEY (E.W). Introduction to approximation theory. Mc Graw Hill, New York, 1966.

[5] KARLIN (S.) - Mathematical methods and theory in games, programming and economics, Addison-Wesley, 1959.

[6] KUHN (H.W.) and TUCKER (A.W.) - Nonlinear programming. Second Berkeley Symposium on Mathematical statistics and Probability. University of California Press (1951).

[7] LEMAIRE (B.) - Problèmes Min-Max et applications au contrôle optimal de systèmes gouvernés par des équations aux dérivées partielles linéaires. Thèse, Paris, 1970.

[8] MOREAU (J.J.) - Théorèmes inf sup, CRAS, Paris 258, 1964, pp. 2720-2722.

[9] MOREAU (J.J.) - Fonctionnelles convexes. Séminaire sur les équations aux dérivées partielles II, Collège de France, 1966/67.

[10] TANTER (A.) - Définition et résolution numérique d'un problème de contrôle optimal pour un système perturbé gouverné par des équations aux dérivées partielles (modèle min-max). Thèse de 3e cycle, Paris, 1972.

[11] VALADIER (M.) - Contribution à l'Analyse convexe. Thèse, Paris, 1970. Publication n°92 du Secrétariat des Mathématiques, Université MONTPELLIER II.

Mathematical Programming in Theory and Practice,
P.L. Hammer and G. Zoutendijk, (Eds.)

Stochastic Programs with Simple Recourse*

William T. Ziemba**

University of British Columbia

ABSTRACT

Consider a situation where a decision maker wishes to plan over a specified time horizon. In each period he chooses a constrained decision vector in such a way that when the random vectors for this and all future periods are observed the constraints will always be satisfied. His objective is to minimize the sum of the expected costs associated with the decisions made and states observed. Such a model is termed a stochastic recourse model. The term recourse refers to the phenomenon that some decisions amount to recourses to previous realizations of random vectors and decision choices. If the recourse

---

* Presented by invitation at the NATO Advanced Study Institute on Mathematical Programming in Theory and Practice, Figueira da Foz, Portugal, June 12-23, 1972. This research has been partially supported by the National Research Council of Canada, Grant NRC A7-7147, the Samuel Bronfman Foundation, and Graduate School of Business, Stanford University, and Atomic Energy Commission, Grant AT 04-3-326-PA #18.

** Without implicating him, I would like to thank Professor Arthur F. Veinott, Jr. for some useful information regarding the material in section three. Thanks are also due to A. Amershi for a careful reading of an earlier version of this paper.

in each period is uniquely determined once the decision and random vectors in that period are known, the recourse is said to be simple. Many business and economic decision problems may be modeled as simple recourse models. Examples include production and inventory planning, economic policy, transportation, marketing allocation, goal programming, portfolio choice, cash management and pension fund management problems. This paper is primarily concerned with static and two-period problems, in which most of the functions involved may be nonlinear. The random problem is shown to have a static deterministic equivalent that often has useful convexity and differentiability properties. Bounds on the optimal objective value may be determined by solving ordinary nonlinear programs. Existing gradient and other algorithms may be modified to solve these problems. The details are given in the context of specific examples for several leading algorithms. A number of special cases, particularly those which simplify the calculation of optimal decision rules, are also illustrated. For the static problem duality and related certainty equivalence results are also developed.

Introduction

Since the early 1940's there has been an intensive study of mathematical programming theory, computations and applications. A basic problem is to

$$\text{(1)} \quad \begin{array}{ll} \text{minimize} & f(x), \\ \text{subject to} & x \in K, \end{array}$$

where $K \equiv \{x | g_i(x) \geqq 0, \ i=1,\ldots,m\}$ is a subset of R^n (Euclidean n-space) and $f, g_1, \ldots, g_m$ are real-valued (scalar) functions of the n-vector x. In applications some or all of the parameters are usually random. A common procedure is to replace these random parameters by their (perhaps estimated) means. This results in a deterministic program. A theoretically more satisfying procedure is to consider (1) to be a probabilistic program and to take account of the risk in the decision making procedure. Suppose that randomness occurs only in the constraints vis-à-vis the random s-sector ξ. Four basic approaches have been suggested.

In the "fat" formulation [42] one considers only those x for which $g_i(x,\xi)$ is $\geqq 0$ for all possible ξ. If the vector ξ may equal $\xi^1, \ldots, \xi^L$ then the problem is to

$$\text{(2)} \quad \begin{array}{ll} \text{minimize} & f(x), \\ \text{subject to} & x \in T, \end{array}$$

where $T \equiv \{x | g_i(x,\xi^{\ell}) \geqq 0, \ i=1,\ldots m, \ \ell=1,\ldots,L\}$. If L is finite, as it will be if each ξ_k has a discrete distribution, (2) is an explicit deterministic program. If L is not finite but there are a finite number of points $\xi^1,\ldots,\xi^{L_1}$ such that $\Pr\{\xi = \xi^1,\ldots, \xi = \xi^{L_1}\} = 1$, an analogous solution procedure is suggested if the constraints are allowed to be satisfied with probability one. A basic difficulty with this approach is that T may be empty or contain (relatively) few elements from which a decision choice must be made.

Charnes, Cooper and Symonds [11] have presented a formulation called "chance-constrained programming," in which each constraint i must be satisfied with a certain tolerance probability γ_i. The problem is to

$$\text{(3)} \qquad \begin{array}{ll} \text{minimize} & f(x), \\ \text{subject to} & x \in U, \end{array}$$

where $U \equiv \{x | P_r\{g_i(x,\xi) \geqq 0\} \geqq \gamma_i,\ i=1,\ldots,m\}$, $0 \leqq \gamma_i \leqq 1$, $i=1,\ldots,m$.

A third approach suggested by Tintner [64] and called, "stochastic linear programming" is to study distribution problems. In the "passive" approach one computes for each given ξ^ℓ

$$\text{(4)} \qquad \begin{array}{lll} Z_\ell & \equiv \text{minimum} & f(x), \\ & \text{subject to} & x \in T_\ell, \end{array}$$

where $T_\ell \equiv \{x | g_i(x,\xi^\ell) \geqq 0,\ i=1,\ldots m\}$ and generates the distribution of Z based on the distribution of ξ. In the active approach [65] one chooses a series of a priori allocation matrices A_k that allocate, say, resources to activities, and calculates the distribution of $\min f$ for each A_k. The decision maker then chooses, based on his preferences, a preferred allocation matrix.

A fourth approach was suggested independently by Beale [3], and Dantzig [15] and was termed "linear programming under uncertainty" or "stochastic programming with recourse." The basic innovation is to amend the problem to allow the decision maker the opportunity to: (a) make corrective actions after ξ is observed; and/or (b) pay penalties for constraint violations. If E_ξ represents mathematical expectation with respect to ξ this approach leads to the problem

$$
\begin{aligned}
&\text{minimize} \quad && e(x) + E_\xi \left[\min_y \ell_1(y) \right], \\
(5) \qquad &\text{subject to} \quad && \ell_2(x,y,\xi) = 0, \\
& && x \in X, \ y \in Y,
\end{aligned}
$$

where ℓ_2 is a real-valued m dimensional vector function of $x \in R^n$, $y \in R^m$, $\xi \in \Xi \subset R^s$, $Y \subset R^m$ and e and ℓ are real-valued scalar functions. The decision procedure is to choose $x \in X$, observe ξ and then to choose y from $Q(x,\xi) \equiv \{y \mid y \in Y, \ \ell_2(x,y,\xi) = 0\}$. The objective is to choose x such that the sum of the preference for the decision x plus the expected preference for the second stage objective $\ell_1(\cdot)$ is minimized. For case (b) the second stage minimization over y is ficitious as $Q(x,\xi)$ has only one member for each x and ξ. In this case the state vector y is unique and the problem is termed a "stochastic program with simple recourse."

The theoretical development and potential and actual applications concerned with these approaches to stochastic programming problems has become very extensive. Technical surveys of the entire field have been given by Parikh [49], Kolbin [38], Vajda [68] and Wets [72]. Walkup and

Wets [69] explore relations between the various models. Fricks [28] and Smith [58] have also provided useful research bibliographies. There have also been surveys of particular subfields such as those by Bereanu [6], Byrne et. al., [8], Dempster [20], Kirby [37], Sengupta and Tintner 57 , Wets 73 , and Ziemba and Vickson 89 .

This paper is concerned with the recent research related to static and two period stochastic programs with simple recourse where most of the functions involved are nonlinear. However, certain features of some of the other models will arise naturally in our development. The discussion will be largely tied to specific applications to business and economic problems.
Examples are given in section two to motivate the general static simple recourse problem. Some basic convexity and differentiability results are presented in section three. Certain useful bounds, duality and certainty equivalent results are given in section four. General algorithmic approaches along with details concerned with the application of a specific algorithm are given in section five. A number of special cases are considered in section six. Section seven is concerned with two period problems. Finally some general comments and historical remarks are made in section eight.

2. Some Examples and the General Static Problem

A. Some Examples

To motivate the general development of the simple recourse model it will be useful to begin by formulating several examples.

Example 2.1: Portfolio Selection

An investor is assumed to have a utility function u over wealth w. Let ξ_j be the random return per dollar invested in alternative $j, j=1,\ldots,n$, and $F(\xi_1,\ldots,\xi_n)$ be the joint cumulative distribution of returns. Assume that x_j, the number of dollars invested in j, does not affect F. Suppose that the investor wishes to allocate his initial wealth w_o in such a way that expected utility of final wealth is maximized. The choice of the x_j is limited by the available budget and also by certain size purchase regulations. The problem is

$$
(6) \qquad
\begin{aligned}
&\max \int \cdots \int u\{\sum_j \xi_j x_j\}\, d\,F\,\{\xi_1,\ldots,\xi_n),\\
&\text{s.t.} \quad \sum_j x_j = w_o,\\
&\qquad \alpha_j \le x_j \le \beta_j,\\
&\qquad x_j \le w_o \gamma, \quad j=1,\ldots,n,
\end{aligned}
$$

where γ and the α_j and β_j are constants.

Example 2.2: Inventory Control

A manager wishes to choose production levels $x \equiv (x_1,\ldots,x_n)$ for final goods whose demand $\xi \equiv (\xi_1,\ldots,\xi_n)$ is random. When the actual demands occur there is either a shortage (y_j^-) or a surplus (y_j^+) in each item j. Assume that the sales prices $(p_1,\ldots,p_n)$ and salvage prices $(r_1,\ldots,r_n)$ do not depend upon the actual demand. That is assume that the buying and selling occurs in perfect markets. Let h be the production cost function and g_j^+ and g_j^- be the surplus and shortage cost functions for item j. Assume that the production technology is represented by K a subset of R^n. If the manager wishes to maximize expected profits his decision problem is

$$\max \quad \sum_j p_j x_j - h(x) - E_\xi \left[\min \left\{ \sum_j \left[g_j^+(y_j^+) + g_j^-(y_j^-) + (p_j - r_j) y_j^+ \right] \right\} \right], \tag{7}$$

$$\text{s.t.} \quad x_j - y_j^+ + y_j^- = \xi_j,$$

$$x \in K,\ y_j^+ \geq 0,\ y_j^- \geq 0 \ \text{ and } \ y_j^+ y_j^- = 0,\ j=1,\ldots,n.$$

Example 2.3: Allocation of Salesmen to Territories

Consider a firm that wishes to allocate m salesmen among n sales districts. Let x_j be the number of salesmen allocated to district j. Let p_j be the price charged in j for the firm's product. Assume that promotional, transportation and other variable expenses in j amount to c_j per sales unit. An econometric analysis has indicated that s_j, sales in district j, are related to price and the number of salesmen in j via

$$s_j = \gamma_j(x_j, p_j) + \xi_j, \quad j=1,\ldots,n. \tag{8}$$

In (8) the γ_j are known functions and ξ_j is the error in equation j. If (8) is correctly specified the distribution of the ξ_j is independent of the x_j and p_j. The firm pays the salesman in j an amount determined by an incentive function $I_j(s_j)$. The firm may determine the allocations and prices that maximize expected profit by solving

$$(9) \qquad \max \quad E_{\xi_1,\ldots,\xi_n} \sum_j \Big\{(\gamma_j(x_j,p_j)) + \xi_j)(p_j-c_j) - I_j(\gamma_j(x_j,p_j)) + \xi_j)\Big\},$$

$$\text{s.t.} \quad \sum_j x_j = m,$$

$$x_j \geq 0 \text{ and integer},$$

$$p_j \geq 0, \qquad j=1,\ldots,m.$$

Example 2.4: Economic Policy Under Uncertainty

Suppose that the over-all behavior of an economic system is described by the linear econometric model

$$\Gamma Y = \beta_1 X + \beta_2 Z + E.$$

The elements of the matrices Γ, β_1 and β_2 are known constants. The observed vectors Y, X, and Z are, respectively, the endogenous, decision and purely exogenous variables of the economic system during any time period. The vector E refers to the (unobserved) random disturbances during the same time period. To formulate a decision problem, following Tinbergen [61], suppose that there exist unique targets Y* and X* that together represent the most preferred levels for the vectors Y and

X, respectively. Deviations away from these targets are defined by

$$y \equiv Y - Y^*, \quad \text{and} \quad x = X - X^*.$$

Suppose that the decision variables have upper and lower bounds described by

$$\alpha_1 \leq X \leq \alpha_2,$$

where α_1 and α_2 are known constant vectors. Upon substitution of the deviation definitions the bounding constraints may be represented as

$$A\,x \geq b,$$

where $A \equiv \binom{I}{I}$, $b \equiv \binom{X^*-\alpha_1}{\alpha_2-X^*}$, and I is an identity matrix.

Similarly the model may be represented as

$$\Gamma y = \beta_1 x + \xi,$$

where $\xi \equiv \beta_2 Z + E + \Gamma Y^* - \beta_1 X^*$ has a joint distribution that is independent of x. Let $h_1(x)$ and $h_2(y)$ be penalty functions associated with deviations x and y, respectively. The goal of choosing the X to minimize the expected penalty cost may be implemented by solving

$$\text{(10)} \quad \begin{aligned} \min \quad & h_1(x) + E_\xi \{h_2(y)\}, \\ \text{s.t.} \quad & Ax \leq b, \\ & \Gamma y - \beta_1 x = \xi. \end{aligned}$$

Example 2.5: Transportation

Suppose a firm has S_i units of a particular product available at warehouse i, $i=1,\ldots,m$. It is desired to ship the product to several outlets. It costs $c_{ij}x_{ij}$ to ship x_{ij} units from warehouse i to outlet j. The outlet demand vector $\xi = (\xi_1,\ldots,\xi_n)$ is assumed to have a known joint distribution independent of the x_{ij}. The shortage and overage cost functions at outlet j are g_j^- and g_j^+, respectively. The firm may find a shipping schedule that minimizes expected costs by solving

$$\begin{aligned}
\text{Min}\quad & \sum_i \sum_j c_{ij}x_{ij} + E_\xi \min \sum_j [g_j^+(y_j^+) + g_j^-(y_j^-)] \\
\text{s.t.}\quad & \sum_j x_{ij} = S_i, \\
& \sum_i x_{ij} - y_j^+ + y_j^- = \xi_j, \\
& x_{ij} \geq 0,\ y_j^+ \geq 0,\ y_j^- \geq 0,\ y_j^+ \cdot y_j^- = 0, \\
& i = 1,\ldots,m, \\
& j = 1,\ldots,n.
\end{aligned}$$

B. The General Static Problem

In the portfolio selection and allocation of salesmen problems the uncertainty occurs only in the objective function. The general form of such problems is

$$(12)\qquad \begin{aligned}
\min\quad & Z(x) \equiv E_\xi\,\{f(x,\xi)\}, \\
\text{s.t.}\quad & x \in K.
\end{aligned}$$

Assume for (12) that[1] $x \in R^n$, $K \subset R^n$, $\xi \in \Xi \subset R^s$, $f: K \times \Xi \to R$ and $Z: K \to R$. A solution of (12) is an x^* that minimizes the expected loss associated with the decision choice x^* and the realization of the random vector ξ. The portfolio problem has form (12) where the objective $f(x,\xi) = u\{\sum_j \xi_j x_j\}$. The univariate property of u lends itself to a special analysis that is considered in detail in "Choosing Investment Portfolios...", in this volume.

The allocation of salesmen problem is also a special case of (12). However the objective in this example may be written as (minus)

$$\text{(13)} \qquad \begin{array}{ll} \min & e(x) + E_\xi[g(x,\xi)], \\ \text{s.t.} & x \in K, \end{array}$$

where $e(x) \equiv -\sum_j [\gamma_j(x_j p_j) + \bar{\xi}_j]\,(p_j - c_j)$, $g(x,\xi) \equiv \sum_j \{I_j[\gamma_j(x_j p_j) + \xi_j]\}$, and $K \equiv [(x_1,\ldots x_m, p_1,\ldots p_m) \mid \sum_j x_j = m,\ p_j \geq 0,\ x_j \geq 0,\ x_j$ integer, $j=1,\ldots,m]$.

It will often be convenient to consider models in form (13) even though they may be written in form (12).

---

[1]The random elements inherent in the decision problem are assumed to belong to the probability space $(\Xi, \mathcal{F}, F)$ which is the space induced in R^m (Euclidean m-space). The set of all possible realizations of the random vector ξ is denoted by Ξ (a subset of R^m), F determines a Lebesgue-Stieltjes measure, and $\mathcal{F}$ is the completion for F of the Borel algebra in R^m. It is assumed that ξ is distributed independently of the decision choice x. The marginal distribution function of F and the density with respect to the random variable ξ_j are denoted by $F_j(\xi_j)$, respectively. $E_\xi[u]$ represents the mathematical expectation of the vector u with respect to the random vector ξ. Bars above random vectors and functions denote their mathematical expectations, e.g., $\bar{\xi} \equiv E_\xi[\xi]$.

A second class of problems is

$$(14)\quad \begin{aligned} \min \quad & e(x) + E_\xi [\min_y \{\ell_1(y)\}], \\ \text{s.t.} \quad & \ell_2(x,y,\xi) = 0, \\ & x \in K,\ y \in Y. \end{aligned}$$

In (14) assume that $y \in R^m$, $Y \subset R^m$, $\ell_1 : Y \to R$ and $\ell_2 : K \times Y \times \Xi \to R^m$. Note that there are as many equations $\ell_{2j} = 0$ as there are elements of y. The vector y constitutes a state vector that must be chosen after x is chosen and ξ is observed. The choice of y is limited to the set $U(x,\xi) \equiv [y \mid y \in Y, \ell_2 (x,y,\xi) = 0]$. The model (14) is a nonlinear stochastic program with recourse a version of which Mangasarian and Rosen [46] have studied. The property of simple recourse requires that U has precisely one element for each $(x,\xi) \in K \times \Xi$. In this case the minimization over y is ficticious and (14) is equivalent to

$$(15)\quad \begin{aligned} \min \quad & e(x) + E_\xi\{\ell_1(y) \mid \ell_2(x,y,\xi) = 0\}, \\ \text{s.t.} \quad & x \in K. \end{aligned}$$

Now formulation (15) is a special case of (13) with $g(x,\xi) \equiv \{\ell_1(y,\xi) \mid \ell_2(x,y,\xi) = 0\}$. In most[2] practical problems the equations $\ell_2 = 0$ may be solved for $y = c(x,\xi)$. In such cases $g(x,\xi)$ is the explicit function $\ell_1 [c(x,\xi),\xi]$. Under minor regularity assumptions examples

---

[2]When the ℓ_2 are linear the c_i exist if the matrix associated with y is non-singular with probability one. Sufficient conditions for the existence of the c_i in the case when the ℓ_2 are non-linear have been investigated by Gale and Nikaidô [30]. Warburton and Ziemba [70] have utilized the results in [30] and present sufficient conditions on ℓ_2 for the c_i to exist and be convex continuously differentiable functions. These conditions, however, are quite strict and are rarely satisfied in practice.

2.2, 2.4 and 2.5 may be reduced to this form. In the economic policy example it is required that Γ be non-singular then $g(x,\xi) = h_2[\Gamma^{-1}(\beta,x + \xi)]$. The non-singularity assumption is a standard one since it means that there are as many independent structural equations as there are variables to be explained. In econometric terminology one says that the model has a unique reduced form. It is easy to put the inventory control and transportation problems into the form of (13) once the $y_j^+ y_j^- = 0$ constraints are dispensed with. These complementary constraints are needed to insure that one does not have a shortage and an overage in any item. A sufficient condition to guarantee that any solution to the transportation problem always satisfies the complementary conditions is

$$\partial g^+/\partial y^+ + \partial g^-/\partial y^- \geqq 0, \text{ for any } \partial y^+ = \partial y^- \geqq 0.$$

Hence under this minor regularity assumption one may delete the complementary conditions from (11). For the inventory control problem the weaker assumption (since presumably $p_j > r_j$)

$$\partial g^+/\partial y^+ + \partial g^-/\partial y^- + (p_j - r_j) \geq 0$$

is sufficient. The device needed to put the inventory control and transportation problems into the form of (13) is to define unconstrained variables $y_j = y_j^+ - y_j^-$, and assume $g^+(0) = g^-(0) = 0$. For the inventory control example, one then lets

$$g(x,\xi) \equiv - \sum_j g_j(x_j,\xi_j)$$

where each
$$g_j(x_j,\xi) \equiv \begin{cases} g_j^+(x_j-\xi_j) + (p_j-r_j)(x_j-\xi_j) & \text{if } x_j \geq \xi_j \\ g_j^-(\xi_j - x_j) & \text{if } x_j \leq \xi_j. \end{cases}$$

In the sequel it will be assumed that the problem may be put into form (12).

Examples 2.1 - 2.5 may be compared in the following tabular form:

	Has e(x) term?	Has Univariate Objective?	Has Separable Objective?	Has Randomness in Constraints
1 Portfolio Selection	No	Yes	No	No
2 Inventory Control	Yes	No	Yes	Yes
3 Allocation of Salesmen	Yes	No	Yes	No
4 Economic Policy	Yes	No	No	Yes
5 Transportation	Yes	No	Yes	Yes

3. Basic Convexity and Differentiability Results

The static simple recourse problem is to find an x^* that solves

$$\text{(12)} \qquad \min \ Z(x) \equiv E_{\xi}[f(x,\xi)], \quad \text{s.t. } x \in K.$$

In (12) it is assumed that the decision variable $x \in R^n$ and that f and Z are real valued scalar functions and $K \subset R^n$. The random s-vector ξ is defined on the probability space $(\Xi, \mathcal{F}, F)$ that is distributed independently of x.

The following Theorem presents sufficient conditions for Z to be continuous on K.

Theorem 1: Suppose that for any $x^o \in K$ and $\epsilon > 0$, $\exists \delta > 0$: $|x - x^o| < \delta$ for $x \in K$ implies $|f(x,\xi) - f(x^o,\xi)| < \epsilon$, for almost all $\xi \in \Xi$, where $Z(x) \equiv \int \cdots \int f(x,\xi) dF(\xi)$ exists and is finite at x^o and exists at all x in any neighborhood of x^o.

Then Z is continuous on K.

Proof:

$$|Z(x) - Z(x^{\circ})| \equiv |\int\cdots\int f(x,\xi)dF(\xi) - \int\cdots\int f(x^{\circ},\xi)dF(\xi)|$$

$$< \int\cdots\int |f(x,\xi) - f(x^{\circ},\xi)|dF(\xi)$$

$$< \int\cdots\int \varepsilon dF(\xi) = \varepsilon.$$

Roughly speaking the theorem indicates that Z will be continuous if with probability one f is continuous.

We say that K is a convex set if $x^1, x^2 \varepsilon K$ implies that $\lambda x^1 + (1-\lambda)x^2 \varepsilon K$ for all $\lambda \varepsilon [0,1]$.

A function e defined on a convex set D is said to be convex if $x^1, x^2 \varepsilon D$ implies that

$$e(\lambda x^1 + (1-\lambda)x^2) \leq \lambda e(x^1) + (1-\lambda)e(x^2) \tag{16}$$

for all $\lambda \varepsilon [0,1]$. The function e is strictly convex if (16) holds with strict equality for all $x^1 \neq x^2$ and $\lambda \varepsilon (0,1)$.

The following theorem relates the convexity or strict convexity of f to that of Z.

Theorem 2: Let K be a convex set.

(a) Suppose $f(x,\xi)$ is convex in $x\varepsilon K$ for all fixed $\xi \varepsilon \Xi$ then Z is convex on K.

(b) Suppose $f(x,\xi)$ is strictly convex in $x\varepsilon K$ for all fixed $\xi \varepsilon \Xi$ then Z is strictly convex on K.

(c) Suppose $f(x,\xi)$ takes the form $u(w) \equiv u(\xi' x)$. Then Z is strictly concave in x if u is strictly concave in w as long as $\xi' x^1 = \xi' x^2$ for all $\xi \varepsilon \Xi$ implies that $x^1 = x^2$.

The proof follows directly from the definitions and the fact that expectation is a non-negative operation. One may note that Z may fail to be strictly concave in (c) if the variance-covariance matrix of ξ is singular.

It is often useful to be able to legitimately interchange the order of differentiation and integration. For example, such an interchange provides a method for the evaluation of partial derivatives. Under suitable regularity conditions the monotone and dominated convergence theorem may be applied to obtain sufficient conditions for the validity of the interchange.

A function $h(x,\xi)$ is said to be integrable with respect to ξ if

$$|E_{\xi}\, h(x,\xi)| \le \int |h(x,\xi)|\, dF(\xi) < \infty.$$

Monotone Convergence Theorem : Assume that e is integrable with respect to ξ. Let $g_n(x,\xi) \ge e(\xi)$ and $g_n(x,\xi) \ge g_{n-1}(x,\xi)$, $\lim_{n\to\infty} g_n(x,\xi)$ exists then $\lim_{n\to\infty}\left[\int g_n(x,\xi)dF(\xi)\right] = \int\left[\lim_{n\to\infty} g_n(x,\xi)\, dF(\xi)\right]$.

Dominated Convergence Theorem:

Suppose $|g_n(x,\xi)| \le h(\xi)$, with probability one, where h is integrable.

If $\lim_{n\to\infty} g_n(x,\xi) = g(x,\xi)$, with probability one,

$$\text{then} \quad \lim_{n\to\infty}\left[\int g_n(x,\xi)dF(\xi)\right] = \int\left[\lim_{n\to\infty} g_n(x,\xi)\right] dF(\xi)$$

$$= \int g(x,\xi)\, dF(\xi)$$

See Loève [39] for proofs of these theorems.

For our purposes it is most appropriate to consider conditions under which Z is continuously differentiable and its partial derivatives may be calculated using the interchange.

Theorem 3:

(a) Suppose f is integrable with respect to ξ and continuously differentiable in x for almost all ξ. Assume that there exist integrable functions $e_1(\xi),\ldots,e_n(\xi)$ such that

$$\left|\frac{\partial f(x,\xi)}{\partial x_j}\right| \le e_j(\xi), \qquad j=1,\ldots,n.$$

Then Z is continuously differentiable on an open subset of R^n and each

$$\frac{\partial Z}{\partial x_j} = E_\xi[\partial f(x,\xi)/\partial x_j].$$

Proof:

By the mean value theorem for given vectors x and x^1 there exists an x^2 such that

$$f(x^1,\xi) - f(x,\xi) = (x_j^1 - x_j)\frac{\partial f(x^2,\xi)}{\partial x_j}.$$

Thus $$\left|\frac{f(x^1,\xi) - f(x,\xi)}{x_j^1 - x_j}\right| = \left|\frac{\partial f(x_j^2,\xi)}{\partial x_j}\right| \le e_j(\xi).$$

By definition

$$\frac{Z(x^1) - Z(x)}{(x_j^1 - x_j)} = E_\xi \left[\frac{f(x^1,\xi) - f(x,\xi)}{x_j^1 - x_j} \right].$$

Since the integral is bounded by e_j the dominated convergence theorem may be applied to verify that

$$\frac{\partial Z(x)}{\partial x_j} = E_\xi \left[\frac{\partial f(x,\xi)}{\partial x_j} \right], \qquad j=1,\ldots,n.$$

Now by Theorem 1 the partials are continuous as long as $\partial f/\partial x_j$ is continuous for almost all ξ.

A function $c(x)$: $D \to R$ is said to be subdifferentiable on an open convex set $D \subset R^n$ if there exists a vector $s(x)$, termed a subgradient, such that

$$c(x^1) \geq s(x)' (x^1 - x) + c(x) \qquad \text{for all } x, x^1 \in D.$$

Let $s(x) = (s_1(x),\ldots, s_n(x))$.
If c is convex then c is differentiable except on a set of measure zero. Then $d_j^- c(x) \leq s_j(x) \leq d_j^+ c(x)$, where d_j^- and d_j^+ are respectively the left and right directional partial derivatives of c with respect to x_j at x. A function is differentiable at x if $s(x)$ is unique. In this case $d_j^-(x) = d_j^+(x)$ for all j. This material is standard, see e.g., Rockafellar [54].

Theorem 4:

Let D be an open convex subset of R^n. Suppose that $f(x,\xi)$: $D \times \Xi \to R$ is convex in x for all fixed $\xi \in \Xi$

and integrable with respect to ξ. Let $d_j^-(x,\xi)$ and $d_j^+(x,\xi)$ be the left and right directional derivatives of f with respect to x_j at x given ξ. Suppose that the d_j^- and d_j^+ are bounded. Assume that for any given $x\in D$ and w_j^- (resp. w_j^+) the number of ξ such that $w_j^- = d_j^-(x,\xi)$ (resp. $w_j^+ = d_j^+(x,\xi)$) has measure zero. Assume that F is absolutely continuous. Then Z is continuously differentiable on D and

$$\frac{\partial Z}{\partial x_j} = E_\xi[s_j(x,\xi)]$$

where $s_j(x,\xi)$ is any number between $d_j^-(x,\xi)$ and $d_j^+(x,\xi)$, $j=1,\ldots,m$.

Proof:

$$\text{Let } g_{tj}^-(x,\xi) = \frac{f(x,\xi) - f(x_1,\ldots, x_j - \frac{1}{t},\ldots,x_n,\xi)}{1/t}$$

$$\text{and } g_{tj}^-(x,\xi) = \frac{f(x_1,\ldots,x_j + \frac{1}{t},\ldots,x_n) - f(x,\xi)}{1/t}.$$

By convexity the g_{tj}^- and g_{tj}^+ are non-decreasing and non-increasing, respectively. Now

$$- e_j(\xi) \le g_{tj}^-(x,\xi) \le g_{tj}^+(x,\xi) \le e_j(\xi),$$

where $e_j \ge 0$ is integrable with respect to ξ. Hence the g_{tj}^- and g_{tj}^+ have limits g_j^- and g_j^+, respectively. Thus by the monotone convergence theorem

$$d_j^+[Z(x)] = E_\xi \left\{d_j^+[f(x,\xi)]\right\} \quad \text{and}$$

$$d_j^-[Z(x)] = E_\xi \left\{d_j^-[f(x,\xi]\right\}.$$

By convexity $d_j^+[f(x,\xi)] = d_j^-[f(x,\xi)]$ except on a set of measure zero. Hence the absolute continuity of F coupled with the assumption on w_j^+ and w_j^- implies that the right and left partial derivatives of Z are equal. Thus partial derivatives of Z exist. Since Z is convex these partials are continuous [54], and

$$\frac{\partial Z(x)}{\partial x_j} = E_\xi \left\{d_j^+[f(x,\xi)]\right\} = E_\xi \left\{d_j^-[f(x,\xi)]\right\}$$

$$= E_\xi \; [s_j(x,\xi)],$$

where $d_j^-[f(x,\xi)] \le s_j(x,\xi) \le d_j^+[f(x,\xi],$ $\quad i=1,\ldots m.$

4. Bounds, Certainty Equivalents and Duality

Consider problem

$$\text{(12)} \qquad \begin{array}{ll} \min & Z(x) \equiv E_\xi\{f(x,\xi)\}, \\ \text{s.t.} & x \in K. \end{array}$$

Let $K \equiv \{x | g_i(x) \ge 0, \; i=1,\ldots,m\}$. Assume that the g_i are differentiable concave functions and that K satisfies a constraint qualification.

Suppose further that Z is differentiable and convex. A Wolfe dual [78] to (12) is

$$\text{(17)} \quad \begin{aligned} &\max \quad W(x,\nu) \equiv Z(x) + \sum \nu_i g_i(x), \\ &\text{s.t.} \quad \nabla Z(x) + \sum \nu_i \nabla g_i(x) = 0, \\ &\qquad\qquad \nu \geq 0, \end{aligned}$$

where $\nu \equiv (\nu_1,\ldots,\nu_m)$ is a vector of Lagrange multipliers. Let $S \equiv \{(x,\nu) | \nabla Z(x) + \sum \nu_i \nabla g_i(x) = 0, \nu \geq 0\}$. Assume that the sets K and S are non-empty. The following theorem summarizes the major relationships between (12) and (17).

Theorem 5:

(a) $Z(x) \geq W(x,\nu) \quad \forall x \in K, (x,\nu) \in S$.

(b) If x^* solves (12) there exists ν^* such that (x^*,ν^*) solves (17) and $Z(x^*) = W(x^*,\nu^*)$.

(c) Suppose further that Z and the g_i are twice differentiable and that the Hessian matrix $\nabla\nabla W(\hat{x},\hat{\nu})$ is non-singular. If $(\hat{x},\hat{\nu})$ solves (17) then $\hat{x}$ solves (12) and $Z(\hat{x}) = W(\hat{x},\hat{\nu})$.

Wolfe proved the weak duality result (a) and the duality theorem (b) in [78]. Huard [36] proved the converse duality theorem (c). Mangasarian [43] has proved a similar converse duality theorem under slightly different hypotheses. See also Mangasarian [45] for a weakening of the convexity-concavity assumptions.

The dual problem often has a meaningful economic interpretation, see [2] and [77]. Consider the inventory control problem 2.2. The model may be written in the form

$$Z^* \equiv \max Z(x) \equiv p'x - h_1(x) - \sum_j E_{\xi_j}\{h_j(\xi_j - x_j)\}. \tag{18}$$

$$\text{s.t.} \qquad x \geq 0,\ g_i(x) \leq b_i,\ i=1,\ldots,m\ .$$

The $g_i(x)$ refer to the usage of the raw materials and other resources that have availability levels b_i. The h_j represent the overage-shortage and net salvage costs. The program (18) has the interpretation: choose that feasible production level that maximizes expected profits.

The dual of (18) is

$$\min\ W(x,\nu) \equiv Z(x) + \sum_i \nu_i(b_i - g_i(x)) + \sum_j x_j \left[\frac{\partial Z}{\partial x_j} - \sum_i \nu_i \frac{\partial g_i}{\partial x_j}\right] \tag{19}$$

$$\text{s.t.} \quad \nu_1 \frac{\partial g_1}{\partial x_1} + \cdots + \nu_n \frac{\partial g_n}{\partial x_1} \geq \frac{\partial Z}{\partial x_1},$$

$$\vdots \qquad\qquad \vdots \qquad\qquad \vdots$$

$$\nu_1 \frac{\partial g_1}{\partial x_m} + \cdots + \nu_n \frac{\partial g_n}{\partial x_m} \geq \frac{\partial Z}{\partial x_m},$$

$$\nu_1 \geq 0,\ \ldots,\ \nu_n \geq 0.$$

Usually each dual variable ν_i may be interpreted as the marginal worth on Z^* of infinitesimal increases in the availability of resource i. Rearranging W gives

$$W(x,\nu) = \sum \nu_i b_i + \{Z(x) - \sum \nu_i g_i(x)\} \tag{20}$$
$$- \sum_j x_j \left[\frac{\partial Z}{\partial x_j} - \sum_i \nu_i \frac{\partial g_i}{\partial x_j} \right].$$

The first term of (20) represents the marginal valuation of the firm's scarce resources. The second term may be interpreted as an economic rent Suppose that by holding the scarce resources the manager considers the possibility of not producing at all. Then he can claim as economic rent the difference between total profits $Z(x)$ were he to sell and the cost of the inputs $g_i(x)$ evaluated at their marginal worth ν_i. In the last term suppose for some j that $\partial Z/\partial x_j < \sum_i \nu_i \partial g_i/\partial x_j$, that is selling good j at level x_j is an unprofitable activity. Then this difference times the activity level x_j should be subtracted from the economic rent. Minimizing W means that the marginal valuation of the scarce resources and economic rent minus marginal profit loss of unprofitable activities is minimized.

Each $\partial g_i/\partial x_j$ is the marginal quantity of input i to obtain an additional unit of good j. Hence $\nu_i \partial g_i/\partial x_j$ represents the marginal value of input i required to obtain an additional infinitesimal amount of good j. Hence each of the inequalities in the constraints of the dual problem mean that the marginal costs for infinitesimal increases in output levels must not be less than the marginal revenue obtained from that increase.

Thus the dual problem is to choose output levels x and imputed marginal prices $\nu \geq 0$ so that the marginal cost of infinitesimal increases in output levels does not exceed the marginal revenue of these increases and the valuation of the scarce resources and economic rent minus marginal profit loss of unprofitable activities is minimized.

Suppose that (12) and $\{\min f(x,\bar{\xi}) | x \in K\}$ have optimal solutions x^* and $\bar{x}$, respectively. The following bounds on $Z(x^*)$ are available.

Theorem 6:

(a) $$Z(x^*) \leq E_\xi \{f(\bar{x},\xi)\}.$$

(b) Assume that f is a continuous convex function of ξ. Then
$$f(\bar{x},\bar{\xi}) \leq Z(x^*).$$

(c) Assume the hypotheses of Theorem 5(a). Then
$$W(x,\nu) \leq Z(x^*).$$

Proof:

(a) $\bar{x}$ is feasible.

(b) Jensen's inequality [54] states that $E_\xi \psi(x,\xi) \geq \psi(x,\bar{\xi})$ if ψ is a continuous convex function of ξ. The result follows because minimization over the set K preserves the inequality.

(c) By Theorem 5(a).

Mangasarian [44] proved (a) and (b); (c) appears in Ziemba [83]. Note that the assumptions made previously about f are not needed in (a).

The pair of bounds (a)-(b) may be obtained by solving one nonlinear program and performing a single numerical integration. One may dispense with the numerical integration by taking a discrete approximation to ξ at the cost of a slightly less tight bound.

Suppose problem (12) takes the form

$$\text{(21)} \qquad \begin{aligned} \min \quad & Z(x) = e(x) + E_\xi[g(x,\xi)], \\ \text{s.t.} \quad & x \in K, \end{aligned}$$

where $g(x,\xi) \equiv \tilde{g}(\xi - Ax)$. Examples 2.2, 2.4 and 2.5 have this form. It is of interest to consider when there exists a certainty equivalent ξ^* such that an optimal solution to

$$(22) \qquad \begin{aligned} \min \quad & e(x), \\ \text{s.t.} \quad & Ax = \xi^*, \\ & x \in K, \end{aligned}$$

solves (21).

Theorem 7:

Consider the assumptions of Theorem 5(c) and suppose that the dual of (22) has an optimal solution. Then x^* solves (21) iff x^* solves (22) and

$$E_{\xi} \nabla_x g(x^*,\xi) = \pi^*,$$

where π^* are optimal dual multipliers for $Ax = \xi^*$ in (22).

A proof of this certainty equivalence result appears in Ziemba [83]. Note that the result indicates that one may solve (21) by merely solving (22) with the correct ξ^*. Finding the correct ξ^* generally involves solving (21). However one may utilize the result to devise algorithms that solve a sequence of problems having formed (22) that home in on ξ^*. Williams [76] proved a version of Theorem 7 in the context of the linear simple recourse problem and devised such an algorithm. Theorem 7 can be used to construct an algorithm to solve the transportation problem (11). In such an application one adjusts ξ via a master problem that utilizes information from a sequence of problems of form (22) that are ordinary deterministic linear transportation problems.

It seems reasonable that the differentiability assumptions in Theorem 7 can be relaxed using a Fenchel-Rockafellar dual problem [54]. In such an extended theorem ξ^* would not generally be unique.

5. Algorithms

The problem is

$$\begin{aligned} \text{(12)} \qquad & \min \; Z(x) \equiv E_\xi \, f(x,\xi), \\ & \text{s.t.} \quad x \in K. \end{aligned}$$

The simplest conceptual case is when one has available an explicit expression for $Z(x)$. Under reasonable assumptions, as outlined in section three, Z will be convex and possibly continuously differentiable. Hence a number of algorithms may be employed to solve the non-linear program (12). See e.g., [1, 25, 27, 55, 56, 80, 90-92]. Consult the cited articles for precise assumptions and iterative procedures for each method. Numerical surveys comparing many of these methods have been given by Colville [12], McCormick [40], Tjian and Zangwill [66], and Zoutendijk [91,92].

The calculation of an explicit expression for Z is generally difficult[3] and often impossible. Hence there is motivation to develop algorithmic procedures that do not require this explicit expression. Let ∇Z and $\nabla\nabla Z$ refer to the gradient vector of Z having typical element $\partial Z/\partial x_j$ and the Hessian matrix having typical element $\partial^2 Z/\partial x_i \partial x_j$, respectively. Most non-linear programming algorithms do not require that one have available explicit expressions for

---

[3] See section six below for some special cases in which explicit expression for Z is relatively easy to obtain.

Z, ∇Z and $\nabla\nabla Z$. They merely require that one can evaluate some or all of these functions at specific points. Hence one may modify such algorithms to apply to the solution of (12) assuming that the appropriate functions may be numerically evaluated. Feasible direction algorithms, see e.g., [89,80,92] generally proceed as follows:

Step 0. Initialization. Find an $x^1 \epsilon K$. Set $k=1$.

Step 1. Find an n dimensional direction vector d^k that is:

a. Feasible. $\exists \lambda^o > 0$: $0 \leqq \lambda \leqq \lambda^o$ $(x^k + \lambda d^k) \epsilon K$;

and

b. Usable. $\nabla Z(x^k)'d^k < 0$.

Step 2: Find a step length λ^k:

$$Z(x^k + \lambda^k d^k) = \min_{\lambda \geqq 0} \{Z(x^k + \lambda d^k) \mid (x^k + \lambda d^k) \epsilon K\}.$$

$$\text{Set } x^{k+1} = x^k + \lambda^k d^k.$$

Step 3. Terminate if the stopping criteria is satisfied. Otherwise go to (1) with $k=k+1$.

Let $K \equiv \{x | g_1(x) \geq 0, \ldots, g_m(x) \geq 0\}$ where each g_i is a real valued scalar function. The initialization step may proceed e.g. by a repeated application of a standard algorithm or by a phase I procedure [16] (if the g_i are linear). Feasible direction vectors

d^k have the property that they do not point out of K, i.e. $\nabla g_i(x^k)'d \geq \epsilon$ $(\epsilon>0)$ for constraints that are active at x^k, i.e. $g_i(x^k) = 0$. The direction finding subproblem can take a number of forms such as a simplex pivot (in the convex simplex method), or a linear or quadratic program (in various first and second order gradient methods, respectively). These calculations may require that Z, ∇Z and the Hessian matrix $\nabla\nabla Z$ be evaluated at any $x^k \epsilon K$. The step length calculation, of which that suggested in step (2) is one of several possibilities, usually proceeds via a search procedure or an explicit calculation (possibly resulting from an approximation along a ray). Evaluations of Z and ∇Z are generally required at a number of points. Topkis and Veinott [67, Thm. 2] and Zangwill [80, p. 281] have proven general convergence theorems for feasible direction methods. Under appropriate assumptions convergence is guaranteed to a stationary point which will be an optimal point if Z is convex (or pseudo-convex) and the g_i are concave (or quasi-concave). The essential conditions that these authors utilize in their convergence proofs that is of relevance here (since K corresponds to a "standard" feasible region) is that Z must be continuously differentiable.

A particularly simple feasible direction algorithm is the Frank-Wolfe (FW) algorithm [27].[4]

The rationale for the FW algorithm is based on the fact that (under appropriate assumptions) x^k solves $\{\min Z(x) | x \epsilon K\}$ if and only if it

---

[4]This adaptation of the Frank-Wolfe algorithm is based on Ziemba [81].

solves $\{\min \nabla Z(x^k)'x \mid x \in K\}$. The procedure is to linearize the objective about some feasible point x^k and then to solve a linear program to obtain a solution vector y^k. All points along the interval $[x^k, y^k]$ are feasible; hence, one may search along the interval for a minimum x^{k+1}. It develops that x^{k+1} will be a better solution than x^k. One then linearizes about x^{k+1} and repeats the process until $x^{k+1} = x^k$, i.e., $y^k = x^k$. The main facts relevant to the present discussion will be summarized as:

Theorem 8:

Suppose that Z is convex and differentiable and K is a a convex polyhedron, then,

(i) if $\bar{x} \in K$ and $\nabla Z(\bar{x})' \ (x - \bar{x}) \geqq 0 \ \forall \, x \in K$, then $\bar{x}$ is a minimum for $Z(x)$ over K;

(ii) suppose x, $\bar{x} \in K$ and $\nabla Z(x)' \ (x - \bar{x}) > 0$, then,

$\exists \lambda^o$: $0 \leqq \lambda^o \leqq \lambda < 1$, $Z\{\lambda x + (1 - \lambda)\bar{x}\} < Z(x)$;

(iii) if $x \in K$ and $Z(x) > Z(x^*) = \min \{Z(x) \mid x \in K\}$, then

$\exists \hat{x}$: $\nabla Z(x)' \ x > Z(x)' \ \hat{x}$; and

(iv) $Z(x^*) = \min \{Z(x) \mid x \in K\}$ if and only if

$\nabla Z(x^*)' \ x^* = \min \{\nabla Z(x^*)' \ x \mid x \in K\}$.

Proof. See e.g. [81].

Consider the linear program

$$(23) \qquad \begin{array}{ll} \min & \nabla Z(x^k)' \ x , \\ \text{s.t.} & x \in K. \end{array}$$

Assume

B1. For each $\hat{x} \in K$ the linear function $\nabla Z(\hat{x})' \ x$ is bounded from below on K.

B1 is satisfied, e.g., if K is bounded or if $\partial Z(x)/\partial x_i$ is nonnegative $\forall i, \forall x \in K$ and K contains the constraints $x_i \geq \ell_i$, where $\ell_i > -\infty \; \forall i$.

Under B1 the continuous objective is bounded from below on a non-empty closed convex set; hence (23) has an optimal solution $\forall \; x^k \in K$. If x^k is a solution to (23), then it follows that x^k is a solution to (12). If x^k is not a solution to (23), it follows that $\exists \; x^{k+1} \in (x^k, y^k]$: $Z(x^{k+1}) < Z(x^k)$, where y^k is a solution to (23).

Lemma.

If Z is convex and differentiable in x: $\nabla Z(x^k)' (y^k - x^k) \leqq Z(x^*) - Z(x^k)$, where x^* is a solution to (12).

Proof:

By convexity, $Z(x^*) - Z(x^k) \geqq \nabla Z(x^k)' (x^* - x^k) \; \forall x^*, \; x^k$. Since y^k is a solution to (23), $\nabla Z(x^k)' x^* \geqq \nabla Z(x^k)' y^k$. Combining these two inequalities provides the result.

Hence, in each iteration the following bounds obtain:

$$Z(x^k) + \nabla Z(x^k)' (y^k - x^k) \leqq Z(x^*) \leqq Z(x^k). \tag{24}$$

The best lower bound is

$$h_k \equiv \max_{\ell=1,\ldots,k} \{Z(x^\ell) + \nabla Z(x^\ell)' (y^\ell - x^\ell)\}. \tag{25}$$

It is shown in [7] and [16] under our assumptions that the algorithm does converge in the sense that $Z(x^k) \to Z(x^*)$. Much is known about the convergence rates of the FW algorithm. It has been shown [27] that there is an $\alpha > 0$ such that, for all sufficiently large k, $Z(x^k) - Z(x^*) \leqq \alpha/k$, and, for all $\epsilon > 0$, $\alpha/k^{1+\epsilon} \leqq Z(x^k) - Z(x^*)$ for an infinite number of k [9]. This is not particularly fast convergence, especially as k get large. However Canon, Cullum, and Wolfe [10] have suggested a modification that improves this convergence of order zero to convergence of order one. The idea is to move away occasionally from a 'bad' vertex that is the solution of $\{\max \nabla Z(x^k)' x | x \epsilon K\}$, instead of always moving toward 'good' vertices. The rationale is that, via the convex combination nature of the sequence of x^k generated, this approach allows one to decay quickly certain variables that do not belong to the optimal facet. A flow chart of the algorithm appears in Figure 1.

In each iteration one must numerically evaluate $\nabla Z(x^k)$ to determine the coefficients for the linear program (23). Strictly speaking one also needs to numerically evaluate ∇Z at several points in the search procedure. However, since these evaluations are generally difficult (see the remarks below) it is advisable to avoid these additional calculations. Now $Z\{\lambda y^k + (1 - \lambda) x^k\} = E_\xi f(\lambda y^k + (1 - \lambda) x^k, \xi) \equiv \phi(\lambda)$ which is a concave function of the scalar λ. A second or third order Taylor series approximation to f yields (see section 6C below) a deterministic quadratic or cubic function of λ, say $\phi_a(\lambda)$. One may then determine an explicit expression for λ_a, a λ that minimizes ϕ_a over $(0, 1]$, see e.g. [25]. No other numerical integration are needed except for occasional $Z(x\ )$ evaluations that one would make to evaluate the bounds h_k. Hence under this scheme at most two numerical integrations are needed in each iteration and most iterations would require only a single numerical integration.

The presentation of the modification of the Frank-Wolfe algorithm illustrates some of the considerations inherent in algorithmic modifications that can be made to solve problems of form (12). Similar

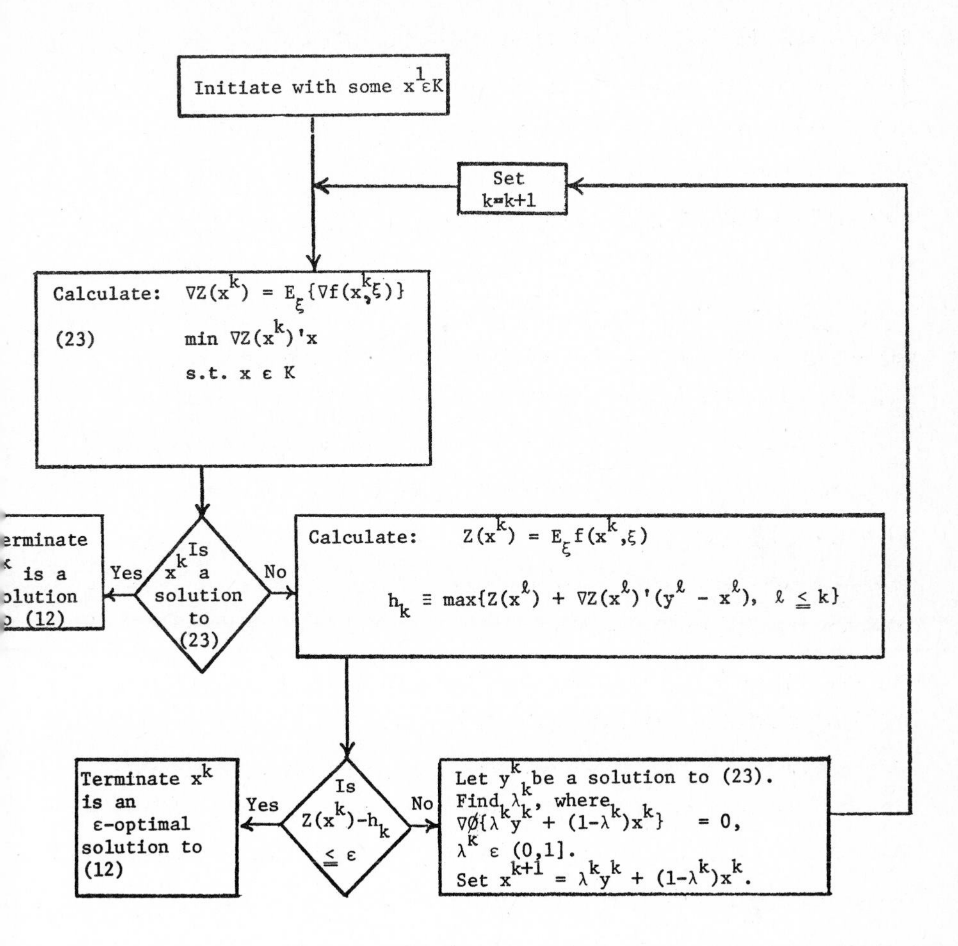

Fig. 1. Flow Chart of the Frank-Wolfe Algorithm Applied to Problem (12).

modifications of the Convex Simplex Method [80], SUMT [24], and the Generalized Programming Algorithm [16] appear in [81], [82] and [49], respectively. Perhaps the central difficulty in actually solving such problems manifests itself in the number and difficulty of the numerical integration calculations.

The general problem of the evaluation of multivariate integrals is considered in detail by Davis and Rabinowitz [19], Gupta [32] and Haber [33]. Univariate calculations may be performed quickly to high accuracy. However, such calculations are very time consuming if there are many dependent random variables. The difficulty of the calculations goes up far more than linearly with the dimension size. Fortunately in many applications one need only perform univariate integrations. A leading special case,(see, in this volume, "Choosing Investment Portfolios..." occurs when the objective is univariate, i.e., $f(x,\xi) = u(\xi'x)$, and ξ is either joint normal or symmetric stable. When f is separable into $\sum f_j(x_j,\xi_j)$ or more generally $\sum f_j(x,\xi_j)$ only marginal distributions are required even if the ξ_j are correlated. Univariate calculations also suffice if the ξ_j are independent. In many problems there is only one random variable of interest and univariate calculations suffice, see e.g., section seven below. Finally we may note that no numerical integrations are needed at all if the ξ_j have discrete distributions or if one is willing to accept a discrete approximation to $F(\xi)$. In the discrete case one performs a summation operation.

6. Special Cases

A. Ξ is finite

Suppose that the vector $\xi \equiv (\xi_1,\ldots,\xi_s)$ has only a finite number of possible realizations say $\xi^1,\ldots,\xi^L$ that occur with probabilities $p_1,\ldots,p_L > 0$ (where $L < \infty$, $\sum_\ell p_\ell = 1$). Then (12) becomes

$$(26)\qquad \begin{aligned} \min \quad & Z(x) = \sum_\ell p_\ell\, f(x,\xi^\ell), \\ \text{s.t.} \quad & x \in K. \end{aligned}$$

Now (26) is an explicit deterministic program that has a large number of similar terms in the objective function. Column generation schemes [31] might be used in solution approaches to solve (26).

When the problem is the linear simple recourse model the expression for (26) is

$$(27)\qquad \begin{aligned} \min \quad & c'x + \sum_\ell p_\ell (q^{+'} y_\ell^+ + q^{-'} y_\ell^-), & \\ \text{s.t.} \quad & Ax \le b, & : \pi \\ & Tx + y_\ell^+ - y_\ell^- = \xi^\ell, & : \mu_\ell \\ & y_\ell^+ \ge 0,\ y_\ell^- \ge 0, \quad \ell=1,\ldots,L. & \end{aligned}$$

In (27) it is assumed that the vectors c, q^+, q^- and b and the matrices A and T are composed of constant elements. The vectors π and μ_ℓ are dual variables. The dual of (27) is

$$\max \quad \pi' b + \sum_{\ell} \mu_{\ell}' \xi^{\ell},$$

$$\text{s.t.} \quad \pi' A + \mu_{\ell}' T \geq c, \qquad \ell = 1,\ldots,L$$

$$- p_{\ell} q^{-} \geq \mu_{\ell} \geq p_{\ell} q^{+}.$$

The dual has far fewer constraints than the primal and its solution would proceed via a generalized upper bounding scheme [18].

B. f is Quadratic

Theil [62] has shown that $\bar{\xi}$ is certainty equivalent for ξ if f is quadratic. The verification is as follows

$$Z(x) \equiv E_{\xi}[f(x,\xi)] \equiv E_{\xi}[a'x + b'\xi + (x',\xi')\begin{pmatrix} A & C \\ C' & B \end{pmatrix}\begin{pmatrix} x \\ \xi \end{pmatrix}]$$

$$= a'x + b'\bar{\xi} + x'Ax + 2\bar{\xi}'Cx + E_{\xi}\{\xi' B \xi\}$$

$$= f(x,\bar{\xi}) + \delta,$$

where δ is the constant, $E_{\xi}\{\xi' B\xi\} - \bar{\xi}' B\bar{\xi}$. Note that the result holds if f is a quadratic function of δ and an arbitrary vector valued function $\phi(x)$. In this case

$$Z(x) \equiv E_{\xi}[f(x,\xi)] \equiv E_{\xi}[a'\phi(x) + b'\xi + (\phi(x)',\xi')$$

$$\begin{pmatrix} A & C \\ C' & B \end{pmatrix}\begin{pmatrix} \phi(x) \\ \xi \end{pmatrix}] = f(x,\bar{\xi}) + \delta.$$

C. f is Polynomial

Suppose $f(x,\xi)$ takes the univariate form $u(w) = u(\sum_{j=1}^{m} \xi_j x_j)$ as in example 2.1 and in all problems where the utility is measured by a single index such as wealth w. Let

$$u(\sum_j \xi_j x_j) = \gamma_0 + \gamma_1 (\sum_j \xi_j x_j) + \gamma_2 (\sum_j \xi_j x_j)^2 + \dots + \gamma_I (\sum_j \xi_j x_j)^I . \tag{28}$$

Upon expansion this expression may be put into the form

$$\sum_{i=0}^{I} \sum_{\nu=0}^{i} \gamma_{\nu i} (x_1^{\alpha_{1\nu i}} \dots x_m^{\alpha_{m\nu i}}) (\xi_1^{\beta_{1\nu i}} \dots \xi_m^{\beta_{m\nu i}}).$$

The α's, β's and γ's are constants with the α's and β's being integers whose sum over ν and j is $2I$. The expected utility is then

$$\sum_i \sum_\nu \gamma'_{\nu i} (x_1^{\alpha_{1\nu i}} \dots x_m^{\alpha_{m\nu i}}) \tag{29}$$

where each $\gamma'_{\nu i} \equiv \gamma_{\nu i} \overline{(\xi_1^{\beta_{1\nu i}} \dots \xi_m^{\beta_{m\nu i}})}$. The expression (29) is a deterministic signomial that is generally non-concave. Geometric programming methods, see e.g. [22], might be used to maximize this function over a given set. In the separable case

$$u(\sum_j \xi_j x_j) = \sum_j u_j(\xi_j x_j) = \sum_j \{\gamma_{0j} + \gamma_{1j}(\xi_j x_j) + \dots + \gamma_{Ij}(\xi_j x_j)^I\}. \tag{30}$$

Then

$$\text{(31)} \qquad Eu\left(\sum_j \xi_j x_j\right) = \sum_{j=1}^{m} \sum_{i=0}^{I} \delta_{ij} x_j^i$$

where each constant $\delta_{ij} \equiv \gamma_{ij} \overline{\xi_j^i}$. The expression (31) is a sum of (generally non-concave polynomial functions of the x_j.

The expressions (28) and (30) might arise from Taylor series approximations [59] to general u's. For example suppose $u_j(\xi_j x_j)$ is expanded about 0 and $u_j^i(0)$ is the derivative of order i of u_j at 0. Then

$$u_j(\xi_i x_j) = \sum_{i=1}^{I} \left\{(\xi_j x_j)^i \frac{U_j^i(0)}{i!}\right\} + R_j,$$

where remainder term $R_j = (\xi_j x_j)^{I+1} \dfrac{U_j^{I+1}(y_j)}{I+1!}$ for some $y_j \in [0, \xi_j x_j]$.

D. Some Separable Asymmetric Quadratic Cases

Obtaining an explicit expression for Z is generally quite difficult unless f and the joint distribution of ξ take a mutually convenient form. When f is separable into $\sum_j f_j(x_j, \xi_j)$

$$Z(x) \equiv E_\xi\{f(x,\xi)\} = \sum_j E_{\xi_j}\{f_j(x_j,\xi_j)\} \equiv \sum_j Z_j(x_j)$$

Hence only the marginal distributions of the ξ_j enter the calculation of Z which reduces considerably the complexity of the calculations. Utilizing this fact we compute the explicit deterministic equivalents and their derivatives for the case when f is separable and each f_j is an asymmetric quadratic function.[5] Let f take the form of a overage-shortage cost function, see examples 2.2 and 2.5. Then $f_j(x_j,\xi_j) = \tilde{f}_j(x_j - \xi_j) \equiv \tilde{f}_j(y_j)$, and

$$\tilde{f}_j(y_j) = \begin{cases} c_j y_j + C_j y_j^2 & \text{if } y_j \geqq 0 \\ -d_j y_j + D_j y_j^2 & \text{if } y_j \leqq 0, \end{cases}$$

The c_j, C_j, d_j and D_j are known constants, $j = 1, \ldots, n$, and the ξ_j have normal, exponential or uniform distributions. Similar calculations for the asymmetric linear case (i.e. $C_j = D_j = 0$) have been given for the uniform [4, 71] and exponential [71] distributions.

For expositional ease we will suppress the j subscripts. Now each Z (j suppressed) is

$$Z(x) = [c(x - \xi) + C(x - \xi)^2]\, dF(\xi) + [d(\xi - x) + D(\xi - x)^2]\, dF(\xi) \tag{32}$$

When the ξ have the (absolutely) continuous densities $g(\xi)$, the Z have the continuous partial derivatives

---

[5]The case when ξ has a joint normal distribution and f is a non-separable, asymmetric quadratic function is considered in [87]. In this case, one may utilize an orthogonal transformation and the separability results presented here. This procedure, because of its complexity, is satisfactory only for small m.

$$\frac{\partial Z(x)}{\partial x} = \quad [c + 2C(x - \xi)]\ g(\xi)\ d\xi \tag{33}$$

$$[d + D(\xi - x)]\ g(\xi)\ d\xi.$$

D1. ξ is Normally Distributed

The marginal density of $F(\xi)$ with respect to ξ is, $g(\xi) \equiv (2\pi\sigma^2)^{-1/2} \exp\{-(\xi - \bar{\xi})^2/2\sigma^2\}$, where $\bar{\xi}$ is the mean of ξ, and σ^2 is the variance of ξ. The quantity $n(x) \equiv (2\pi)^{-1/2} e^{-1/2\, t^2}|_{t=x}$ refers to a standard normal density evaluated at the point x, and $N(x) \equiv \int_{-\infty}^{x} n(t)dt$ is the cumulative distribution function of a standard normal variable evaluated at the point x. The first two incomplete moments of a standard normal variable are [53]:

$$\int_{-\infty}^{x} t.n(t)dt = n(x), \quad \int_{x}^{\infty} t\cdot n(t)dt = n(x), \quad \int_{-\infty}^{x} t^2 n(t)dt =$$

$-nx(x) + N(x)$, and $\int_{x}^{\infty} t^2 n(t)dt = xn(x) + 1 - N(x)$.

Equation (32) may be written as

$$Z(x) = \int_{\infty}^{x} [(C\,x^2 + c\,x) - (c + 2C\,x)\xi$$

$$+ C\,\xi^2]\ f(\xi)d\xi + \int_{x}^{\infty} [(D\,x^2 - d\,x)$$

$$+(d - 2D\,x)\xi + D\,\xi^2]\ g(\xi)d\xi$$

$$= (C x^2 + c x) \int_{-\infty}^{x} n(t)dt + (D x^2 - d x) \int_{x}^{\infty}$$

$$n(t)dt - (c + 2C x) \int_{-\infty}^{x} (t\sigma + \bar{\xi})$$

$$n(t)dt + (d - 2D x) \int_{x}^{\infty} (t\sigma + \bar{\xi}) n(t)dt$$

$$+ \quad C \int_{-\infty}^{x} (t^2\sigma^2 + \bar{\xi}^2 - 2t\,\bar{\xi}) n(t)dt + D \int_{x}^{\infty}$$

$$((t^2\sigma^2) + \bar{\xi}^2 - 2t\,\bar{\xi}) n(t)dt = \{D[(x - \bar{\xi})^2 + \sigma^2]$$

$$- \quad d(x - \bar{\xi})\} + \{\mathcal{D}[(x + \bar{\xi})^2 + \sigma^2] + \mathcal{S}(x + \bar{\xi})\} N(x)$$

$$+ \quad \{\mathcal{S}\sigma + (2x + 2\bar{\xi} - \sigma z)\,\sigma\,\mathcal{D}\}\, n(x),$$

where

$$t \equiv (x - \bar{\xi})/\sigma, \mathcal{D} \equiv C - D, \mathcal{S} \equiv c + d$$

Using (33) the partial derivatives are then

$$\frac{\partial Z(x)}{\partial x} = \{\mathcal{S} + 2\mathcal{H}(x - \bar{\xi})\}\, N(z) + 2\mathcal{D}\sigma\, n(z)$$

$$- \quad \{d + 2D\,(x + \bar{\xi})\},$$

where $\mathcal{H} \equiv C + D$.

D2. ξ is Exponentially Distribution

The marginal density of ξ is

$$g(\xi) \equiv \begin{cases} \lambda e^{-\lambda \xi} & \text{if } \xi \in (0, \infty),\ \lambda > 0 \\ 0 & \text{otherwise} \end{cases}$$

where $1/\lambda$ is the mean of ξ. The expressions required to calculate $\breve{Z}$ are

$$\int_0^x (x - \xi) g(\xi) d\xi = -\left(\frac{1}{\lambda}\ 1 - e^{-\lambda x}\right),$$

$$\int_x^\infty (\xi - x) g(\xi) d\xi = \frac{1}{\lambda} e^{-\lambda x},$$

$$\int_0^x (x - \xi)^2 g(\xi) d\xi = \frac{2}{\lambda^2}\left(1 - e^{-\lambda x}\right)$$

$$\int_x^\infty (\xi - x)^2 g(\xi) d\xi = \frac{2}{\lambda^2} e^{-\lambda x}$$

Utilizing these results one obtains

$$Z(x) = E\, g(x) + G, \quad \text{and} \quad \frac{\partial Z^{\cdot}(x)}{\partial x} = -\lambda\, E\, g(x),$$

where

$$E \equiv \{\lambda(c + d) + 2C - D)\}/\lambda,$$

$$G \equiv \{2C - \lambda\, c\}/\lambda^2.$$

D3. ξ is Uniformly Distributed.

The marginal density of ξ is

$$g(\xi) \equiv \begin{cases} \dfrac{1}{\beta - \alpha} & \text{if } \xi \in [\alpha, \beta], \quad -\infty < \alpha < \beta + \infty \\ 0 & \text{otherwise,} \end{cases}$$

where α and β are constants. The expressions required to calculate Z are

$$\int_{\alpha}^{x} (x - \xi)^k \, g(\xi) d\xi = \frac{(x - \alpha)^{k+1}}{(k+1)\,(\beta - \alpha)},$$

$$\int_{x}^{\beta} (\xi - x)^k \, g(\xi) d\xi = \frac{(\beta - x)^{k+1}}{(k+1)\,(\beta - \alpha)}, \qquad k = 1,2.$$

Utilizing these results one obtains

$$Z(x) = H\,x^3 + I\,x^2 + J\,x + K, \quad \text{and}$$

$$\frac{\partial Z(x)}{\partial x_j} = 3H\,x^2 + 2I\,x + J.$$

where

$$H \equiv 2(C - D),$$

$$I \equiv 3(c + d) + 6(C\,\beta - D\,\alpha),$$

$$J \equiv -6(c\,\beta + d\,\alpha + D\,\beta^2 - C\,\alpha^2,$$

$$K \equiv 3(d\,\beta^2 + c\,\alpha^2) + 2(D\,\beta^3 - C\,\alpha^3),$$

7. Two Period Problems[6]

Each of the decision situations depicted in examples 2.1-2.5 has a natural generalization to a two period decision problem. For example the transportation problem 2.5 may be formulated as Example 7.1: Transportation.

Suppose a firm will have S_{1i} and S_{2i} units of a particular product available at warehouse $i(i=1,\ldots,m)$ in periods one and two, respectively. It costs $c_{tij}x_{tij}$ to ship x_{tij} units from warehouse i to outlet j in period t where $t=1,2$. The demand vectors at the destinations are $\xi_1 \equiv (\xi_{11},\ldots,\xi_{1n})$ and $\xi_2 \equiv (\xi_{21},\ldots,\xi_{2n})$, which are distributed independently of the x_{tij}. Let the shortage cost functions at outlet j be $g_{1j}^-(y_{1j}^-)$ and $g_{2j}^-(y_{2j}^-)$. The overage-storage cost functions are $g_{1j}^+(y_{1j}^+)$ and $g_{2j}^+(y_{2j}^+)$. Let β be the firm's one period discount factor. The firm may find a shipping policy that minimizes expected costs by solving

$$\text{(34)} \qquad \min \sum_i \sum_j c_{1ij}x_{1ij} + E_{\xi_1}\{\min \sum_j [g_{1j}^+(y_{1j}^+)$$

$$+ g_{1j}^-(y_{1j}^-)] + \beta(\min [\sum_i \sum_j c_{2ij}x_{2ij} + E_{\xi_2|\xi_1} \min$$

$$\{\sum_j [g_{2j}^+(y_{2j}^+) + g_{2j}^-(y_{2j}^-)]\})]\},$$

[6] This section is based largely on Ziemba [88].

$$\text{s.t.} \quad \sum_j x_{1ij} = S_{1i},$$

$$\sum_i x_{1ij} - y_{1j}^{+} + y_{1j}^{-} = \xi_{1j},$$

$$\sum_j x_{2ij} = S_{2i},$$

$$\sum_i x_{2ij} - y_{2j}^{+} + y_{2j}^{-} + y_{1j}^{+} = \xi_{2j},$$

$$x_{tij} \geq 0,\ y_{tij}^{+} \geq 0,\ y_{tij}^{-} \geq 0,\ y_{tij}^{+} \cdot y_{tij}^{-} = 0,$$

$$t = 1,2, \quad i = 1,\ldots,m \quad \text{and} \quad j = 1,\ldots,n.$$

A second example is Example 7.2: Pension Fund Management[7]

We consider the management activities of a firm's employee pension plan. These activities consist of paying benefits, investing the plan's resources and making contributions to accumulate a fixed terminal worth for the plan. Suppose that there are two decision periods and that the fund is initially endowed with resources of value m_0 and the goal is m_2. The firm must decide upon contribution levels c_1 and c_2 at the beginning of each of these periods. A contribution must also be made at the end of period two to achieve the goal m_2, taking into account that benefits of b_1 and b_2 must be paid. The firm's contribution cost function $v(c)$ will be assumed to be strictly increasing and convex.

---

[7] This problem was described to me by my colleague Irwin Tepper. He is, considering among other things, the economic and computational aspects of finite horizon problems of this nature in his dissertation [60]. The formulation given here was adapted from Ziemba [88].

The convexity property reflects the situation that the firm's contributions at any given time become more and more costly. Small contributions must be obtained in lieu of investment in inventories and other short term assets in which sales and other revenues decrease more than in proportion to the funds taken from these activities. If β_1 and β_2, are the firm's discount factors. The problem is

$$\min_{c_1 \epsilon K_1} \left\{ v(c_1) + \beta_1 E_{\xi_1} \min_{c_2 \epsilon K_2(c_1,\xi_1)} \left[v(c_2) \right.\right.$$

$$\left.\left. + \beta_2 E_{\xi_2|\xi_1} \{v[m_2 - \{(m_1 + c_1 - b_1)\xi_1 + c_2 - b_2\} \xi_2]\} \right] \right\}.$$

In addition to upper boundedness it is assumed that c_1 and c_2 are large enough so that: (i) b_1 and b_2 may be paid out; and (ii) the accrual benefits for the participants of the plan are covered to maintain the plan's tax exempt status. This latter constraint means that $c_t \geqq \alpha_t \equiv$ accrual benefits in t plus the capitalized value of all past accruals which have not been funded times the appropriate rate of interest. Hence

$$K_1 \equiv \{c_1 | \alpha_1 \leqq c_1 \leqq c_1^u,\ b_1 - m_0 \leqq c_1\}, \quad \text{and}$$

$$K_2(c_1,\xi_1) \equiv \{c_2 | \alpha_2 \leqq c_2 \leqq c_2^u,\ b_2 - \xi_1(m_0 + c_1 - b_1) \leqq c_2\}.$$

A natural generalization of problem (12) is[8]

---

[8] In [88] there is a discussion of a more general class of problems, similar in spirit to (14), that can be reduced to form (23) in an exactly analogous way that problems of type (14) may be reduced to form (12). Utilizing the Simon-Theil Theorem [63] and the direct reduction results in Ziemba [84] one can reduce certain problems of type (34) and its generalization to either form (12) or an explicit deterministic equivalent.

$$Z^* \equiv \min_{x_1 \in K_1} E_{\xi_1} \left[f_1(x_1,\xi_1) + \min_{x_2 \in K_2(x_1,\xi_1)} E_{\xi_2|\xi_1} f_2(x_1,x_2,\xi_1,\xi_2) \right] \tag{34}$$

In (34) the x_t are decision vectors chosen at the beginning of period t. The ξ_t are random vectors, belonging to the probability spaces $(\Xi_t, \mathcal{F}_t, F_t)$ that are observed after the x_t are chosen but before the x_{t-1} are chosen. The K_t are constraints on the choice of the decision vectors and the f_t are cost functions. Problem (34) may be interpreted as follows. Choose an initial decision x_1 so that a second decision $x_2 = x_2(\xi_1)$ may be chosen after ξ_1 is observed and total expected costs are minimized.

The transportation problem may be put into form (34) as follows. As in the static problem the complementary conditions may be deleted if

$g^+_{tj}(0) = g^-_{tj}(0) = 0$, and

$$\partial g^+_{tj} / \partial y^+_{tj} + \partial g^-_{tj} / \partial y^-_{tj} \geq 0, \text{ for any } \partial y^+_{tj} = \partial y^-_{tj} \geqq 0.$$

Let $y_{tj} = y^+_{tj} - y^-_{tj}$,

and $x_t \equiv (x_{t1}, \ldots, x_{tmn})$, for $t=1,2$.

Then
$$f_1(x_1,\xi_1) = \sum_i \sum_j c_{1ij} x_{1ij} + \sum_j \begin{cases} g^+_{1j} \left(\sum_i x_{1ij} - \xi_{1j} \right) & \text{if } \sum_i x_{1ij} \geqq \xi_{1j} \\ g^+_{2j} \left(\xi_{1j} - \sum_i x_{1ij} \right) & \text{if } \sum_i x_{1ij} \leqq \xi_{1j} \end{cases}$$

$$f_2(x_1,x_2,\xi_1,\xi_2) = \beta \left\{ \sum_i \sum_j c_{2ij} x_{2ij} \right.$$

$$\left. + \sum_j \begin{bmatrix} g_{2j}^+ (\overbrace{\sum_i x_{2ij} - \xi_{2j} - \max (\sum x_{1ij} - \xi_{ij}, 0)}^{y_{2j}}) \\ \text{if } y_{2j} \geq 0 \\ g_{2j}^- (\xi_{2j} + \max (\sum x_{1ij} - \xi_{1j}, 0) - \sum_i x_{2ij}) \\ \text{if } y_{2j} \leq 0, \end{bmatrix} \right\}$$

and $K_t \equiv \{x_t | x_t \geq 0, \sum_j x_{tij} = S_{ti}, i=1,\dots m\}, \quad t=1,2.$

The pension fund management problem has the form of (34) with

$$x_1 \equiv c_1, \; x_2 \equiv c_2,$$

$$f_1(x_1) \equiv v(c_1), \text{ and}$$

$$f_2(x_1,x_2,\xi_1,\xi_2) \equiv v(c_2) + \beta_1\beta_2 \, v[m_2 - \{(m_1 + c_1 - b_1)\xi_1 + c_2 - b_2\}\xi_2].$$

It is possible to develop bounds on Z^* analogous to the bounds given in Theorem 5 a, b. Let $\bar{\xi}_1$ be the mean of ξ_1 and $\bar{\xi}_2(\xi_1)$ be the mean of ξ_2 conditional on ξ_1. Let $(\bar{x}_1,\bar{x}_2)$ be a solution, assumed to exist, of

$$(35) \qquad \min_{x_1 \epsilon K_1, \; x_2 \epsilon (x_1,\bar{\xi}_1)} \left[f_1(x_1,\bar{\xi}_1) + f_2(x_1,x_2,\bar{\xi}_1,\bar{\xi}_2(\bar{\xi}_1)) \right] \equiv Z_L.$$

Theorem 9 [88].

(a) Assume that f_1 and f_2 are finite continuous convex functions of (x_1,ξ_1) and (x_1,x_2,ξ_1,ξ_2) respectively. Suppose that $\bar{\xi}_2(\cdot)$ is either linear or that it is convex and f_2 is non-decreasing in ξ_2. Finally assume that K_1 and $K_2(\cdot)$ are convex sets.

(b) Suppose that $\bar{x}_2 \epsilon K_2(\bar{x}_1,\xi_1)\ \forall \xi_1 \epsilon \Xi$.

Then the following bounds obtain

$$Z_L \overset{(a)}{\leqq} Z^* \overset{(b)}{\leqq} E_{\xi_1,\xi_2}\left[f_1(\bar{x},\xi_1) + f_2(\bar{x}_1,\bar{x}_2,\xi_1,\xi_2)\right].$$

Inequality (b) follows directly because $(\bar{x}_1,\bar{x}_2)$ is feasible for (34). Note however that it is not generally true that $\bar{x}_2 \epsilon K_2(\bar{x}_1,\xi_1)\ \forall \xi_1 \epsilon \Xi_1$. One may, of course, generate an upper bound utilizing $\bar{x}_1$ and any x_2 that belongs to $K_2(\bar{x}_1,\xi_1)\ \forall \xi_1 \epsilon \Xi_1$. Such an x_2 will not always exist. The proof of inequality (a) and some related discussion appears in [88] and utilizes Jensen's inequality and other well known properties of convex functions. Note that one may obtain the lower bound Z_L by solving a deterministic convex program.

Ziemba [88] has developed a modified feasible direction algorithm to solve (34). The algorithm is a consequence of the fact that under certain moderate assumptions Z is convex and continuously differentiable and it is relatively easy to numerically evaluate the partial derivatives of Z. The proof utilizes a result similar to those in Danskin [14] regarding the calculation of directional and partial derivatives of functions of the form $\psi(x) = \{\min h(x,y) \mid y \epsilon D(x)\}$, and Theorem 3.

Let $s_2(x_1,x_2,\xi_1) \equiv E_{\xi_2|\xi_1}[f_2(x_1,x_2,\xi_1,\xi_2)]$, and

$$t_1(x_1,\xi_1) \equiv \min_{x_2 \in K_2(x_1,\xi_1)} [s_2(x_1,x_2,\xi_1)].$$

Then (34) is equivalent to

(36) $$\min_{x_1 \in K_1} Z_1(x_1) \equiv \min_{x_1 \in K_1} E_{\xi_1}[f_1(x_1,\xi_1) + t_1(x_1,\xi_1)].$$

Theorem 10 [88]:

Suppose K_1 is a convex set and that $K_2(\cdot)$ is a convex set $\forall\ (x_1,\xi_1) \in K_1 x \Xi$.

Assume that for each $x_1 \in K_1$ and $\xi_1 \in \Xi_1$ there exists a unique $x_2^*(x_1,\xi_1) \in K_2(x_1,\xi_1)$ continuously differentiable in x_1 that minimizes s_2 and that there exists an x_1^* that minimizes Z_1. Suppose also that for each fixed $\xi_1 \in \Xi$, $\{f_1(x_1,\xi_1) + t_1(x_1,\xi_1)\}$ is finite on K_1.

Then Z_1 is a continuously differentiable convex function and

$$\frac{\partial Z_1}{\partial x_{1j}} = E_{\xi_1}\left\{\frac{\partial f_1(x_1,\xi_1)}{\partial x_{1j}} + \frac{\partial s_2[x_1,x_2^*(x_1,\xi_1),\xi_1]}{\partial x_{1j}}\right\}.$$

Suppose $K_1 \equiv \{x_1 | g_{1r}(x_1) \leqq 0,\ r=1,\ldots,m_1\}$ and $K_2(x_1,\xi_1) \equiv \{x_2 | g_{2r}(x_1,x_2,\xi_1) \leqq 0,\ r=1,\ldots,m_2\}$ satisfy a constraint qualification, see [80]. Assume that the f_t and g_{tr} are real-valued continuously differentiable convex functions of x_1 and (x_1,x_2) respectively. Theorem 10 then leads to the following feasible direction algorithm.

Step 0: Initialization: Solve (35) for $(\bar{x}_1,\bar{x}_2)$.

Set $x_1^1 = \bar{x}_1$, $\epsilon^1 > 0$, $k=1$.

Step 1: For x_1^k and $\forall \xi_1 \epsilon \Xi_1$ solve

$$\min_{x_2 \epsilon K_2(x_1,\xi_1)} s_2(x_1^k, x_2, \xi_1) \quad \text{for} \quad x_2^*(x_1^k, \xi_1) \quad \text{[subproblem]}$$

Calculate $$E_{\xi_1}\left[\frac{\partial f_1\ (x_1^k, \xi_1}{\partial x_{1j}} + \frac{\partial s_2[x_1^k, x_2^*(x_1^k,\xi_1),\xi_1]}{\partial x_{1j}}\right] \equiv c_j^k$$ [summation of functions]

Step 2: Direction Finding Problem: find γ_j^k, α^k:

$$\min_{\alpha,\gamma} \quad \alpha$$

$$\text{s.t.} \quad \nabla g_{1r}(x_1^{\ k})'\gamma \leqq \alpha, \quad r \ \epsilon\ R_k, \quad \text{[linear program]}$$

$$\sum_j c_j^{\ k} \gamma_j \leqq \alpha$$

$$-1 \leqq \gamma_j \leqq 1, \quad j=1,\ldots,n_1,$$

where $R_k \equiv \{r \mid g_{1r}(x_1^{\ k}) \geqq -\epsilon^k\}$.

Step 3: Step Size Determination: Find θ^k:

$$\min_{\theta \geqq 0} \psi(\theta) \equiv \min_{\theta \geqq 0} E_{\xi_1}\left[f_1(x_1^{\ k} + \theta\gamma^{\ k}, \xi_1) + \min_{x_2 \epsilon K_2(x_1,\xi_1)} \left\{ s_2(x_1^{\ k} + \theta\ \gamma^k, x_2, \xi_1) \right\}\right]$$

Step 4: Set $x_1^{k+1} = x_1^k + \theta^k \gamma^k$.

STOP if x_1^{k+1} satisfies the Kuhn-Tucker Conditions, or a tolerance bound. Otherwise set

$$\epsilon^{k+1} = \begin{cases} \epsilon^k & \text{if } -\alpha^k > \epsilon^k \\ \epsilon^k/2 & \text{otherwise} \end{cases}$$

Go to Step 1 with $k = k+1$.

The procedure is a primal algorithm since each x_1^k is feasible for (34). The subproblems in step 1 may be thought of as one step dynamic programming calculations. Hence the procedure is a forward-backward algorithm. The primal approach has the distinct advantage over a dynamic programming approach that one does not have to perform the calculations for all feasible x_1.

It is necessary to assume that Ξ_1 has finite cardinality for there to be a finite number of subproblems to be solved in step 1 of each iteration. (Sampling approximations could, of course, be used if Ξ_1 does not have finite cardinality). The finiteness of the cardinality of ξ_2 is not of crucial importance. Indeed the case when ξ_2 has a continuous density generally would be computationally less difficult than when it is discrete. In the case when both ξ_1 and ξ_2 are discrete the algorithm presented is only one of many algorithms that could be applied to solve (34), since using an argument similar to that in section 6A (34) has an explicit deterministic equivalent.

In the direction finding problem c_γ^k represents the directional derivative at x_1^k. Hence any γ such that $c^k\gamma < 0$ is a good direction and the linear program finds the best γ subject to the stipulation that the direction is usable in the sense that there exists a $\tau > 0$ such that all $\theta \in [0,\tau]$ are feasible step sizes. Column generation schemes, see [31], may thus be used in step 2 to reduce the sampling requirements in step 1. The algorithm is an application of Zoutendijk's ϵ-pertubation method [89]. If $x_2^*(\cdot)$ is not differentiable then one may utilize a similar algorithm that utilizes steep subgradients as direction vectors, see [88].

The Kuhn-Tucker conditions for (34) are that $g_{1r}(x_1) \leqq 0$, there exist $\lambda_r \geq 0$ such that $\lambda_r g_{1r}(x_1) = 0$, and $\nabla Z_1(x_1) + \sum \lambda_r \nabla g_{1r}(x_1) = 0$, where $r=1,\ldots,m_1$.

In general the computational burden of the algorithm is quite prohibitive. However it is believed that the algorithm would be economically feasible to solve problems where only univariate (or perhaps bivariate) numerical integrations are required. It can be shown that only univariate integrations are needed for example 7.1 and 7.2. The decision variable in each period of the Pension Fund Management Problem is constrained to lie in a bounded interval. For such problems it is possible to devise a modified embedded bisecting algorithm. The precise steps are presented in [88]. In Ziemba [86] the feasible direction algorithm is applied to a portfolio revision problem.

8. Comments and Historical Remarks

The original formulation of the simple recourse model dates to the papers by Beale [3] and Dantzig [15]. They formulated the linear simple recourse model

$$\text{(37)} \qquad \min \quad c'x + E_\xi \{\min [q^{+'} y^{+} + q^{-'} y^{-}],$$

$$\begin{aligned} \text{s.t.} \quad & Ax & & = b, \\ & Tx + y^+ - y^- & & = \xi, \\ & y^+ \geq 0,\ y^- \geq 0, & y^{+'} y^- & = 0. \end{aligned}$$

There has been an extensive study of models of form (37). Williams [75] and Wets [71] have given thorough accounts of existence and solution questions, respectively, for (37). A presentation of most of the important results for the linear simple recourse model appears in Parikh [49]. Many of the results presented here are generalizations of results that were initially developed for (37). Walkup and Wets [69] studied the case when q^+, q^- and T are also random. The analysis here applies directly to this problem as well since (37) may easily be put in form (12) if $q^+ + q^- \geq 0$ with probability one.

An important special case of (12) occurs when some constraints of K are of the integer variety as in example 2.3. One may modify implicit enumeration schemes see e.g. [48] in this case to solve (12). See Hillier [34, 8 Ch. 1] for such an algorithm applied to a capital budgeting problem. In this case the objective is univariate, i.e., $f(x,\xi) = u(\xi'x)$. Hillier develops an efficient algorithm assuming that the ξ's are normally distributed and that all the decision variables must be integers. The scheme in section five points to more general applications particularly in cases when f is not univariate and some but not all variables must be integer valued as in example 2.3.

Most of the analysis related to problem (12) has been made under certain convexity assumptions. The assumptions on the basic functions required to guarantee that examples 2.1-2.5 and 7.1-7.2 have the required convexity properties generally have natural realistic interpretations.

However it is of interest to consider more general problems. Unfortunately a number of major difficulties arise when the convexity assumptions are weakened. For example pure strategies are no longer generally optimal. Standard algorithms may no longer converge to global optima, etc. See Fromowitz [29] for some first results.

Versions of Theorems 1-4 have been presented by numerous authors see e.g. [15, 44, 52, 69, 81, 82]. The results given here are the most general that the author is aware of.

In addition to the applications mentioned in sections two and seven there are numerous other problems that have forms (12) or (34). Examples pertaining to goal programming [13], inventory and transportation problem [5, 23, 71, 74] cash balance [88], consumers lifetime income allocation [79], consumption-investment planning problems [21], product mix [35], scheduling of aircraft [17, 47] team theory [50, 51] and agricultural planning [26] have been considered.

REFERENCES

[1] J. Abadie and J. Carpentier, "Generalization of the Wolfe Reduced Gradient Method to the Case of Nonlinear Constraints," in Optimization, R. Fletcher (ed.), Academic Press, New York, 1969.

[2] M.L. Balinsky and W.J. Baumol, "The Dual in Nonlinear Programming and its Economic Interpretation," Review of Economic Studies, XXXV (1968), 237-256.

[3] E.M.L. Beale, "On Minimizing a Convex Function Subject to Linear Inequalities," J. Royal Stat. Soc. B, XVII (1955), 173-184.

[4] ___________, "The Use of Quadratic Programming in Stochastic Linear Programming," Santa Monica: The RAND Corp., P-2404 1961.

[5] M. Bellmore et. al., "A Multiperiod Transportation Model with Stochastic Demands," mimeo, Dept. of O.R., Johns Hopkins University, 1969.

[6] B. Bereanu, "The Distribution Problem in Stochastic Linear Programming," in Operations Research-Verfahren, ed. by R. Henn et. al., New York: Springer-Verlag, 1970.

[7] C. Berge and A. Ghouli-Houri, Programming, Games and Transportation Networks, Wiley, New York, 1965.

[8] R.F. Byrne et. al., Studies in Budgeting, Amsterdam: North Holland Publishing Co., Inc., 1971.

[9] M.D. Canon and C.D. Cullum, "A Tight Upper Bound on the Rate of Convergence of the Frank-Wolfe Algorithm," SIAM J. Control, VI (1968), 509-516.

[10] ___________, ___________, and P. Wolfe, "On Improving the Rate of Convergence of the Frank-Wolfe Algorithm," unpublished, IBM Research, Yorktown Heights, N.Y., 1968.

[11] A. Charnes, W.W. Cooper and G.H. Symonds,"Cost Horizons and Certainty Equivalents: An Approach to Stochastic Programming of Heating Oil," Management Science, IV (1958), 235-263.

[12] A.R. Colville, "A Comparative Study of Nonlinear Programming Codes," in Proceeding of the Princeton Symposium on Mathematical Programming, ed. by H.W. Kuhn, Princeton University Press, 1970.

[13] B. Contini, "A Stochastic Approach to Goal Programming," Operations Research, XVI (1968), 576-586.

[14] J.M. Danskin, The Theory of Max-Min, New York: Springer-Verlag, Inc., 1967.

[15] G.B. Dantzig, "Linear Programming Under Uncertainty," Management Science, III-IV (1955), 197-206.

[16] ___________, Linear Programming and Extensions, Princeton: Princeton University Press, 1963.

[17] ___________ and A.R. Ferguson, "The Allocation of Aircraft to Routes, Management Science, III, (1954), 45-73.

[18] ___________ and R.M. van Slyke, Generalized Upper Bounding Techniques, I, J. Comput. Systems Sci,, I (1968), 213-226.

[19] P.J. Davis and P. Rabinowitz, Numerical Integration, Waltham, Mass: Blaisdell Publishing Co., 1967.

[20] M. A. H. Dempster, "On Stochastic Linear Programming: I. Static Linear Programming Under Risk", J. Math. Anal. and Appl., XXI (1968), 304-343.

[21] J.H. Dreze and F. Modigliani, "Consumption Decisions Under Uncertainty," Journal of Economic Theory, V (1972), 308-335.

[22] R.J. Duffin, "Lineariz ing Geometric Programs," SIAM Review, XII (1970), 211-227.

[23] M. El - Agizy, "Two Stage Programming Under Uncertainty with Discrete Distribution Function," Operations Research, XV (1967), 55-70.

[24] A.V. Fiacco and G.P. McCormick, "The Sequential Unconstrained Minimization Technique for Non-linear Programming: A Primal-Dual Method," Management Science, X (1964), p. 360-366.

[25] R. Fletcher, ed., Optimization, New York: Academic Press, 1969.

[26] K.A. Fox et. al., The Theory of Quantitative Economic Policy, Chicago: Rand McNally, 1966.

[27] M. Frank and P. Wolfe, "An Algorithm for Quadratic Programming," Naval Research Logistics Quarterly, III (1956), 95-110.

[28] R.E. Fricks, "Probabilistic Programming 1950-1963: An Annotated Bibliography" and "A Research Bibliography on Two Stage Probabilistic Programming," Tech. Memos. 155 and 160, Dept. of O.R., Case Western Reserve Univ., 1969.

[29] S. Fromowitz, "Nonlinear Programming with Randomization", Management Science, XI (1965), 831-846).

[30] D. Gale and H. Nikaidô, "The Jacobian Matrix and Global Univalence of Mappings," Mathematics Annalen, CLIX (1965), 81-93.

[31] A.M Geoffrion, "Elements of Large-Scale Mathematical Programming," Management Science, XVI (1970), Part I, 652-675, Part II, 676-691.

[32] S.S. Gupta, "Probability Integrals of Multivariate Normal and Multivariate t," Ann. Math. Stat., XXXIV (1963), 792-828, and "Bibliography on the Multivariate Normal Integrals and Related Topics," Ann. Math. Stat., XXXIV (1963), 829-838.

[33] S. Haber, "Numerical Evaluation of Multiple Integrals," SIAM Review, XII (1970), 481-526.

[34] F.S. Hillier, The Evaluation of Risky Interrelated Investments Amsterdam: North-Holland Publishing Co., 1969.

[35] S.D. Hodges and P.G. Moore, "The Product-Mix Problem under Stochastic Seasonal Demand," Management Science, XVII (1970), 107-114.

[36] P. Huard, "Dual Programs," in Recent Advances in Mathematical Programming, edited by R.L. Graves and P. Wolfe, McGraw-Hill, 1963.

[37] M.J.L Kirby, "The Current State of Chance-Constrained Programming," in Proceedings of the Princeton Symposium on Mathematical Programming, H.W. Kuhn (ed) 1970.

[38] V.V. Kolbin, "Stochastic Programming," in Progress in Mathematics, vol. 11, ed. by R.V. Gamkrelidze, New York: Plenum Press, 1971.

[39] M. Loeve, Probability Theory, Princeton: D. Van Nostrand and Co., 3rd Ed., 1963.

[40] G.P. McCormick, "Penalty Function Versus Non-Penalty Function Methods for Constrained Nonlinear Programming Problems," Mathematical Programming, I (1971), 217-238.

[41] A. Madansky, "Bounds on the Expectation of a Convex Function of a Multivariate Random Variable," Ann. Math. Stat., XXX (1959), 743-746.

[42] ___________, "Methods of Solutions of Linear Programs Under Uncertainty" Operations Research, X (1962), 165-176.

[43] O.L. Mangasarian, "Duality in Nonlinear Programming," Quar. Appl. Math., XX (1962), 300-302.

[44] ___________, "Nonlinear Programming Problems with Stochastic Objective Functions," Management Science, X (1964), 353-359.

[45] ___________, "Optimality and Duality in Nonlinear Programming," in Proceedings of the Princeton Symposium on Mathematical Programming, ed. by H.W. Kuhn, Princeton University Press, 1970.

[46] __________ and J. B. Rosen, "Inequalities for Stochastic Non-linear Programming Problems," Operations Research, XII (1964), 143-154.

[47] J.L. Midler and R.D. Wollmer, "Stochastic Programming Models for Scheduling Airlift Operations," Naval Research Logistics Quarterly, XVI, (1969), 315-330.

[48] L.G. Mitten, "Branch and Bounds Methods: General Formulation and Properties," Operations Research, XVIII (1970), 24-34.

[49] S.C. Parikh, "Lecture Notes on Stochastic Programming," Unpublished University of California, Berkeley, 1968.

[50] R. Radner, "The Linear Team: An Example of Linear Programming Under Uncertainty," in H. Antosiewicz, ed., Second Symposium on Linear Programming, Washington: National Bureau of Standards (1955), 381-396.

[51] __________, "The Application of Linear Programming to Team Decision Problems," Management Science, V (1959), 143-150.

[52] __________, "Team Decision Problems," Ann. Math. Stat., XXXIII (1962), 857-881.

[53] H. Raiffa and R. Schlaifer, Applied Statistical Decision Theory, Cambridge: Division of Research, Graduate School of Business Administration, Harvard University, 1961.

[54] R.T. Rockafellar, Convex Analysis, Princeton: Princeton University Press, 1970.

[55] J. B. Rosen, "The Gradient Projection Method for Nonlinear Programming Part I, Linear Constraints," SIAM J. Appl. Math. 8, 181-217 (1960).

[56] __________ and J. Krouser, "A Gradient Projection Method for Nonlinear Constraints," unpublished research paper, Dept. of Computer Science, University of Wisconsin, 1972.

[57] J.K. Sengupta and F. Tintner, "A Review of Stochastic Linear Programming," Review of the International Statistical Institute (1969).

[58] D. Smith, "Programming Under Risk: A Bibliography," Center for Population Studies, Harvard University, 1969.

[59] Stone, M.H., "The Generalized Weirstrass Approximation Theorem," Mathematics Magazine, XXI (1948), 167-184.

[60] I. Tepper, "Economics of Pension Plans: A Study in Dynamic Financial Planning Under Risk", unpublished Ph.D. dissertation. U. of Penn., 1972.

[61] J. Tinbergen, On the Theory of Economic Policy, Amsterdam: North-Holland Publishing Co., 2nd ed., 1966.

[62] H. Theil, "Econometric Models and Welfare Maximization," Weltwirtschaftliches Archiv, LXXII (1954), 60-83.

[63] __________, Optimal Decision Rules for Government and Industry, Chicago: Rand-McNally, 1964.

[64] G. Tintner, "Stochastic Linear Programming with Applications to Agricultural Economics," in Proc. 2nd Symp. L.P., ed. by H.A. Antosiewicz, 1955.

[65] __________, "A Note on Stochastic Linear Programming," Econometrica, XXVIII (1960), 490-495.

[66] T.Y. Hans Tjian and W.I. Zangwill, "Analysis and Comparison of the Reduced Gradient and the Convex Simplex Method for Convex Programming," Center for Research in Management Science Working Paper No. 273, University of California, Berkeley, July 1969.

[67] D.M. Topkis and A.F. Veinott, Jr., "On the Convergence of Some Feasible Direction Algorithms for Nonlinear Programming," J. SIAM Control, V (1967), 268-279.

[68] S. Vajda, Probabilistic Programming, New York: Academic Press, 1972.

[69] D.W. Walkup and R.J.B. Wets, "Stochastic Programs with Recourse: Special Forms," in Proceedings of the Princeton Symposium on Mathematical Programming, H.W. Kuhn (ed.), 1970.

[70] A. Warburton and W.T. Ziemba, "Convex Inversion," Journal of Mathematical Analysis and Applications, XXXV (1971), 58-66.

[71] R. Wets, "Programming Under Uncertainty: The Complete Problem," Z. Warscheinlichkeitstheorie und Verw. Gebiete, IV (1966), 319-339.

[72] __________, "Notes on Stochastic Programming" unpublished, Dept. of I.E. and O.R. , University of California, Berkeley, 1967.

[73] __________, "Stochastic Programs with Fixed Recourse: the Equivalent Deterministic Programs," SIAM Review, forthcoming.

[74] A.C. Williams, "A Stochastic Transportation Problem," Operations Research, XI (1963), 759-770.

[75] __________, "On Stochastic Linear Programming," J. SIAM Appl. Math., XIII (1965), 927-940.

[76] __________, "Approximation Formulas for Stochastic Linear Programming," J. SIAM Appl. Math. XIV (1966), 668-677.

[77] __________, "Nonlinear Activity Analysis and Duality" in Proc. of the Princeton Symposium on Mathematical Programming, ed. by H.W. Kuhn, Princeton University Press, 1970.

[78] P. Wolfe, "A Duality Theorem for Nonlinear Programming," Quart. Appl. Math., XIX (1961), 239-244.

[79] M.E. Yaari, "On the Consumers Lifetime Allocation Process," International Economic Review, V (1964), 304-317.

[80] W.I. Zangwill, Nonlinear Programming: A Unified Approach Prentice-Hall, Inc., Englewood Cliffs, N.J., 1969.

[81] W.T. Ziemba, "Computational Algorithms for Convex Stochastic Programs with Simple Recourse," Operations Research, XVIII (1970), 414-431.

[82] ___________, "Solving Non-Linear Programming Problems with Stochastic Objective Functions," Journal of Financial and Quantitative Analysis, VII (1972), 1809-1827.

[83] ___________, "Duality Relations, Certainty Equivalents and Bounds for Convex Stochastic Programs with Simple Recourse," Cahiers du Centre d'Etudes de Recherche Operationnelle, XIII (1971), 85-97.

[84] ___________, "Transforming Stochastic Dynamic Programming Problems into Nonlinear Programs," Management Science, XVII (1971), 450-462.

[85] ___________, "Stochastic Programming Models of Dynamic Planning Problem" Economic Computation and Economic Cybernetics: Studies and Research, No. 3 (1970), 67-84.

[86] ___________, "Revising Investment Portfolios," paper in process.

[87] ___________, "Essays on Stochastic Programming and the Theory of Economic Policy," unpublished Ph.D. dissertation, Berkeley: University of California, 1969.

[88] ___________, "Two Period Stochastic Programs with Simple Recourse," GSB, Stanford University, Research Paper no. 82, 1972.

[89] ___________ and R. G. Vickson, Stochastic Optimization Models in Finance, Academic Press, Inc., forthcoming.

[90] G. Zoutendijk, Method of Feasible Directions, New York: Elsevier Publishing Co., 1960.

[91] ___________, "Nonlinear Programming: A Numerical Survey," SIAM J. Control, IV (1966), 194-210.

[92] ___________, "Nonlinear Programming Computation Methods," in Integer and Nonlinear Programming, J. Abadie (ed.). Amsterdam: North-Holland Publishing Co., 1970.

Mathematical Programming in Theory and Practice,
P.L. Hammer and G. Zoutendijk, (Eds.)

A CONSTRAINT-ACTIVATING OUTER POLAR METHOD FOR SOLVING PURE OR MIXED INTEGER 0-1 PROGRAMS

by

Egon Balas

Management Sciences Research Group
Graduate School of Industrial Administration
Carnegie-Mellon University

Abstract

This paper discusses a procedure for solving pure and mixed integer 0-1 programs, based on the properties of outer polar sets introduced in [1]. Rather than generating cutting planes from outer polars, here we use the latter in a different way. Starting with a subset of the problem constraints, we activate as many of the remaining constraints as are needed to produce a convex polytope that is contained in the outer polar of the convex hull of feasible integer points. When this is achieved, the algorithm terminates and the best solution found in the process is optimal.

This report was prepared as part of the activities of the Management Sciences Research Group, Carnegie-Mellon University, under Grant #GP 31699 of the National Science Foundation and Contract N0014-67-A-0314-0007 NR 047-048 with the U. S. Office of Naval Research.

A CONSTRAINT-ACTIVATING OUTER POLAR METHOD FOR SOLVING PURE OR MIXED INTEGER 0-1 PROGRAMS

by

Egon Balas

1. Introduction: Outer Polars

The approach discussed in this paper is based on the outer polar theory of [1]. It uses the fact that the outer polar of the feasible set contains the latter (and is usually considerably larger), and contains in its boundary all feasible 0-1 points. Based on these properties, we generate a polytope defined by a subset of the problem constraints, which has relatively few vertices (the fewer the constraints defining the set, the smaller the number of its vertices), and is contained in the outer polar. When this state is achieved, no further problem constraints need to be activated, i.e., no further vertices of the feasible set need to be generated.

We will discuss the pure 0-1 case and then show how the procedure extends to the mixed case.

Consider the problem

(P) $\max\{cx \mid x \in X,\ x \text{ integer}\}$

where

$$X = \{x \in R^n \mid Ax \leq b,\ 0 \leq x \leq e\}$$

and where A is $m \times n$, $e = (1,\ldots,1)$, c is integer.

Let $K = \{x \in R^n \mid 0 \leq x \leq e\}$ and $\hat{X} = \text{conv}(X \cap \text{vert } K)$, i.e., let $\hat{X}$ denote the convex hull of feasible 0-1 points. Further, for any set S, let the boundary, the interior, the convex hull and the set of vertices of S, be denoted by bd S, int S, conv S, and vert S respectively.

In [1] we introduced the concept of an outer polar and studied its properties. Some of these properties will be restated below in a slightly different form; namely, in [1] we found it convenient to translate X to a coordinate system centered at $\frac{1}{2}$ e, whereas here we prefer to work with the original coordinate system (computations become easier this way).

Apart from X, we will also have to do with polyhedral sets obtained from X by adding or removing some inequalities. For the purposes of this paper, the outer polar of any such set $S \subset R^n$ will be defined as

$$S^* = S^o(\tfrac{n}{4}) = \{y \mid (y - \tfrac{1}{2} e)(x - \tfrac{1}{2} e) \leq \tfrac{n}{4}, \ \forall\, x \in S\}$$

i.e., as the (translated) scaled polar of S, with scaling factor $\frac{n}{4}$. In other words, S* is defined with reference to the euclidean ball with radius $\sqrt{n}/2$, centered at $\frac{1}{2}$ e.

The following properties of outer polars can then be shown to hold (see [1]).

1. $$K^* = \{y \mid (y - \tfrac{1}{2} e)x \leq \tfrac{1}{2} ye, \ \forall\, x \in \text{vert } K\}$$
$$= \{y \mid |y - \tfrac{1}{2} e| \leq \tfrac{n}{2}\}$$

2. $$X^* = \{y \mid (y - \tfrac{1}{2} e)x \leq \tfrac{1}{2} ye, \ \forall\, x \in \text{vert } X\}$$
$$= \left\{ y \;\middle|\; \begin{array}{l} y \leq A^T u + v + \tfrac{1}{2} e \\ \tfrac{1}{2} ey \geq bu + ev \\ \text{for some } u, v \geq 0 \end{array} \right\}$$

3. $\hat{X} \subset X \subset K \subset K^* \subset X^* \subset \hat{X}^*$

4. $X \cap \text{bd } X^* = X \cap \text{bd } \hat{X}^* = X \cap \text{vert } K$

5. $$\text{bd } X^* = \{y \mid \max\{(y - \tfrac{1}{2} e)x \mid x \in X\} = \tfrac{1}{2} ey\}$$
$$\text{bd } \hat{X}^* = \{y \mid \max\{(y - \tfrac{1}{2} e)x \mid x \in \hat{X}\} = \tfrac{1}{2} ey\}$$

6. $\hat{X}^*$ is the intersection of the halfspaces defined by those facets of X^* which contain a facet of K^*.

7. There exists a (duality) mapping φ from vertices of conv $(X \cup \{\frac{1}{2}e\})$ to facets of X^*, and from facets of conv $(X \cup \{\frac{1}{2}e\})$ to vertices of X^*, with the following properties. A point $x \neq \frac{1}{2}e$ is a vertex of conv $(X \cup \{\frac{1}{2}e\})$, contained in n (extended) facets $H_i = \{x | \beta^i x = \beta_i\}$ of conv $(X \cup \{\frac{1}{2}e\})$, (where the signs of β^i, β_i are chosen so that $X \subset H_i^+ = \{x | \beta^i x \leq \beta_i\}$), if and only if

$$\varphi(x) = \{y | (x - \frac{1}{2}e)y = \frac{1}{2}ex\}$$

is an (extended) facet of X^* containing the n vertices $\varphi(H_i)$ of X^*, where $\varphi(H_i)$ is an ordinary vertex if $\beta_i - \frac{1}{2}\beta^i e \neq 0$, and a "vertex at infinity" (a direction) otherwise, namely

$$\varphi(H_i) = \begin{cases} \dfrac{n}{4(\beta_i - \frac{1}{2}\beta^i e)}\beta^i + \frac{1}{2}\beta^i e & \text{if } \beta_i - \frac{1}{2}\beta^i e \neq 0 \\ \beta^i & \text{if } \beta_i - \frac{1}{2}\beta^i e = 0 \end{cases}$$

Furthermore, for any $x \in X$,

$$x \in \text{vert } K \Leftrightarrow x \in \varphi(x)$$

$$\Leftrightarrow \left\{ \begin{array}{l} \text{the (extended) facet } \varphi(x) \text{ of } X^* \text{ is also} \\ \text{an (extended) facet of } \hat{X}^* \text{ and } K^* \end{array} \right\}.$$

2. Outline of the Approach

The basic idea of our approach is the following. We choose a subset of n+1 inequalities of X, such that the intersection of the corresponding half-spaces defines a simplex S. For each vertex α^i of S, we check whether $\alpha^i \epsilon X^*$. If so, we go to the next vertex. Otherwise, we cut off α^i by attaching to the constraint set of S an inequality of X violated by α^i. This replaces S by a smaller set, and α^i by a new set of vertices. Whenever a feasible 0-1 point is found, it is stored as the current best one. When all vertices of the current set have been examined, we are done.

The feasible set X, the simplex S with vertices 1, 2, 3, and the outer polar X* of X are shown for two 2-dimensional examples in fig. 1. In fig. 1a, vertex 2 of S is contained in X*, whereas vertices 1 and 3 are not; therefore they have to be cut off by adding a constraint of X ($x_1 \geq 0$ in the first case, $x_1 \leq 1$ in the second). In fig. 1b, vertex 3 of S belongs to X*; vertex 1 belongs to bd X*, and is a feasible 0-1 point; whereas vertex 2 has to be cut off by imposing the inequality $x_2 \leq 1$.

In the following, $H_i^+ = \{x|\beta^i x \leq \beta_i\}$ will denote an inequality of X (of the form $x_i \leq 1$, or $-x_i \leq 0$, or $a^i x \leq b_i$, where a^i is a row of A), and the closed halfspace defined by the inequality. Also, $H_i = \{x|\beta^i x = \beta_i\}$ will denote the equation as well as the corresponding hyperplane. An extended facet (i.e., the affine hull of a facet) will simply be called a facet. Furthermore, $H_o^+(z^*) = \{x|- cx \leq - z^* - 1\}$ will be considered one of the constraints of X, with z* set sufficiently small at the start, and updated whenever a new feasible 0-1 point is found. (Thus, X itself is also updated.)

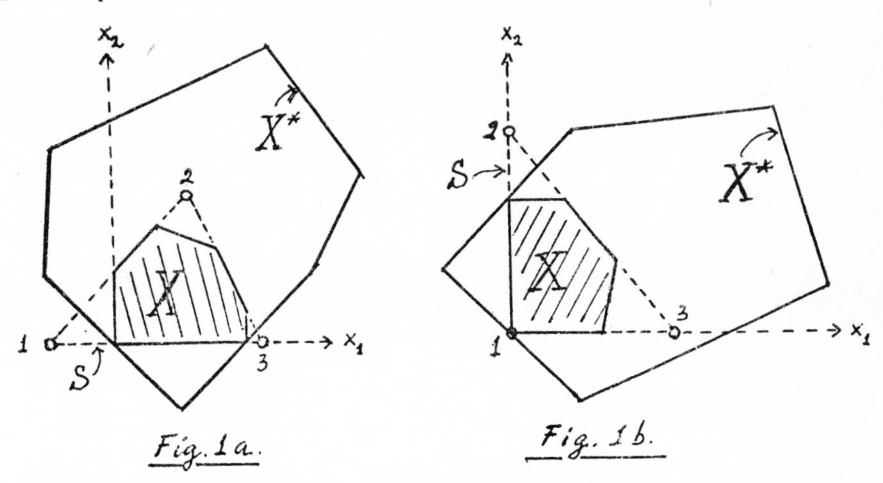

Fig. 1a. Fig. 1b.

Since X is bounded in all directions, it seems that one can always choose n+1 of the 2n+m inequalities of X so as to obtain a simplex. This, however, is not true, as illustrated by the following 3-dimensional example:

$$X = \{(x_1, x_2, x_3) | x_1 + x_2 \leq 1,\ 0 \leq x_i \leq 1,\ i=1,2,3\}$$

No matter how one chooses 4 of the 7 inequalities, one cannot obtain an ordinary simplex: the intersection of the corresponding halfspaces is unbounded in at least one direction.

However, the following property is true and sufficient for our purposes.

Theorem 1. There always exists a subset of n inequalities of X which, together with a nonnegative combination of the remaining inequalities, defines a simplex.

Proof. Let v be any vertex of X, and let $t_j \geq 0$, $j \in J$, be n inequalities of X that are tight for v. Further, let w be a vertex at which the linear function $\sum_{j \in J} t_j$ attains its maximum over X, and let this maximum be b_0. Then the inequality $\sum_{j \in J} t_j \leq b_0$ is a nonnegative combination of those inequalities of X that are tight for w. Also, the intersection of the $n+1$ inequalities

$$t_j \geq 0 \quad , \quad j \in J$$

$$\sum_{j \in J} t_j \leq b_0$$

is clearly a simplex. Q.E.D.

The procedure for solving the 0-1 program (P) can now be outlined as follows:

Procedure 1.

0. Choose $n+1$ inequalities H_i^+, $i=i_o, i_1, \ldots, i_n$, of X, such that $S = \bigcap_{i=i_o,\ldots,i_n} H_i^+$ is a simplex. Define $S_o = S$, set $k = 0$, and go to 1.

1. Choose a vertex α^i of S_k that has not yet been examined. If there is none, stop. Otherwise,

   (a) If $\alpha^i \in X^* \cap \text{bd}\, \hat{X}^*$, then the facet of $\hat{X}^*$ containing α^i contains a feasible 0-1 point x^i. Store x^i as the current best feasible solution, and update $H_o^+(z^*)$ by setting $z^* = cx^i$. Then go to (b).

   (b) If $\alpha^i \in X^* \cap \text{int}\, \hat{X}^*$, define $S_{k+1} = S_k$, set $k=k+1$, and go to 1.

   (c) If $\alpha^i \notin X^*$, choose an inequality H_j^+ of X, such that $\alpha^i \in H_j^-$; define $S_{k+1} = S_k \cap H_j^+$, set $k=k+1$, and go to 1.

The procedure terminates when all vertices of the current set S_k have been examined.

Theorem 2. Procedure 1 is well defined and terminates in a finite number of iterations either by finding an optimal solution or by establishing that no feasible solution exists.

Proof. According to Theorem 1, one can always choose n+1 inequalities (or combinations of inequalities) of X that define a simplex. For any vertex α^i of the current set S_k, situations (a), (b) and (c) of step 1 are easily seen to be mutually exclusive and jointly exhaustive. In case (a), α^i is contained in a facet of $\hat{X}^*$, which according to Property 6 of section 1, contains a vertex of $\hat{X}$, i.e., a feasible 0-1 point x^i. Since $\hat{X}$ is the convex hull of the currently feasible 0-1 points (i.e., feasible also with regard to $cx \geq z^* + 1$), x^i is always better than the 0-1 points found earlier in the procedure, and therefore becomes the current best. The updating of z^* diminishes $\hat{X}$ and increases $\hat{X}^*$, so that α^i is now in case (b). Whenever $\alpha^i \in X^* \cap \text{int}\, \hat{X}^*$ [case (b)], the procedure leaves α^i and goes to another vertex. Whenever, on the other hand, $\alpha^i \notin X^*$, α^i is cut off by adding to the set of inequalities defining S_k, a new inequality H_j^+ of X, i.e., by replacing S_k with $S_{k+1}=S_k \cap H_j^+$. The existence of such an inequality (i.e., one that cuts off α^i) follows from the fact that $\alpha^i \notin X^*$ and $X \subset X^*$. Thus, when all vertices of the current set S_k have been examined (and this is bound to happen in a finite number of iterations, since X has a finite number of facets), then $S_k \subset \text{int } \hat{X}^*$, i.e., S_k contains no more feasible integer points.

Q.E.D.

Remark 1. Step 1(c) can be modified so as to choose an inequality H_j^+ of K (rather than X), such that $\alpha^i \varepsilon H_j^-$. The procedure still remains well defined and finite, since from $K \subset X^*$, $\alpha^i \notin X^*$, there always exists an inequality H_j^+ of K that cuts off α^i.

3. An Alternative View

Before discussing further details, it will be useful to give an alternative view of the above approach. In [1] we have shown how the dual correspondence between vertices of a set and facets of its polar makes it possible to establish further and more elaborate correspondences between the two sets, to the extent that every operation performed on a polyhedral set (like cutting off a vertex, passing a hyperplane through some vertices, intersecting some extended facets, etc.), can be put in correspondence with a certain operation on the polar set.

Thus, instead of considering the simplex S, one can work with its (translated) polar (scaled by $\frac{n}{4}$):

$$S^* = S^o(\tfrac{n}{4}) = \{y \varepsilon R^n \mid (y - \tfrac{1}{2} e)x \leq \tfrac{1}{2} ye, \ \forall\, x \varepsilon S\}.$$

Since $X \subset S$, we have $S^* \subset X^*$. Actually, since S is the intersection of a subset of the halfspaces defining X, S* is the convex hull of a subset of the points and directions of X*. Therefore, from property 4 of section 1,

$$(X \cap \text{vert } K) \subset (X \sim \text{int } S^*),$$

i.e., no feasible 0-1 point is contained in int S^*. To state the same thing in a different way, let η be the index set for those vertices α^i of conv $(S \cup \{\frac{1}{2}e\})$, that are different from $\frac{1}{2}e$, and for each $i \in \eta$, let F_i^+ be the half-space defined by the (extended) facet $F_i = \varphi(\alpha^i)$ of S^*, i.e.,

$$F_i^+ = \{x \mid (\alpha^i - \frac{1}{2} e)x \leq \frac{1}{2} \alpha^i e\}.$$

Then

$$S^* = \{x \mid (\alpha^i - \frac{1}{2}e)x \leq \frac{1}{2} \alpha^i e, \ i \in \eta\}$$

$$= \bigcap_{i \in \eta} F_i^+.$$

and any feasible 0-1 solution satisfies at least one of the $|\eta|$ conditions

$$x \in F_i \cup F_i^-, \quad i \in \eta$$

where F_i and F_i^- are obtained from F_i^+ by replacing $\leq$ with $=$ and $>$ respectively. In other words, for any feasible 0-1 solution x, the inequality

$$(\alpha^i - \frac{1}{2}e)x \geq \frac{1}{2} \alpha^i e$$

holds for at least one $i \in \eta$. (Here $|\eta| = n$ or $n+1$).

We could then proceed as follows. For each (extended) facet F_i of S^*, we check whether $X \cap F_i^- = \emptyset$. If so, we go to the next facet. Otherwise, we add a vertex v^j of X^*, such that $v^j \in F_i^-$, to the set of points and directions whose convex hull is S^*. This replaces S^* by a larger set, and F_i by a set of new facets. Whenever F_i contains a feasible 0-1 point, it is stored as the current best. When all facets of the current set have been examined, we are done.

This approach is illustrated in fig. 2 for the same two examples shown in fig. 1. The facets 1′, 2′, 3′ of S* correspond to (are the images, under the polarity mapping, of) the vertices 1, 2, 3 of S in fig. 1.

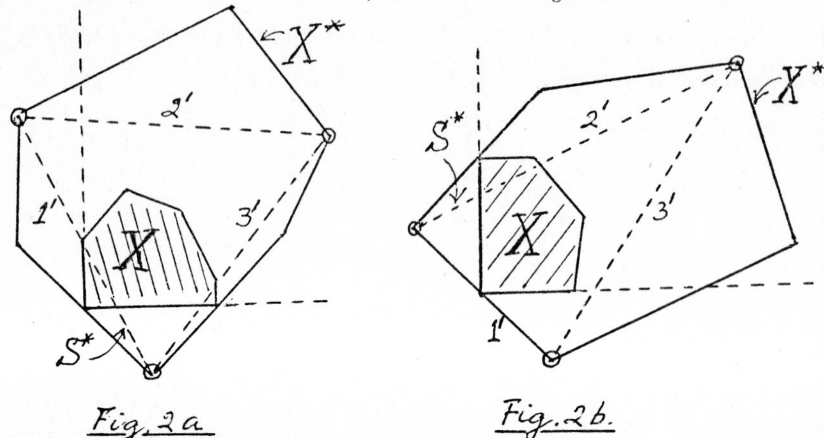

Fig. 2a Fig. 2b.

An outline of this procedure is as follows.

Procedure 2.

0. Choose n+1 inequalities H_i^+, $i=i_o, i_1,\ldots,i_n$ of X, such that $S = \bigcap_{i=i_o,\ldots,i_n} H_i^+$ is a simplex. Define $\mathcal{S}_o = S^*$, set k=0, and go to 1.

1. Choose a facet F_i of $\mathcal{S}_k$ that has not yet been examined. If there is none, stop. Otherwise,

   (a) If $X \cap F_i^- = \emptyset$ but $\hat{X} \cap F_i \neq \emptyset$, then $\hat{X} \cap F_i = x^i$ is a feasible 0-1 point. Store x^i as the current best feasible solution, and update $H_o(z^*)$ by setting $z^* = cx^i$. Then go to (b).

   (b) If $X \cap F_i^- = \emptyset$ and $\hat{X} \cap F_i = \emptyset$, define $\mathcal{S}_{k+1} = \mathcal{S}_k$, set k=k+1, and go to 1.

   (c) If $X \cap F_i^- \neq \emptyset$, choose a vertex v^j of X* such that $v^j \epsilon F_i^-$, define $\mathcal{S}_{k+1} = \text{conv}(\mathcal{S}_k \cup \{v^j\})$, set k=k+1, and go to 1.

The procedure terminates when all facets of the current set $\mathscr{S}_k$ have been examined.

The next Theorem shows that Procedure 2 is essentially the same as Procedure 1. In order to have a perfect correspondence between the steps of the two procedures, we will assume that $\{\frac{1}{2}\ e\}\epsilon S$, and that each inquality H_j^+ chosen in step 1(c) is such that H_j is a facet of X, with $\{\frac{1}{2}\ e\}\epsilon H_j^+$; i.e., each set of the sequence generated under Procedure 1 contains $\frac{1}{2}$ e. It is always possible to choose S and the inequalities H_j^+ in this manner: it suffices, for instance, to generate S from a vertex of X that is also a vertex of K, (see the proof of Theorem 1), and to choose the inequalities H_j^+ from among the facets of K (see Remark 1). If one rejects this assumption, then minor differences appear between the two procedures, but they are unessential and not worthy of a detailed discussion. With the assumption, the following statement is true.

Theorem 2. Procedures 1 and 2 are equivalent: there is a one-to-one correspondence between the sequence of sets S_k generated under Procedure 1, and the sequence of sets $\mathscr{S}_k$ generated under Procedure 2, namely $\mathscr{S}_k = S_k^*$ for all k; also, between the vertices α^i of each set S_k, and the facets F_i of the corresponding set $\mathscr{S}_k$, namely $F_i = \varphi(\alpha^i) = \{x | (\alpha^i - \frac{1}{2}\ e)x = \frac{1}{2}\ \alpha^i e\}$.

Proof. For k = 0, $S_k = S$ and $\mathscr{S}_k = S_k^* = S^*$; further, since $\{\frac{1}{2}\ e\}\epsilon S$ by assumption, each vertex α^i of S corresponds (Property 8 of section 1) to a (unique) facet

$$F_i = \varphi(\alpha^i) = \{x | (\alpha^i - \frac{1}{2}\ e)x = \frac{1}{2}\ \alpha^i e\}$$

of S* .

Suppose now that r is the number of sets S_k generated under Procedure 1, s is the number of sets $\mathcal{S}_k$ generated under Procedure 2, and the correspondence stated in the Theorem is true for $k = 0, 1, \ldots, p$, with $p < \min\{r,s\}$. We will then show that it is also true for $k = p+1$. Indeed, under step 1 of Procedure 1 a vertex α^i of S_p is selected, whereas under step 1 of Procedure 2 a facet F_i of $\mathcal{S}_p$ is chosen, with $\mathcal{S}_p = S_p^*$. But there is a one-to-one correspondence between the vertices of S_p and the facets of S_p^*, given by

$$F_i = \varphi(\alpha^i) = \{x \mid (\alpha^i - \frac{1}{2}e)x = \frac{1}{2}\alpha^i e\}$$

Also, there is a one-to-one correspondence between the 3 cases (a), (b), (c) in step 1 of Procedure 1, and the corresponding 3 cases of Procedure 2, namely:

(a) $\alpha^i \in X^* \cap \mathrm{bd}\,\hat{X}^* \Leftrightarrow (\alpha^i - \frac{1}{2}e)x^i = \frac{1}{2}\alpha^i e$, where

$$(\alpha^i - \frac{1}{2}e)x^i = \max\{(\alpha^i - \frac{1}{2}e)x \mid x \in X\}$$

$$= \max\{(\alpha^i - \frac{1}{2}e)x \mid x \in \hat{X}\}$$

$$\Leftrightarrow X \cap F_i^- = \emptyset,\ \hat{X} \cap F_i \neq 0, \text{ where}$$

$$F_i = \varphi(\alpha^i) = \{x \mid (\alpha^i - \frac{1}{2}e)x = \frac{1}{2}\alpha^i e\},\ F_i^- = \{x \mid (\alpha^i - \frac{1}{2}e)x > \frac{1}{2}\alpha^i e\}.$$

(b) $\alpha^i \in X^* \cap \mathrm{int}\,\hat{X}^* \Leftrightarrow \max\{(\alpha^i - \frac{1}{2}e)x \mid x \in X\} \leq \frac{1}{2}\alpha^i e$, but

$$\max\{(\alpha^i - \frac{1}{2}e)x \mid x \in \hat{X}\} < \frac{1}{2}\alpha^i e$$

$$\Leftrightarrow X \cap F_i^- = \emptyset,\ \hat{X} \cap F_i = \emptyset$$

(c) $\alpha^i \notin X^* \Leftrightarrow \exists x \in X: \quad (\alpha^i - \frac{1}{2} e)x > \frac{1}{2} \alpha^i e$

$\Leftrightarrow X \cap F_i^- \neq \emptyset$

$$\left\{ \begin{array}{l} X \text{ has a facet } H_j, \\ \text{with } \{\frac{1}{2} e\} \epsilon H_j^+, \ \alpha^i \epsilon H_j^- \end{array} \right\} \Leftrightarrow \left\{ \begin{array}{l} X^* \text{ has a vertex } v^j = \varphi(H_j) \\ \text{with } v^j \epsilon F_i^- \end{array} \right\}$$

Further, since $\{\frac{1}{2} e\} \epsilon S_p \cap H_j^+$, and $\mathscr{S}_p = (S_p)^*$, from basic properties of polar sets (see [1], section 1)

$$\mathscr{S}_{p+1} = \text{conv}(\mathscr{S}_p \cup \{v^j\}) = (S_p \cap H_j^+)^*$$

where $H_j^+ = \{v^j\}^*$, $H_j = \text{bd}(H_j^+) = \varphi^{-1}(v^j)$ Q.E.D.

In the sequel we will discuss our approach basically in terms of Procedure 1.

4. Some Details

(a) Choosing the starting simplex

One can follow several different strategies in choosing the starting simplex S. We shall discuss three possible choices, each of them having some advantages and some drawbacks.

1. Solve the linear program

$$\max\{cx | x \in X\} .$$

Let $t = (x,y)$, where $y \in R^m$ is a slack vector, let $t = \bar{a}_0$ be an optimal (basic) solution with nonbasic index set J, and let

$$z = \bar{a}_{00} + \sum_{j \in J} \bar{a}_{0j}(-t_j)$$

$$t = \bar{a}_0 + \sum_{j \in J} \bar{a}_j (-t_j)$$

Further, let z^* be the objective function value associated with a known feasible solution (if there is one) or otherwise, let

$$z^* = \sum_{i \in N} \min\{0, c_i\}$$

The constraint, $H_0^+(z^*) = \{x \mid -cx \leq -z^* - 1\}$ becomes, when expressed in terms of the nonbasic variables, $\sum_{j\varepsilon J} \bar{a}_{0j} t_j \leq \bar{a}_{00} - z^* - 1$, or, introducing a slack variable s_0,

$$s_0 = \bar{a}_{00} - z^* - 1 + \sum_{j\varepsilon J} \bar{a}_{0j}(-t_j).$$

Since the solution $t = \bar{a}_0$ is optimal, $\bar{a}_{0j} \geq 0,\ \forall\ j\varepsilon J.$

Whenever $\bar{a}_{0j} > 0,\ \forall\ j\varepsilon J$, the n+1 inequalities $t_j \geq 0$, $j\varepsilon J$, $s_0 \geq 0$, define a simplex. When this is not the case, one can temporarily replace each $\bar{a}_{0j} = 0$ by a positive number γ smaller than any of the nonzero coefficients $\bar{a}_{0h}$, for instance

$$\gamma = \frac{1}{2} \min_{h\varepsilon J \mid \bar{a}_{0h} > 0} \{\bar{a}_{0h}\}.$$

As soon as another inequality is activated which has a positive coefficient in the column containing γ (i.e. which cuts off the vertex obtained from $t = \bar{a}$ by pivoting on γ), one can replace γ by the original coefficient $\bar{a}_{0j} = 0.$

2. Define S as the intersection of $H_o^+(z^*)$ with the n halfspaces

$$x_i \leq 1 \quad \text{if} \quad c_i \geq 0,$$

$$x_i \geq 0 \quad \text{if} \quad c_i < 0.$$

This is clearly a simplex whenever $c_i \neq 0,\ \forall\ i$, and whenever this is not the case, one can temporarily replace $c_i = 0$ by $\gamma = \min_{j\varepsilon N \| c_j \| \neq 0} \{|c_j|\}$, as in the previous instance. The advantage of such a choice lies in the fact that every vertex of S and, if only facets of K are used in step 1(c), every vertex of all sets S_k subsequently generated, has at most one noninteger component. Thus, whenever a vertex of the current set S_k is cut off, at least one of the newly created vertices is a 0-1 point (though, of course, not necessarily feasible). The drawback of this approach is, on the other hand, the fact that the vertices lying on $H(z^*)$ will often tend to be far outside X^*.

3. Choose S so that $H_o^+(z^*)$ and r inequalities of the system $Ax \leq b$, where $r = \text{rank}(A)$, are among the facets of S. This means that whenever $r \leq n$, we shoose S so that every nonredundant inequality of $Ax \leq b$, as well as the inequality $cx \geq z^* + 1$, defines a facet of S. Such a choice has the advantage that every vertex of K contained in S, and in any of the subsequently generated sets S_k, is feasible; this enhances the chances of finding feasible 0-1 points early in the procedure. The drawback of this choice is that, like in case 2, the vertices of S will often tend to lie farther from the boundary of X^* than in case 1 (though less so than in case 2). A set S with the above properties can be constructed in several ways, and unless $r = m = n$, it is by no means unique.

(b) Choosing a vertex of the current set

We have no choice criterion that could be called best on theoretical grounds. Choosing the vertex with best objective function value is of little use here, since $H_o^+(z^*)$ is among the facets of the starting simplex S, and hence n of the n+1 vertices of S lie on $H_o^+(z^*)$ and thus yield the same objective function value. A similar situation prevails for subsequent sets S_k.

There is, however, a choice rule which looks good on empirical grounds. The 2^n orthants of the coordinate system centered at $\frac{1}{2}e$ have the property that each orthant contains exactly one vertex of the cube K, i.e. exactly one 0-1 point (in the original coordinate system). Since every vertex α^i of S_k (like any arbitrary point) clearly lies in (at least) one orthant of the same coordinate system, we may associate to each vertex α^i a unique 0-1 point x^i, defined by

$$x_j^i = \begin{cases} 1 & \text{if } \alpha_j^i > \frac{1}{2} \\ 0 & \text{if } \alpha_j^i \leq \frac{1}{2} \end{cases}$$

Let $\beta^h x + y_h = \beta_h$ denote row h of the system

$$\begin{aligned} -cx + y_0 &= -z^* - 1 \\ Ax \quad +y &= b \end{aligned}$$

where $h \epsilon M \cup \{0\}$. Then the amount of infeasibility associated with a 0-1 point x^i is

$$v(x^i) = \sum_{h \epsilon M \cup \{0\}} |\min\{y_h, 0\}|$$

where the bars mean absolute value.

The rule that we recommend is then the following. Choose the vertex α^j o S_k whose associated vertex x^j of K is "least infeasible", i.e. the one for which

$$v(x^j) = \min_{\alpha^i \epsilon \text{vert } S_k} v(x^i)$$

The rationale behind this choice rule is that the "closer to feasibility" the 0-1 point x^i associated with the vertex α^i that is selected, the more likely it is that cutting off α^i will produce a vertex whose associated 0-1 point is feasible. The discussion of the next topic [see (c) below] should yield further insight into the rationale behind this rule.

(c) Testing the selected vertex

Having chosen a vertex α^i of S_k, we have to check whether $\alpha^i \in X^*$, and if so (which then implies $\alpha^i \in \hat{X}^*$), whether $\alpha^i \in \text{bd}\, \hat{X}^*$. This can be done by solving the linear program

$$\text{(LP)} \qquad \max\{(\alpha^i - \tfrac{1}{2}e)\, x \mid x \in X\}$$

where

$$X = \{x \in R^n \mid Ax \leq b,\ -cx \leq -z^*-1,\ 0 \leq x \leq e\}$$

(with z^* updated whenever a new feasible 0-1 point is found).

Since X is bounded, (LP) always has a finite optimum. Let x^i be an optimal solution to (LP). Then, from the properties of outer polars listed in section 1, $\alpha^i \in X^*$ if and only if

$$(\alpha^i - \tfrac{1}{2}e)\, x^i \leq \tfrac{1}{2}\alpha^i e\ ;$$

and, if $\alpha^i \in X^*$ (which implies $\alpha^i \in \hat{X}^*$), then $\alpha^i \in \text{bd}\, \hat{X}^*$ if and only if x^i is a 0-1 point and

$$(\alpha^i - \tfrac{1}{2}e)\, x^i = \tfrac{1}{2}\alpha^i e.$$

There are two advantages in solving (LP) by the dual rather than the primal simplex method, starting with the basic solution $\tilde{x}$ defined by

$$\tilde{x}_j = \begin{cases} 1 & \text{if } \alpha^i_j - \tfrac{1}{2} > 0 \\ 0 & \text{if } \alpha^i_j - \tfrac{1}{2} \leq 0 \end{cases}$$

which is easily seen to be dual-feasible, since it is an optimal solution to

$$\max\{\alpha^i - \tfrac{1}{2}e)x \mid x \in K\}\ .$$

Here, as before, $K = \{x | 0 \leq x \leq e\}$.

The first advantage consists in the fact that the 0-1 point $\tilde{x}$ may be feasible (i.e. $A\tilde{x} \leq b$, $c\tilde{x} \geq z^* + 1$), even though $(\alpha^i - \frac{1}{2} e)\tilde{x} > \frac{1}{2} \alpha^i e$, in which case one can store $\tilde{x}$ and update X by setting $z^* = c\tilde{x}$. The second one comes from the fact that whenever the value of the objective function gets below $\frac{1}{2} \alpha^i e$, one can stop: $\alpha^i \notin X^*$.

It should be mentioned that the algorithm under discussion can be sharpened by slightly redefining cases (b) and (c) of step 1 as follows:

(b) If one can show that $\alpha^i \in \hat{X}^*$, define $S_{k+1} = S_k$, set $k = k+1$, and go to 1.

(c) If one cannot show that $\alpha^i \in \hat{X}^*$, choose an inequality H_j^+ of X, such that $\alpha^i \in H_j^-$; define $S_{k+1} = S_k \cap H_j^+$, set $k = k+1$, and go to 1.

From the proof of Theorem 2 it should be clear that this change does not affect the validity of the procedure, since when it stops then $S_k \subset \hat{X}^*$, and $S_k \cap X \cap \mathrm{bd}\hat{X}^* = \emptyset$, i.e. S_k contains no more feasible 0-1 points. The reason for drawing the dividing line between cases (b) and (c) in terms of whether one can show or not that $\alpha^i \in \hat{X}^*$ (rather than in terms of the actual situation), lies in the difficulty of establishing the actual situation: one wants to limit the effort of checking whether $\alpha^i \in \hat{X}^*$ to some relatively easy test representing a condition that is sufficient but not necessary.

Thus, for any dual-feasible fractional solution x^i to (LP), one may calculate a penalty $\pi(x^i)$, i.e. some valid lower bound on the amount by which $(\alpha^i - \frac{1}{2} e)x^i$ must decrease if x^i is to be made integer and primal-feasible. Then, clearly,

$$(\alpha^i - \frac{1}{2} e)x^i - \pi(x^i) \leq \frac{1}{2} \alpha^i e$$

is a sufficient condition for $\alpha^i \in \hat{X}^*$, and if this condition holds, one goes to case (b) above; whereas otherwise one continues. If the solution gets primal feasible (fractional) and $(\alpha^i - \frac{1}{2} e)x - \pi(x) > \frac{1}{2} \alpha^i e$, one goes to (c).

The penalty $\pi(x^i)$ can be calculated in the same way as in the conventional branch and bound procedures; i.e. if P_j^U and P_j^D stand for the "up penalty" and "down penalty" associated with the variable x_j (where x_j^i is fractional), then

$$\pi_1(x^i) = \max_j \min\{P_j^U, P_j^D\}$$

is a legitimate penalty. Also, if $I^- = \{i \in I | t_i < 0\}$ is the index set for the basic variables which are negative, and P_i^- is the penalty for making t_i nonnegative, then

$$\pi_2(x^i) = \max_{i \in I^-} \{P_i^-\}$$

is another valid penalty, and

$$\pi(x^i) = \max\{\pi_1(x^i), \pi_2(x^i)\}$$

can be used in the above context.

It should be mentioned, that whenever

$$\sum_{j \in N} | \alpha_j^i - \frac{1}{2} | < \frac{n}{2}$$

there is no need to solve (LP), since $\alpha^i \in K^*$, which implies $\alpha^i \in X^*$, and we are in case (b) above.

(d) Updating the objective function constraint

Whenever a new feasible 0-1 point x^i is found, the constraint $H_o^+(z^*) = \{x | cx \geq z^* + 1\}$ is updated by setting $z^* = cx^i$; and since many of the active vertices of the current set lie on $H_o(z^*)$, they also have to be updated. Now each vertex α^i generated under the procedure is given by the x-components of a basic feasible solution $t = (x,y)$ to the system

$$\begin{aligned} Ax + y \quad &= b \\ -cx \quad + y_0 &= -z^* - 1 \end{aligned}$$

and the basic components of t can be represented as

$$t_B = B^{-1}\begin{pmatrix} b \\ \hline -z^*-1 \end{pmatrix}$$

where B is the basis defining α^i (and t).

When z^* is replaced by z_1^*, the basic components of t have to be changed by

$$\Delta t_B = B^{-1}\begin{pmatrix} 0 \\ \hline z^*-z_1^* \end{pmatrix}$$

Since only the last column of B^{-1} is needed to compute Δt_B, it is worth storing this column along with α^i rather than reinverting the basis.

(e) Activating a problem constraint

When a vertex $\alpha^i \notin X^*$ is cut off by activating a problem inequality H_j^+, α^i is replaced by the set of newly created vertices. This poses the problem of generating all vertices of the (n-1)-dimensional polyhedron $S_k \cap H_j$. There are several known methods for doing this, like the ones proposed by Motzkin, Raiffa, Thompson and Thrall [3], Balinski [4], and more recently Shefi [5], Pollatschek and Avi-Itzhak [6], Mattheis [7], Burdet [8]. One should also mention Chernikova's method [9],[10] for generating all edges of a polyhedral cone (a problem equivalent to that of finding all vertices of a polytope that generates the cone), which was recently restated in the simplex format by Rubin [11]. However, none of these methods deals specifically with the problem facing us here, namely that of finding all the new vertices generated by intersecting a polytope S_k whose vertices are known, with a half-space H_j^+. Theorem 3 below provides a simple and straightforward way of obtaining the new vertices from the vertices of S_k cut off by H_j^+, i.e. from $H_j^- \cap \text{vert } S_k$.

Theorem 3. A point $v \varepsilon R^n$ is a vertex of $S_k \cap H_j^+$, but not of S_k, if and only if it is of the form

$$v = (\alpha^i,\alpha^h) \cap H_j,$$

where α^i and α^h are vertices of S_k such that $\alpha^i \varepsilon H_j^-$, $\alpha^h \varepsilon H_j^+ - H_j$, and (α^i,α^h) is the open edge joining α^i and α^h, i.e.

$$(\alpha^i,\alpha^h) = \{x | x = \lambda\alpha^i + (1-\lambda)\alpha^h, \ 0 < \lambda < 1\}.$$

Proof. v is a vertex of $S_k \cap H_j^+$ but not of S_k, if and only if v is the intersection of exactly n facets of $S_k \cap H_j^+$, one of which is H_j. The remaining n-1 facets of S_k define an edge of S_k, say $[\alpha^i,\alpha^h]$, where α^i and α^h are vertices of S_k, and the vertex v of $S_k \cap H_j^+$, lying on this edge, is not a vertex of S_k, if and only if it lies on the open edge (α^i,α^h). Further, $(\alpha^i,\alpha^h) \cap H_j \neq \emptyset$ if and only if $\alpha^i \varepsilon H_j^-$ and $\alpha^h \varepsilon H_j^+ - H_j$, or $\alpha^h \varepsilon H_j^-$ and $\alpha^i \varepsilon H_j^+ - H_j$. Q.E.D.

To simplify the discussion, we first assume that all vertices of S_k are nondegenerate ; we then briefly discuss degeneracy at the end. Then each vertex α^i of S_k has exactly n adjacent vertices, which will be denoted by $\alpha^{i1},\ldots\alpha^{in}$.

Corollary 3.1. Given the vertices of S_k, the following procedure generates the vertices of $S_k \cap H_j^+$:

0. Remove from vert S_k all vertices of S_k contained in H_j^-, and put them on a list L. Put the remaining vertices on a list V, and go to 1.

1. If $L \neq \emptyset$, choose the first element of L, call it α^i, remove α^i from L, and go to 2.

   If $L = \emptyset$, stop: $V = \text{vert } S_k \cap H_j^+$.

2. For each of the n open edges (α^i, α^{ih}), $h=1,\ldots,n$, of S_k, adjacent to α^i, check whether H_j intersects (α^i, α^{ih}), and if so, add the vertex $v = (\alpha^i, \alpha^{ih}) \cap H_j$ to the list V.

   When all n edges (α^i, α^{ih}) adjacent to α^i have been checked, return to 1.

Proof. Step 0 removes from vert S_k all vertices cut off by H_j^+. The starting list V then consists of $H_j^+ \cap \text{vert } S_k$. To make the list of vertices of $S_k \cap H_j^+$ complete, V has to be augmented by addition of the newly created vertices. According to Theorem 3, each such vertex is of the form $v = (\alpha^i, \alpha^{ih}) \cap H$ where $\alpha^i \varepsilon H_j^- \cap \text{vert } S_k$ and α^{ih} is a vertex of S_k adjacent to α^i. Hence by applying step 2 to each $\alpha^i \varepsilon H_j^- \cap \text{vert } S_k$, we obtain all newly created vertices.

Q.E.D.

Identifying the vertices of S_k contained in H_j^- is a trivial matter: if H_j^+ is defined by the inequality $t_j \geq 0$, all one has to do is to check for each vertex α^i, which is the x-component of a solution $t^i = (x^i, y^i)$, whether $t_j^i < 0$.

Checking whether H_j intersects an edge (α^i, α^{ih}) adjacent to some $\alpha^i \in H_j^- \cap \text{vert } S_k$ is again an easy task, which can be performed as follows. Let the simplex tableau associated with the vertex α^i be

$$t_i = \bar{a}_{i0} + \sum_{j \in J} \bar{a}_{ij}(-t_j), \qquad i \in I$$

and let H_j^+, expressed in equation form and in terms of the nonbasic variables t_j, $j \in J$, be

$$s_* = \bar{a}_{*0} + \sum_{j \in J} \bar{a}_{*j}(-t_j)$$

where $\bar{a}_{*0} < 0$ (since α^i does not satisfy H_j^+).

Each of the n vertices α^{ih} of S_k adjacent to α^i can be obtained from α^i by pivoting in one of the columns $j \in J$. The pivot row has to be selected according to the rules of the modified simplex method for variables with upper bounds, but only those rows (and those bounds) are to be considered, which are active for S_k. Suppose that, according to these rules, the vertex α^{ih}, adjacent to α^i, can be obtained from α^i by pivoting into the basis the variable t_h with value ρ_h. Then checking whether H_j intersects (α^i, α^{ih}) amounts simply to checking whether

(i) $\bar{a}_{*h} < 0$

and, if so, whether

(ii) $\dfrac{\bar{a}_{*0}}{a_{*h}} < \rho_h$

If (i) and (ii) hold, then pivoting on $\bar{a}_{*h}$ yields the new vertex $v = (\alpha^i, \alpha^{ih}) \cap H_j$ of $S_k \cap H_j^+$. Otherwise $(\alpha^i, \alpha^{ih}) \cap H_j = \emptyset$.

We now briefly discuss degeneracy. The vertices of the starting simplex S are of course nondegenerate. Whenever activating an inequality H_j^+ produces degeneracy, i.e. whenever H_j intersects a vertex α^i, one can perturb the solution by replacing the zero component with $\varepsilon > 0$ sufficiently small. This changes the situation $\alpha^i \in H_j$ into $\alpha^i \in H_j^+$, and the procedure of Corollary 3.1 remains valid with the following modification. In step 2, we check whether H_j intersects $(\alpha^i, \alpha^{ih}]$ rather than (α^i, α^{ih}); i.e., whether $(\bar{a}_{*o} / \bar{a}_{*h}) \leq \rho_h$, rather than $(\bar{a}_{*o} / \bar{a}_{*h}) < \rho_h$. If H_j intersects $(\alpha^i, \alpha^{ih}]$ in the point α^{ih}, i.e. if $(\bar{a}_{*o} / \bar{a}_{*h}) = \rho_h$, then we pivot on $\bar{a}_{*h}$, just like in the case when $(\bar{a}_{*o} / \bar{a}_{*h}) < \rho_h$. This creates one or several zero entries in the constant column, which we then replace by $\varepsilon > 0$. The vertex obtained in this way is just a new representation for α^{ih} (in terms of a new basis); nevertheless, it has to be treated as a new vertex and added to the set V. Apart from this multiple representation for some of the vertices, the perturbation does not cause any problem. Since we never pivot in a row i for which $\bar{a}_{io} = \varepsilon$, and the problem of cycling does not arise at all, the ε values may all be the same.

Since the procedure is based on the idea of activating only part of the problem constraints, and we also have a choice as to which constraints to activate, one may hope that massive degeneracy can usually be avoided. (A highly degenerate vertex of X may be generated as a nondegenerate vertex of some set S_k, which uses only a subset of the inequalities of X, and dropped from the list before other inequalities are activated). However, for cases where massive degeneracy does occur, one may want to use some procedure which generates all the edges adjacent to a degenerate vertex and thus does not require multiple representation (see for instance [9], [10], [11]).

In the light of the above developments, Procedure 1 of section 2 can be implemented in a branch and bound framework as follows. The vertices of the successive sets S_k generated under the procedure can be associated with the nodes of a rooted tree T. The root of T is a fictitious node α^0, the successors of α^0 are the n+1 vertices of the starting simplex S_0. All vertices generated in the procedure are put on a list V. Whenever a vertex is chosen under step 1, it is marked. Also, whenever a constraint H_j^+ is activated, all vertices that violate H_j^+ are transferred from the list V to a second list L; then each vertex on the list L is replaced by the set of its successors, which are put on the list V. When this is accomplished (i.e., $L=\emptyset$), one reverts to step 1. Under these rules, each node of T has at most n successors. The procedure ends when $L = \emptyset$ and all vertices on the list V are marked.

5. The Mixed Integer Case

The method discussed in this paper is perfectly applicable to mixed-integer 0-1 programming. In that case we partition the problem à la Benders [12] by generating a set of constraints on the integer variables only, and use the outer polar of this set in much the same way as in the pure integer case. Most steps of the procedure are then carried out on this smaller problem in the subspace of the integer-constrained variables.

Let us state the mixed-integer program as

(MP) $$\max\{cx + dy \mid Ax + Dy \leq b,\ y \geq 0,\ x_j = 0 \text{ or } 1,\ \forall j\}$$

where $x \in R^n$, $y \in R^p$, $b \in R^m$.

For any given 0-1 vector $x \in R^n$, let P(x) and D(x) denote the pair of dual linear programs

P(x) $$\max_y\{dy \mid Dy \leq b - Ax,\ y \geq 0\}$$

and

$$D(x) \qquad \min_{u}\{u(b-Ax) \mid uD \geq d,\ u \geq 0\}$$

If $P(x)$ and $D(x)$ have optimal solutions $y(x)$ and $u(x)$ respectively, let

$$\begin{aligned} z(x) &= cx + d\cdot y(x) \\ &= cx + u(x)\cdot(b-Ax) \end{aligned}$$

If, on the other hand, $P(x)$ is infeasible, i.e. $D(x)$ has no finite minimum, let $u(x)$ be a vertex, and $t(x)$ a direction vector, of the feasible set of $D(x)$, such that $u(x) + \lambda t(x)$ is a feasible solution to $D(x)$ for all $\lambda > 0$, with $t(x)\cdot(b-Ax) < 0$.

Now let $\{x^1,\ldots,x^k\}$ be an arbitrary finite set of 0-1 points in R^n, denote $\{1,\ldots,k\} = Q$, and let

$$Q_1 = \{h \in Q \mid P(x^h) \text{ is feasible}\},$$

$$Q_2 = Q - Q_1.$$

Further, denote

$$z^* = \max_{h \in Q_1} z(x^h).$$

Then any $x \in R^n$ such that $P(x)$ is feasible and $z(x) > z^*$, must satisfy the inequalities

$$[u(x^h)\cdot A-c]\cdot x < u(x^h)\cdot b-z^* \qquad , \ h \in Q$$

$$[t(x^h)\cdot x \leq t(x^h)\cdot b \qquad , \ h \in Q_2$$

Inequalities of this type were first derived by Benders [12]. For their justification in the form stated here, and a discussion of their use in the context of implicit enumeration, see [13], section 4; also, see [14].

We will start our procedure by generating a set of constraints on the x variables, by solving $D(x^i)$ [or $P(x^i)$, whichever is more convenient] for several given 0-1 vectors x^i. The latter should be chosen so as to be as "good" as possible, i.e. as "close" as possible to the optimal x vector. To achieve this, one may solve the linear program associated with (MP) and then choose various combinations of 0-1 values close to the fractional values of the optimal solution, or one may use any information provided by the physical characteristics of the problem that is being modelled. The number of constraints to be generated is a matter of convenience, in that one would like to obtain a set as tightly constrained as possible, but without having so many constraints as to make pivoting too expensive.

Let $k > n+1$ be the number of inequalities generated this way, and call X the constraint set defined by these k inequalities, plus the 2n inequalities of $0 \leq x \leq e$. Procedure 1 of section 2 can then be applied to our problem with the following modification of case (a) in step 1:

(a) If $\alpha^i \in X^* \cap \mathrm{bd}\,\hat{X}^*$, then the facet of $\hat{X}^*$ containing α^i contains a 0-1 point x^i satisfying the current constraint set X. Among the constraints of X which are not active (i.e. are not among the inequalities defining the current set S_k), let (σ) be the one that is most "oversatisfied" by x^i, i.e. satisfied with the largest slack. Remove (σ), and replace it by a new constraint as follows.

Solve $D(x^i)$ [either directly, or by solving $P(x^i)$]. If $D(x^i)$ has an optimal solution $u(x^i)$, then replace (σ) by the constraint

$$[u(x^i)\cdot A-c]\cdot x < u(x^i)\cdot b - z^* .$$

If, in addition, $z(x^i) > z^*$, then update X by setting $z^* = z(x^i)$ in the right hand side of each constraint containing z^*.

If, on the other hand, $D(x^i)$ has no finite optimum, but $u(x^i) + \lambda t(x^i)$ is a feasible solution for all $\lambda > 0$, with $t(x^i)\cdot(b-Ax^i) < 0$, then replace (σ) by the constraint

$$[t(x^i)\cdot A]\ x \leq t(x^i)\cdot b\ .$$

In each case, go to (b).

The rest of the procedure remains unchanged.

It follows that, if we have a mixed integer 0-1 program in n integer and p continuous variables, we start by creating a constraint set $X \subset R^n$ on the integer variables alone, and the then we essentially apply the procedure of section 2 to the pure 0-1 program defined by the constraints of X; i.e., we first choose $n+1$ inequalities of X that define a generalized simplex S, then check whether the vertices of S belong to X^* etc.; except that, whenever a 0-1 point x^i is found which satisfies the constraints of the current set X, a linear program $P(x^i)$ in the p continuous variables [or its dual, $D(x^i)$] is solved, and thereby a new constraint is generated, which then replaces an old constraint of X. Besides this periodic updating of X (which does not affect the current set S_k), there is a second type of updating which affects both X and S_k: namely, the updating of z^* in the right hand side of the constraint set. Since most currently active vertices are likely to lie on some of the hyperplanes whose defining inequalities contain z^* in their right hand side, they also have to be updated. This can be done economically as follows. Let the basic components of the solution t, whose x-components yield α^i, be expressed as

$$t_B = B^{-1}\rho\ ,$$

where, say, the first q components of β contain z^*, whereas the remaining components are independent of z^*. When z^* is replaced by z_1^*, the basic components of t have to be changed by

$$\Delta t_B = (z^* - z_1^*)B^{-1} \cdot \begin{pmatrix} e_q \\ 0 \end{pmatrix}$$

where $e_q = (1,\ldots,1) \in R^q$.

Since only the sum of the first q columns of B^{-1}, i.e. a single column, is needed to compute Δt_B, one can avoid reinverting the basis, by storing this column.

6. Numerical Example

Consider the problem

$$\max \ \{cx \mid Ax \leq b,\ 0 \leq x \leq e,\ x \text{ integer}\}$$

where

$$c = (5, 6, 3, 5, -4, -3)$$

$$A = \begin{pmatrix} 9 & -3 & -6 & 8 & -7 & 0 \\ 6 & 0 & 9 & 5 & 1 & -1 \\ -3 & 9 & 7 & 6 & -7 & 3 \end{pmatrix} \qquad b = \begin{pmatrix} -13 \\ 10 \\ 5 \end{pmatrix}$$

The optimal linear programming solution is

$\bar{x} = (0.121, 0.745, 0.809, 0, 1, 0)$, $z = c\bar{x} = 3.504$,

with the associated simplex tableau shown below as Table 1.

	1	$-x_4$	$-x_5'$	$-x_6$	$-y_1$	$-y_2$	$-y_3$
z	3.504	7.422	7.652	5.643	0.777	0.131	0.926
x_1	0.121	1.152	0.854	0.057	0.096	0.039	0.032
x_2'	0.255	1.216	-1.645	-0.467	-0.082	0.053	-0.138
x_3'	0.191	0.213	0.766	0.148	0.064	-0.085	0.021

Table 1.

Here $x'_i = 1 - x_i$, and y_i are the slack variables. To start, we set

$$z^* = \sum_i \min\{0, c_i\} = -7$$

Then the first row of Table 1, with 3.504 replaced by 3.504 + 6 = 9.504, is the constraint $cx \geq z^* + 1$, or $-cx \leq 6$ expressed terms of the nonbasic variable

We take the optimal solution $\bar{x} = \alpha^o$ to be the first vertex of the starting simplex S, and obtain the 6 remaining vertices α^i of S by pivoting into the basis each of the non-basic variables in turn, in place of the basic variable z. Table 2 below shows the vertices α^i, along with the 0-1 points x^i defined by

$$x^i_j = \begin{cases} 1 & \text{if } \alpha^i_j > \frac{1}{2} \\ 2 & \text{if } \alpha^i_i \leq \frac{1}{2} \end{cases}$$

The last column contains the amounts of infeasibility

$$v(x^i) = \sum_k |\min\{y^i_k, 0\}|$$

associated with each point x^i.

	α^i_j						x^i_j						$v(x^i)$
i \ j	1	2	3	4	5	6	1	2	3	4	5	6	
0	0.121	0.745	0.809	0	1	0	0	1	1	0	1	0	5
1	-1.353	-0.811	1.081	1.28	1	0	0	0	1	1	1	0	15
2	-0.938	-1.306	1.759	0	-0.24	0	0	0	1	0	0	0	9
3	0.026	-0.035	1.056	0	1	1.67	0	0	1	0	1	1	0
4	-1.054	-0.259	1.592	0	1	0	0	0	1	0	1	0	0
5	-2.708	4.590	-5.298	0	1	0	0	1	0	0	1	0	3
6	-0.207	-0.671	1.024	0	1	0	0	0	1	0	1	0	0

Table 2.

The points x^3 and x^4 ($= x^6$) are feasible, with $cx^3 = -4$, and cx^4 ($=cx^6$)= -1. We set $z^* = -1$, and replace in Table 1 the constant 9.504 (= 3.504 + 6), by 3.504 (= 3.504 + 0). Since all vertices of S, except for α^o, lie on the hyperplane $cx = z^* + 1$, they now have to be updated. The outcome of this shown in Table 3.

i \ j	$\bar{\alpha}\alpha_j^i$ 1	2	3	4	5	6	x_j^i 1	2	3	4	5	6	$v(x^i)$
1	-0.387	0.203	0.007	0.440	1	0	0	0	0	0	1	0	6
2	-0.278	0.014	1.129	0	0.540	0	0	0	1	0	1	0	1
3	0.086	0.469	0.900	0	1	1.562	0	0	1	0	1	1	4
4	-0.316	0.371	1.100	0	1	0	0	0	1	0	1	0	1
5	-0.920	2.166	-1.404	0	1	0	0	1	0	0	1	0	3
6	0.003	0.223	0.886	0	1	0	0	0	1	0	1	0	1

Table 3.

For i = 1, 2, 3, 6, $\sum_j |\alpha_j^i - \frac{1}{2}| < \frac{6}{2}$, and of course the same is true for i = 0.

Thus 5 of the seven vertices of S are at this stage contained in K^*, hence in X^*, and they can be removed from our list of active vertices [case (b) of step 1]. The remaining vertices are α^4 and α^5.

We choose α^4 and check whether $\alpha^4 \in X^*$. This is done in Table 4 below, featuring the problem

$$\max \{(\alpha^4 - \tfrac{1}{2}e)x \mid x \in X\}$$

with starting (dual-feasible) solution $x^4 = (\ 0\ \ 0\ \ 1\ \ 0\ \ 1\ \ 0\ )$

	1	$-x_1$	$-x_2$	$-x'_3$	$-x_4$	$-x'_5$	$-x_6$
$(\alpha^4 - \frac{1}{2}e)x^4$	1.100	0.816	0.129	0.6	0.5	0.5	0.5
z	-1	-5	-6	3	-5	-4	3
y_1	0	9	-3	6	8	7	0
y_2	0	6	0	-9	5	-1	-1
y_3	5	-3	9	-7	6	7	3

Table 4.

We have to check whether

$$(\alpha^4 - \frac{1}{2}e)x - \pi(x) \leq \frac{1}{2}\alpha^4 e$$

for some feasible x. We calculate

$$\frac{1}{2}\alpha^4 e = 1.077, \quad \pi(x^4) = \min\{0.816, 0.129, 0.5, 0.5\} = 0.5$$

and

$$(\alpha^4 - \frac{1}{2}e)x^4 - \pi(x^4) = 1.100 - 0.5 < 1.077$$

Thus α^4 can be dropped [case (b) step 1] and we choose the one remaining vertex of S, namely α^5.

Testing α^5 in the same way as α^4 shows that

$$\max\{(\alpha^5 - \frac{1}{2}e)x \mid x \in X\} > \frac{1}{2}\alpha^5 e$$

and no obvious penalty makes the left hand side small enough. Thus α^5 has to be cut off by activating a problem constraint other than the facets of S [case (b) of step 1]. Table 5 shows the simplex-tableau associated with α^5 obtained from th optimal simplex Tableau by pivoting y_2 into the basis in place of z. The constrai active for S are $x_4 \geq 0$, $x_5 \leq 1$, $x_6 \geq 0$, $y_1 \geq 0$, $y_2 \geq 0$, $y_3 \geq 0$, and $z \geq 0$. To cut off α^5, we have to activate one of the 3 constraints of x that are violated by α^5: $x_1 \geq 0$, $x_2 \leq 1$, $x_3 \geq 0$. We choose $x_2 \leq 1$, or $x_2' \geq 0$.

	1	$-x_4$	$-x_5'$	$-x_6$	$-y_1$	$-y_3$	$-z$
y_2	26.77	57.09	59.0	43.0	5.98	7.12	7.7
x_1	-0.920	-1.075	-1.447	-1.620	-0.137	-0.246	-0.:3
x_2'	-1.166	-4.242	-4.781	-2.746	-0.399	-0.515	-0.40
x_3'	2.404	5.066	5.781	3.803	0.572	0.626	0.65

Table 5.

The successors of α^5 are then those vertices that can be obtained by pivotin on a negative coefficient in the x_2'-row, without violating $y_2 \geq 0$. Since each coefficient of that row is negative and none of the pivots violates $y_2 \geq 0$, α^5 has 6 successors, shown in Table 6.

i \ j	α^i_j 1	2	3	4	5	6	x^i_j					
7	-0.63	1	-0.036	0.27	1	0	0	1	0	0	1	0
8	-0.567	1	0.007	0	0.756	0	0	1	0	0	1	0
9	-0.234	1	0.208	0	1	0.424	0	1	0	0	1	0
10	-0.520	1	0.266	0	1	0	0	1	0	0	1	0
11	-0.364	1	0.013	0	1	0	0	1	0	0	1	0
12	-0.063	1	0.468	0	1	0	0	1	0	0	1	0

Table 6.

For i = 9 and 12, we have $\sum_j |\alpha^i_j - \frac{1}{2}| < 3$, hence α^9 and α^{12} can be removed from the active set [case (b) of step 1].

For the remaining 4 vertices, Table 7 shows the starting (dual-feasible) simplex tableau with the 4 different objective functions (the starting solution is in each case the same).

	1	$-x_1$	$-x_2'$	$-x_3$	$-x_4$	$-x_5'$	$-x_6$	$\frac{1}{2}\alpha^i e$
$(\alpha^7 - \frac{1}{2}e)x^7$	1.000	1.130	0.5	0.536	0.473	0.5	0.5	0.640
$(\alpha^8 - \frac{1}{2}e)x^8$	0.756	1.067	0.5	0.493	0.5	0.256	0.5	0.598
$(\alpha^{10} - \frac{1}{2}e)x^{10}$	1.000	1.020	0.5	0.234	0.5	0.5	0.5	0.873
$(\alpha^{11} - \frac{1}{2}e)x^{11}$	1.000	0.864	0.5	0.487	0.5	0.5	0.5	0.824
z	2	-5	6	-3	-5	-4	3	-
y_1	-3	9	3	-6	8	7	0	-
y_2	9	6	0	9	5	-1	-1	-
y_3	3	-3	-9	7	6	7	3	-

Table 7.

The last column contains the amounts $\frac{1}{2}\alpha^i e$. Since -6 is the only negative entry in the y_1-row (for which this solution is infeasible), the first 4 entries in the x_3-column can serve as penalties for the corresponding objective function. Thus we have

$$1.000-0.536 < 0.640$$
$$0.736-0.493 < 0.598$$
$$1.000-0.234 < 0.873$$
$$1.000-0.487 < 0.824$$

which shows that $\alpha^i \in \hat{X}^*$ for i = 7, 8, 10, 11, and we are done: the optimal solution is x = (0, 0, 1, 0, 1, 0).

7. Concluding Remarks

The procedure described in this paper uses the properties of outer polar sets in a new context: rather than generating cutting planes, it seeks to replace the constraint set X with a set S_k defined by just a few of the constraints of X, and having relatively few vertices, all of which are contained in the outer polar of the integer hull. When this is achieved, an optimal solution is at hand. The method is finite, but its efficiency depends on the number of problem constraints that have to be activated in order to generate the desired set. No computational experience is available at this time.

However, the approach discussed here has a significance beyond that of solving linear programs with 0-1 variables. In [1], section 6, we have shown how the concepts of polar sets and outer polars can be generalized so as to provide convex sets that can be used for concave quadratic programming (i.e., minimizing a concave or maximizing a convex quadratic function subject to linear constraints), in much the same way as the ordinary outer polars are used for integer programming. An approach of the type discussed in this paper looks quite promising for concave quadratic programming. This, along with other uses of generalized polars for concave quadratic programming, will be discussed in [15]. Furthermore, the same approach can be used to solve concave quadratic (pure or mixed) integer (0-1) programs. Finally, in [16] we discuss a similar approach for solving general (nonconvex) linearly constrained quadratic programs.

References

[1] E. Balas: "Integer Programming and Convex Analysis: Intersection Cuts from Outer Polars." M.S.R.R. No. 246, Carnegie-Mellon University, April 1971, revised October 1971. Published in Mathematical Programming, 2 (1972), No. 3, p. 329-382.

[2] A. Ülküçü: "Eligible Cuts in Simple Polytopes." Seminar given at the NATO Advanced Study Institute on Mathematical Programming in Theory and Practice, June 12-23, Figueira da Foz.

[3] T. S. Motzkin, H. Raiffa, G. L. Thompson and R. M. Thrall: "The Double Description Method." Contributions to the Theory of Games, vol. II., H. W. Kuhn and A. W. Tucker (editors), Princeton University Press, Princeton, 1953, paper No. 3.

[4] M. L. Balinski: "An Algorithm for Finding All Vertices of Convex Polyhedral Sets." J. Soc. Ind. Appl. Math., 9, 1961, No. 1, p. 72-88.

[5] A. Shefi: "Reduction of Linear Inequality Constraints and Determination of All Feasible Extreme Points." Ph.D. Dissertation, Department of Engineering-Economic Systems, Stanford University, June 1968.

[6] M. A. Pollatschek and B. Avi-Itzhak: "Sorting Feasible Basic Solutions of a Linear Program." OR, Stat. and Econ. Mimeo. Series No. 44, Technion, Haifa, May 1969.

[7] T. H. Mattheiss: "An Algorithm for the Determination of Irrelevant Constraints in Systems of Linear Inequalities, or All Vertices of Convex Polytopes." Department of Business Administration, University of Maryland, September 1970.

[8] C. A. Burdet: "The Facial Decomposition Method." W.P. 104-71-2, G.S.I.A., Carnegie-Mellon University, May 1972.

[9] N. V. Chernikova: "Algorithm for Finding a General Formula for the Nonnegative Solutions of a System of Linear Equations." USSR Computational Mathematics and Mathematical Physics, IV, 1964, No. 4, p. 151-158.

[10] N. V. Chernikova: "Algorithm for Finding a General Formula for the Nonnegative Solutions of a System of Linear Inequalities." USSR Computational Mathematics and Mathematical Physics, V, 1964, No. 2, p. 228-233.

[11] D. S. Rubin: "Neighboring Vertices on Convex Polytopes." Graduate School in Business Administration, University of North Carolina at Chapel Hill, March 1972.

[12] J. F. Benders: "Partitioning Procedures for Solving Mixed-Variables Programming Problems." Numerische Mathematik, 4, 1962, 238-252.

[13] E. Balas: "Minimax and Duality for Linear and Nonlinear (Mixed-) Integer Programming." Integer and Nonlinear Programming, J. Abadie (editor), North-Holland Publishing Company, 1970, p. 385-418.

[14] C. E. Lemke and K. Spielberg: "Direct Search Algorithm for Zero-One and Mixed-Integer Programming. Operations Research, 15, 1967, p. 892-915.

[15] E. Balas and C. A. Burdet: "On Concave Quadratic Programming," in preparatio

[16] E. Balas: "Nonconvex Quadratic Programming via Generalized Polars," in preparation.

Mathematical Programming in Theory and Practice,
P.L. Hammer and G. Zoutendijk, (Eds.)

BOOLEAN PROCEDURES FOR BIVALENT PROGRAMMING

Peter L. Hammer

University of Waterloo

Department of Combinatorics and Optimization

Waterloo, Ontario, Canada.

This research was supported by NRC Grant number A8552

INTRODUCTION

The importance of 0-1 programs has lead in the last years to the elaboration of numerous techniques for their solution. Ingenious methods have been obtained making full use of the advantages of powerful computers and of the art of their users. We shall not attempt here to single out any of the existing algorithms, primarily because of the difficulty - or perhaps impossibility - of ranking them in any meaningful order.

The approaches we know about are all iterative, each iteration consisting of two phases, one of analysis of the given problem (aiming to the discovery of infeasibility, of fixed values for certain variables or of good bounds for the optimum), and one of synthesis of a new problem (by branching, by backtracking, by introducing cutting planes, etc.). Attention is usually focused on the sythesis aspect, while the analysis is carried out on a relaxed continuous problem (where $x \varepsilon \{0,1\}$ is replaced by $0 \leq x \leq 1$). If any more is done at the analysis phase, that is the merit of the skill and especially of the art of the performer.

There is, however, a way of fashioning the art of logical deductions into a technique which involves only elementary, mechanically repeated steps, and though these steps are small and hence many, they do not require any more human genius or even presence, than do the usual arithmetical operations. It was George Boole who discovered this technique well over a century ago.

The aim of this paper is to show that Boolean algebra can successfully be applied for the analysis, and to some extent also to the synthesis of 0-1 programs.

A word of caution. Boolean algebra can be applied for the solution of <u>any</u> linear or nonlinear 0-1 program. but being widely applicable does not mean to be universally recommended. There are large classes of 0-1 programs where the solution space is so vast that little structure can be hoped to be found in it, and where thus, it might be perferable to attempt a more straightforward solution technique. To this class belong in particular the problems where all the constraints

$$\sum_{j=1}^{n} a_{ij}x_j \leq b_i$$

involve only nonnegative coefficients, or where the values $|a_{ij}|$ are "small" in comparison to b_i. But even in these cases, the knowledge of a "good" feasible solution allows the introduction of a supplementary constraint (expressing the fact that the objective function is bounded by its value in the particular known good solution) which might so drastically reduce the solution space that a Boolean search for structure might become motivated.

The particular case involving only nonnegative coefficients a_{ij} will be dealt with in a forthcoming paper [6], where the emphasis will be shifted to the Boolean analysis of the objective function rather than that of the constraints.

The structure of this paper is the following. After a short introductory Chapter I outlining basic elements of Boolean algebra, Chapter II will describe a Boolean equivalent of a system of constraints appearing in bivalent programs. Chapter III will be devoted to the question of discovering order relations in the solution space of a bivalent program, and utilizing them for simplifying a linear problem; it also contains a concrete method for solving linear bivalent programs, based on a particularization of the results given before. Similarly, Chapters IV and V concentrate on quadratic 0-1 programs, while Chapter VI is devoted to polynomial 0-1 programs.

Computational experience is available for the method described in Chapters III and VI and is encouraging. It is, however, to be pointed out that the computational experience is limited, that numerous alternative options have not yet been experienced and that improvements can be anticipated. Details are given in Chapter VII.

I. ELEMENTS OF BOOLEAN ALGEBRA

Let $B = \{0,1\}$. For $x \varepsilon B$ we shall denote by $\bar{x} = 1-x$ its complement or negation. We shall also write frequently

(1)
$$x^{\alpha} = \begin{cases} x & \text{if } \alpha = 1 \\ \bar{x} & \text{if } \alpha = 0 \end{cases}$$

This notation can cause no confusion, because the regular powers of $x \varepsilon B$ being all equal to x (idempotency of multiplication) we shall never use them.

For any $x, y \varepsilon B$, we shall define their union $x \vee y$ by

(2)
$$x \vee y = x + y - x y$$

or by the tableau

∨	0	1
0	0	1
1	1	1

Some of the most commonly utilized properties of the above defined operations are the following:

(3) $x \vee y = y \vee x$ (commutativity)

(4) $x \vee (y \vee z) = (x \vee y) \vee z$ (associativity)

(5) $x \vee x = x$ (idempotency)

(6) $x \vee y = 0$ if and only if $x = y = 0$

(7) $x \vee 0 = x$

(8) $x \vee 1 = 1$

(9) $x \vee \bar{x} = 1$

(10) $x \vee y\,z = (x \vee y)(x \vee z)$
(11) $x(y \vee z) = x\,y \vee x\,z$ } (distributivities)

(12) $x \vee x\,y = x$ (absorption)

(13) $x \vee \bar{x}\,y = x \vee y$

(14) $\overline{x \vee y} = \bar{x}.\bar{y}$
(15) $\overline{x\,y} = \bar{x} \vee \bar{y}$ } (De Morgan's Laws)

(16) $\bar{\bar{x}} = x$ (double negation)

(17) $x \leq y$ if and only if $x\,y = x$

(18) $x \leq y$ if and only if $x\,\bar{y} = 0$

(19) $x = y$ if and only if $x\,\bar{y} \vee \bar{x}\,y = 0$

A function $f(x_1,\ldots,x_n)$ whose variables and values belong to B, will be called a Boolean function. Examples of such functions are $x \vee y\,z$, $x \vee y\,z \vee \bar{x}\,\bar{z}$, $(x \vee \bar{y})(\overline{y \vee x\,z})$, etc. The algebraic expression of a Boolean function is not unique, e.g. the expressions $x \vee y$ and $x \vee y\,z \vee \bar{x}\,\bar{z}$ define the same function (this can be seen either by giving to x, y, z all 2^3 possible combinations of values, or noticing that $x \vee y\,z \vee \bar{x}\,\bar{z} = x \vee y\,z \vee \bar{z}$ [due to (13)] $= x \vee y$ [due again to (13)]).

A variable x, or its negation $\bar{x}$ will be called a literal X. A finite product of literals will be called an elementary conjunction.

$$C = \prod_{j \in S} x_j^{\alpha_j} \quad ; \tag{20}$$

by convention, we shall consider sometimes also the constant 1 as being an elementary conjunction (with $S = \emptyset$). A finite union of elementary conjunctions

$$E = C_1 \vee C_2 \vee \ldots \vee C_m \tag{21}$$

will be called a disjunctive form. It can be shown easily that every Boolean function can be expressed in a disjunctive form.

We shall say that an elementary conjunction C is <u>contained</u> in the elementary conjunction C' if every literal appearing as a factor in C is also a factor of C'. E.g. $x\bar{y}$ is contained in $x\bar{y}zu$, also in $x\bar{y}$, but is not contained in xz or in xyz.

An elementary conjunction I is said to be an <u>implicant</u> of the Boolean function $f(x_1,\ldots,x_n)$, if $I = 1$ implies $f(x_1,\ldots,x_n) = 1$. For example, $x\bar{y}$ is an implicant of $x\bar{y} \vee \bar{y}z(x \vee \bar{z})$. Also, $x\bar{y}$ is an implicant of $xz \vee \bar{y}\bar{z}$ (indeed, if $x\bar{y} = 1$, then $x = 1$, $y = 0$, and hence $xz \vee \bar{y}\bar{z}$ becomes $z \vee \bar{z}$ which is equal to 1).

An implicant P of a Boolean function $f(x_1,\ldots,x_n)$ is said to be a <u>prime implicant</u> if there is no other implicant P' of f contained in P. For example, $x\bar{y}$ is a prime implicant of $f = xz \vee \bar{y}\bar{z}$, but $x\bar{y}z$ is a non-prime implicant of f. If all the prime implicants of a Boolean function f are $P_1,\ldots,P_t$, then it is easy to see that

$$f = P_1 \vee \ldots \vee P_t \tag{22}$$

We shall see later that the knowledge of the prime implcants of a given Boolean function is extremely useful. A way of finding all the prime implicants is offered by the so-called <u>consensus</u> method.

Given two elementary conjunctions C and C', such that there is precisely one variable (x_o) appearing unnegated (x_o) is one of them, and negated ($\bar{x}_o$) in the other, then the elementary conjunction obtained from the juxtaposition CC' of C and C' after deleting x_o, $\bar{x}_o$ and repeated literals, will be called the <u>consensus</u> of C and C'. For example, let $C = x\bar{y}\bar{z}u$ and $C = \bar{y}zu\bar{w}$; then their consensus is $C'' = x\bar{y}u\bar{w}$.

The consensus method consists in applying as many times as possible the following two operations to a disjunctive form of a Boolean function:

(i) eliminate any elementary conjunction which contains another one;

(ii) add as a new elementary conjunction the consensus of two elementary conjunctions, provided this consensus does not include any of the listed (undeleted) elementary conjunctions.

All the different expressions obtained along this process represent the same Boolean function, and the elementary conjunctions appearing in the final form at the end of this (finite, but long) process are exactly the prime implicants of the given functions.

Example 1. Let

(23) $$\phi_1 = x_1 x_2 \bar{x}_5 \vee x_1 x_2 \bar{x}_6$$

(24) $$\phi_2 = x_3 \bar{x}_4 x_6 \vee \bar{x}_1 x_3 \bar{x}_4 \vee x_3 \bar{x}_4 x_5 \vee \bar{x}_1 x_3 x_6 \vee x_3 x_5 x_6 \vee x_1 \bar{x}_4 x_6 \vee \bar{x}_4 x_5 x_6$$

(25) $$\phi_3 = x_1 \bar{x}_2 x_6 \vee \bar{x}_2 x_3 x_6 \vee \bar{x}_2 x_4 x_6 \vee x_1 \bar{x}_2 x_3 \vee x_1 x_3 x_6$$

(26) $$\phi_4 = \bar{x}_1 \bar{x}_6 \vee x_3 \bar{x}_4 \bar{x}_6 ,$$

and let us find by the consensus method the prime implicants of the Boolean function

(27) $$\Phi = \phi_1 \vee \phi_2 \vee \phi_3 \vee \phi_4 .$$

The calculations are shown in Table 1 below. The conjunctions are represented in a shorter form, by putting $12\bar{5}$ instead of $x_1 x_2 \bar{x}_5$, etc. The first sixteen conjunctions of the tableau are the originally given ones.

The consensus method is applied as follows:

1). Check whether any absorption occurs (i.e. if there is a conjunction

C_h and one C_k such that C_h is included in C_k, then C_k is redundant because $C_h \vee C_k = C_h$), and if yes delete from our list the absorbed conjunctions;

2). Consider the i-th conjunction (starting with i: = 2) and examine the pairs $(C_1 C_i)$, $(C_2, C_i), \ldots, (C_{i-1}, C_i)$; if such a pair (C_j, C_i) (briefly (j,i)) defines a consensus C_{ji} then check if C_{ji} is absorbed by one of the conjunctions in our list; if not, enter C_{ji} in our list and repeat 1).

3). Repeat steps 1) and 2) as many times as possible.

TABLE I

No.	Conjunction	Given/consensus of	Absorbed by No.
1	$12\bar{5}$	Given	
2	$12\bar{6}$	Given	34
3	$3\bar{4}6$	Given	33
4	$\bar{1}3\bar{4}$	Given	35
5	$3\bar{4}5$	Given	38
6	$\bar{1}36$	Given	33
7	356	Given	33
8	$\bar{1}\bar{4}6$	Given	37
9	$\bar{4}56$	Given	53
10	$1\bar{2}6$	Given	43
11	$\bar{2}36$	Given	33
12	$\bar{2}46$	Given	43
13	$1\bar{2}3$	Given	46
14	136	Given	33
15	$\bar{1}\bar{6}$	Given	
16	$3\bar{4}\bar{6}$	Given	38

Table I (continued)

No.	Conjunction	Given/consensus of	Absorbed by No.
17	$123\bar{4}$	(2,3)	38
18	$23\bar{4}\bar{5}$	(1,4)	38
19	$23\bar{4}\bar{6}$	(2,4)	34
20	$23\bar{5}6$	(1,6)	33
21	1235	(2,7)	39
22	$2\bar{4}\bar{5}6$	(1,8)	40
23	$12\bar{4}6$	(1,9)	40
24	$12\bar{4}5$	(2,9)	41
25	$1\bar{5}6$	(1,10)	
26	$\bar{2}\bar{4}6$	(8,10)	43
27	$\bar{1}\bar{2}6$	(8,12)	43
28	$\bar{2}56$	(9,12)	43
29	$13\bar{5}$	(1,13)	46
30	$13\bar{6}$	(2,13)	45
31	$\bar{2}3\bar{4}$	(4,13)	38
32	123	(2,14)	46
33	36	(6,14)	54
34	$2\bar{6}$	(2,15)	
35	$\bar{1}\bar{4}$	(8,15)	
36	$\bar{1}\bar{2}4$	(12,15)	51
37	$\bar{2}3\bar{6}$	(13,15)	45
38	$3\bar{4}$	(3,16)	54
39	135	(13,21)	46
40	$2\bar{4}6$	(9,22)	52
41	$12\bar{4}$	(1,24)	52

Table I (continued)

No.	Conjunction	Given/consensus of	Absorbed by No.
42	$1\bar{4}6$	(9,25)	53
43	$\bar{2}6$	(10,27)	
44	$3\bar{5}\bar{6}$	(15,29)	45
45	$3\bar{6}$	(15,30)	54
46	13	(13,32)	54
47	$\bar{1}3$	(15,33)	54
48	$2\bar{4}5$	(9,34)	52
49	23	(33,34)	54
50	$\bar{4}\bar{5}6$	(25,35)	53
51	$\bar{1}\bar{2}$	(35,36)	
52	$2\bar{4}$	(34,40)	
53	$\bar{4}6$	(35,42)	
54	3	(33,45)	

Thus, we see that Φ becomes

(28) $$\Phi = x_1x_2\bar{x}_5 \vee \bar{x}_1\bar{x}_6 \vee x_1\bar{x}_5x_6 \vee x_2\bar{x}_6 \vee \bar{x}_1\bar{x}_4 \vee \bar{x}_2x_6 \vee \bar{x}_1\bar{x}_2 \vee x_2\bar{x}_4 \vee \bar{x}_4x_6 \vee x_3$$

If we are interested in the equation $\Phi = 0$, we see that its solutions are characterized by:

1) $x_3 = 0$

2) $x_2 = x_6$ (because $x_2\bar{x}_6 = \bar{x}_2x_6 = 0$)

3) the variables x_1, x_2, x_4, x_5 are characterized by the equation

(29) $$x_1x_2\bar{x}_5 \vee \bar{x}_1\bar{x}_2 \vee \bar{x}_1\bar{x}_4 \vee x_2\bar{x}_4 = 0$$

It is likely that in practical problems finding all the prime implicants of a Boolean function might require an excessive amount of computation. Therefore, in the more practical procedures described in Chapters III - VI we shall work with implicants which will not be necessarily prime, but which will allow a rapid solution of many 0-1 programs.

II. A BOOLEAN EQUIVALENT OF THE CONSTRAINT SET

Consider first a linear inequality

$$\sum_{j=1}^{n} a_j' x_j' \leq b' \tag{30}$$

with $x_j \varepsilon B(j=1,\ldots,n)$. This can be rewritten as

$$\sum_{j=1}^{n} |a_j'| \, x_j^{\alpha_j} \leq b' - \sum_{j=1}^{n} \min(a_j', 0) \tag{31}$$

with

$$\alpha_j = \begin{cases} 1 \text{ if } a_j' \geq 0 \\ 0 \text{ if } a_j' < 0 \end{cases}$$

Let $a_j = |a_j'|$, $X_j = x_j^{\alpha_j}$, $b = b' - \sum_{j=1}^{n} \min(a_j', 0)$; then (31) becomes

$$\sum_{j=1}^{n} a_j X_j \leq b, \tag{32}$$

where now all $a_j \geq 0$. We can also reindex the variables so that $a_1 \geq a_2 \geq \ldots \geq a_n \geq 0$.

For example $-2x_1 + 3x_2 - x_3 - 3x_4 + 4x_5 \leq 2$ becomes in this way $4X_1 + 3X_2 + 3X_3 + 2X_4 + X_5 \leq 8$, where $X_1 = x_5$, $X_2 = x_2$, $X_3 = \bar{x}_4$, $X_4 = \bar{x}_1$, $X_5 = \bar{x}_3$.

A subset S of $\{1,\ldots,n\}$ will be called a <u>cover</u> of (32) if $\sum_{j\varepsilon S} a_j > b$.

It will be called a <u>minimal cover</u> of (32) if it is a cover of it and does not contain (properly) any other cover of (32). E.g. in the above example $\{1, 2, 5\}$ is not a cover, $\{1, 2, 4, 5\}$ is a non-minimal cover, $\{1, 2, 4\}$

is a minimal cover. An algorithm for determining all minimal covers of an inequality (32) will be given at the end of this chapter.

Let $\mathcal{S} = \{S_1,\dots,S_t\}$ be the family of all minimal covers of (32). it is easy to see that any 0-1 solution of (32) will satisfy the conditions

$$\prod_{j\in S_r} X_j = 0 \qquad (r = 1,\dots,t),$$

or simply

$$\bigvee_{r=1}^{t} \prod_{j\in S_r} X_j = 0 \tag{33}$$

The Boolean function $\psi(X_1,\dots,X_n)$ appearing in the left-hand side of (33) will be called the <u>resolvent of (32)</u>. The Boolean function $\phi(x_1,\dots x_n)$ obtained from ψ by substituting $X_j = x_j^{\alpha_j}$ $(j = 1,\dots,n)$ will be called the <u>resolvent of (30)</u>.

It has been shown in [2] that a vector $(x_1,\dots x_n)$ with $x_j \varepsilon B (j = 1,\dots,n)$ is a solution of (30) if and only if $\phi(x_1,\dots,x_n) = 0$.

In the above example, the minimal covers of $4X_1 + 3X_2 + 3X_3 + 2X_4 + X_5 \le 8$ are {1, 2, 3} {1, 2, 4}, {1, 3, 4} and {2, 3, 4, 5}. Hence, the resolvent of it is $X_1 X_2 X_3 \vee X_1 X_2 X_4 \vee X_1 X_3 X_4 \vee X_2 X_3 X_4 X_5$. Substituting here the relations $X_1 = x_5$, $X_2 = x_2$, $X_3 = \bar{x}_4$, $X_4 = \bar{x}_1$, $X_5 = \bar{x}_3$, we see that all the solutions of the original inequality $-2x_1 + 3x_2 - x_3 - 3x_4 + 4x_5 \le 2$ are characterized by the Boolean equation

$$\bar{x}_1 x_2 \bar{x}_5 \vee \bar{x}_1 x_2 \bar{x}_3 \bar{x}_4 \vee \bar{x}_1 \bar{x}_4 x_5 \vee x_2 \bar{x}_4 x_5 = 0 .$$

The main advantage of the resolvent is that it allows to discover the structure of the solution space of systems of constraints. While little information can be expected to be found from the resolvent of a single constraint, the resolvent of a system of constraints might give a considerably better insight into the solution space of the problem. Given a system of constraints

(34) $\sum a_{ij}\, x_j \le b_i \qquad (i = 1,\ldots m)$

if $\phi_i(x_1,\ldots,x_n)$ denotes the resolvent of the i-th individual constraint, then the Boolean function

(35) $\Phi(x_1,\ldots x_n) = \phi_1(x_1,\ldots x_n) \vee\ldots\vee \phi_m(x_1,\ldots,x_n)$

will be called the resolvent of the system. It is clear now that a point $(x_1,\ldots,x_n)(x_j \varepsilon B,\ j = 1,\ldots,n)$ satisfies the system (34) if and only if it satisfies the Boolean equation

(36) $\Phi(x_1,\ldots,x_n) = 0$

Example 2. Consider the system

(37) $8x_1 + 6x_2 + 3\bar{x}_5 + 2\bar{x}_6 \le 14$

(38) $5x_3 + 4\bar{x}_4 + 4x_6 + 2\bar{x}_1 + 2x_5 \le 9$

(39) $3\bar{x}_2 + 3x_6 + 2x_1 + 2x_3 + x_4 \le 6$

(40) $4\bar{x}_6 + 3\bar{x}_1 + 2x_3 + \bar{x}_4 \le 6$

where $x_j \varepsilon B$ $(j = 1,\ldots,6)$. It can be seen that the resolvents of the above four inequalities are exactly the functions $\phi_1,\ldots,\phi_4$ given in (23) - (26), and hence the resolvent of this system is $\Phi(x_1,\ldots,x_6) = \phi_1 \vee \phi_2 \vee \phi_3 \vee \phi_4$. A simplified expression of Φ (using prime implicants) was given in (28). It follows now that every 0-1 solution of our system is characterized by the fact that $x_3 = 0$, $x_6 = x_2$ and by the Boolean equation $x_1 x_2 \bar{x}_5 \vee \bar{x}_1 \bar{x}_2 \vee \bar{x}_1 \bar{x}_4 \vee x_2$

(For those who like it linear, let us notice as in [3] that the last equation simply means that $x_1 x_2 \bar{x}_5 = \bar{x}_1 \bar{x}_2 = \bar{x}_1 \bar{x}_4 = x_2 \bar{x}_4 = 0$. Further $x_1 x_2 \bar{x}_5 = 0$ means that $\overline{x_1 x_2 \bar{x}_5} = 1$, or $\bar{x}_1 \vee \bar{x}_2 \vee x_5 = 1$, or finally, $\bar{x}_1 + \bar{x}_2 + x_5 \geq 1$. Reasoning in the same way on the other conjunctions, we see that our original system is equivalent to: $x_3 = 0$, $x_2 = x_6$, $\bar{x}_1 + \bar{x}_2 + x_5 \geq 1.$, $x_1 + x_2 \geq 1$, $x_1 + x_4 \geq 1$, $\bar{x}_2 + x_4 \geq 1$).

A second important use of the resolvent is in handling nonlinear constraints. Indeed let us consider a polynomial *) constraint

(41)
$$\sum_{i=1}^{m} a_i \prod_{j \in M_i} x_j \leq b ,$$

where $M_i \subseteq \{1,\ldots,n\}$ (here we have taken into account the fact that for $x \in B$, $x = x^2 = x^3 = \ldots = x^k$). If we put now

(42)
$$y_i = \prod_{j \in M_i} x_j \qquad (i = 1,\ldots m) ,$$

then (41) will become a linear inequality

(41')
$$\sum_{i=1}^{m} a_i y_i \leq b$$

in the 0-1 variables $y_i (i = 1,\ldots,m)$. Let $\phi(y_1,\ldots y_m)$ be the resolvent of (41'). Substituting (42) into ϕ we shall obtain a Boolean function $\chi(x_1,\ldots,x_n)$, which we shall call the <u>resolvent of (41)</u>. It was shown [2] that a point $(x_1,\ldots,x_n)(x_j \in B, j = 1,\ldots,n)$ satisfies (41) if and only if

* It was shown in [10] that the most general constraints in 0-1 variables can be represented in such a form.

$\chi(x_1,\ldots,x_n) = 0.$

Example 3. Consider the inequality

$$-2x_1 x_2 + 3x_1 x_3 - x_3 - 3x_2 x_4 x_5 + 4x_2 x_5 x_6 \leq 2 ;$$

if we put $x_1 x_2 = y_1$, $x_1 x_3 = y_2$, $x_3 = y_3$, $x_2 x_4 x_5 = y_4$, $x_2 x_5 x_6 = y_5$, then we know that the resolvent of the corresponding linear inequality in y's, as determined previously, will be

$$\phi = \bar{y}_1 y_2 y_5 \vee \bar{y}_1 y_2 \bar{y}_3 \bar{y}_4 \vee \bar{y}_1 \bar{y}_4 y_5 \vee y_2 \bar{y}_4 y_5 .$$

Substituting, we find

$$\begin{aligned}\chi &= (\overline{x_1 x_2})\, x_1 x_3 x_2 x_5 x_6 \vee (\overline{x_1 x_2})\, x_1 x_3 \cdot \bar{x}_3 \cdot (\overline{x_2 x_4 x_5}) \vee \\ &\quad \vee (\overline{x_1 x_2})(\overline{x_2 x_4 x_5})\, x_2 x_5 x_6 \vee x_1 x_3 (\overline{x_2 x_4 x_5})\, x_2 x_5 x_6 \\ &= \bar{x}_1 x_2 \bar{x}_4 x_5 x_6 \vee x_1 x_2 x_3 \bar{x}_4 x_5 x_6\end{aligned}$$

or simply

$$\chi = x_2 \bar{x}_4 x_5 x_6 (\bar{x}_1 \vee x_3) .$$

A simple branching-type algorithm for the construction of minimal covers runs as follows. We construct at each step of the algorithm a sequence S_t of indices $\{s_1,\dots,s_k\}$ (where $1 \le s_1 < \dots\dots < s_k \le n$); the sequence will be labelled if and only if $\sum_{h=1}^{k} a_{s_h} > b$. The starting sequence is the empty one; if it is labelled (i.e. if $b < 0$) then the process ends at this stage; otherwise denoting by S_t the last of the built sequences we continue to build a new sequence according to the following three rules:

1). If $s_k < n$ and the sequence $S_t = \{s_1,\dots,s_k\}$ is not labelled, then $S_{t+1} = \{s_1,\dots s_k, s_k + 1\}$.

2). If $s_k < n$ and the sequence S_t is labelled, then $S_{t+1} = \{s_1,\dots,s_{k-1},$

3). If $s_k = n$, let ℓ be the greatest index (if any) in $\{1,\dots,k\}$ such that $s_{\ell+1} - s_\ell > 1$. Then $S_{t+1} = \{s_1,\dots,s_{\ell-1}, s_\ell + 1\}$.

The building of sequences stops when we arrive to a sequence with $s_k = n$ where rule 3 cannot be applied for building a next sequence (i.e. $S_t = \{q, q+1, q+2,\dots, n\}$).

It was shown in [2] that the labelled sequences in this list are all the minimal covers of the given inequality.

Example 4. The above algorithm applied to the inequality

$$4X_1 + 3X_2 + 3X_3 + 2X_4 + X_5 \le 8$$

produces the following sequences (the labelled ones are underlined):

$\emptyset$, {1}, {1, 2}, {1, 2, 3}, {1, 2, 4}, {1, 2, 5}, {1, 3}, {1, 3, 4}, {1, 4}, {1, 4, 5}, {2}, {2, 3}, {2, 3, 4}, {2, 3, 4, 5}. Hence the minimal covers are {1, 2, 3}, {1, 2, 4}, {1, 3, 4}, {2, 3, 4, 5}.

One of the major informations which can be derived from the knowledge of all the prime implicants of the resolvent of a system of linear or nonlinear inequalities is the answer to the question of feasibility of the given system. Indeed, if ϕ is the resolvent of the system, then the equation $\phi(x) = 0$ (characterizing all the solutions of the given system) is feasible if and only if the constant 1 is not a prime implicant of ϕ. Based on this remark we could device (similarly to [2]) a procedure for optimization in 0-1 variables. Let us denote by $f_o(X)$ the objective function, let us assume that it has only integer values when the variables are 0 or 1, and let b^o and B^o represent a lower and an upper bound of $f_o(X)$ in the unitcube (e.g. b^o can be taken equal to the sum of the negative coefficients in a polynomial representation of f_o, and B^o the sum of the positive ones). An algorithm could now run as follows:

Algorithm. 1. Produce the resolvent $\phi^o(X)$ of the system of constraints, and all its prime implicants. If the constant 1 is a prime implicant of ϕ^o, then the problem is infeasible. Otherwise, put $i := 0$ and go to

2. If $b^i < B^i$, put $\beta^i = \left\lceil \frac{b^i + B^i}{2} \right\rceil$ and let ψ^i be the resolvent of the constraint $f_o(X) \geq \beta^i$. Let further $\chi^i = \phi^i \vee \psi^i$. If 1 is not a prime implicant of χ^i (then it is possible to find 0-1 solutions to our problem, for which the objective function takes values at least β equal to β^i) then put $b^{i+1} = \beta^i$, $B^{i+1} = B^i$, $\phi^{i+1} = \chi^i$ and repeat step 2 for $i: = i + 1$. If 1 is a prime implicant of χ^i, put $b^{i+1} = b^i$, $B^{i+1} = \beta^i$, $\phi^{i+1} = \phi^i$ and repeat step 2 for $i: = i + 1$. If $b^i = B^i$, then $v^* = b^i = B^i$ is the optimal value of the objective function; go to 3.

3. Solve the Boolean equation $\chi^i(X) = 0$; its solution set will be exactly the same as the set of optimal solutions to the original problem.

Finally, if f_o is linear, we can replace step 3 by

by

3'. Find all the optimal solutions to the linear program consisting in the maximization of f_o subject to all the original constraints, supplemented by $0 \le x_j \le 1$ $(j = 1,\ldots,n)$ and $f_\sigma(x_1,\ldots,x_n) \le v^*$. All the optimal solutions of our discrete problem will be optimal solutions of the associated linear program.

III. THE LINEAR CASE

In spite of - or rather because of - their mathematical merit (showing that 0-1 programming is of the same degree of difficulty as the tautology problem of logic) the results of Chapter II are of not much immediate practical importance; it is not difficult to see that finding (a) the resolvent, (b) its prime implicants, and repeating then (a) and (b) a (very small!) number of times, is time consuming and hence not recommended.

It is thus natural to try to weaken to a certain extent the Boolean representation of the original problem, and possibly also the procedures for processing the available Boolean informations.

By examining elementary conjunctions appearing in the resolvent we see that the most important ones, in decreasing order of importance, are the following:

0) The conjunction of length zero (i.e. the constant 1), which indicates infeasibility;

1) The conjunctions of length one, which show that certain variables have a fixed (0 or 1) value in every feasible solution of the problem;

2) The conjunctions of length two, which indicate the presence of a partial order in the solution space (indeed, $x\,\bar{y} = 0$ is equivalent to $x \le y$, $x\,y = 0$ is equivalent to $x \le \bar{y}$, $\bar{x}\,y$ is equivalent to $y \le x$, $\bar{x}\,\bar{y} = 0$ is equivalent to $\bar{x} \le y$). Moreover, the presence of two conjunctions of length two relating the same two variables gives an even deeper insight into the solution space of our problem. Indeed,

- if $x\,y = x\,\bar{y} = 0$ then $x = 0$
- if $x\,y = \bar{x}\,y = 0$ then $y = 0$
- if $x\,y = \bar{x}\,\bar{y} = 0$ then $x = \bar{y}$
- if $x\,\bar{y} = \bar{x}\,y = 0$ then $x = y$
- if $x\,\bar{y} = \bar{x}\,\bar{y} = 0$ then $y = 1$
- if $\bar{x}\,y = \bar{x}\,\bar{y} = 0$ then $x = 1$

Hence, the conjunctions of order two establish order relations and can show the equality of two variables, their complimentarity, or can fix the value of certain variables.

3) The conjunctions of length three, might be useful in obtaining new conjunctions of length two by the consensus method. Indeed, $x\,y\,z = x\,y\,\bar{z} = 0$ produces $x\,y = 0$; similarly $x\,y = x\,\bar{y}\,z = o$ produces $x\,z = 0$, etc.

It is hence natural to restrict the attention on the conunctions of length ℓ not exceeding a certain value, e.g. 3. By the <u>ℓ-resolvent</u> of a linear inequality we shall mean the union of those conjunctions which correspond to the minimal-covers containing at most ℓ elements. Similarly, the ℓ-resolvent ψ of a system of linear inequalities is the union of the ℓ-resolvents of the inequalities appearing in the system. Obviously, for $\ell = n$, the resolvent and the ℓ-resolvent have the same meaning. However, for $\ell < n$, the condition $\psi(x) = 0$ while remaining a necessary condition for X to be a solution to our original problem, ceases to be a sufficient one. In particular, the infeasibility of $\psi(X) = 0$ implies the infeasibility of the original problem, but the feasibility of $\psi(X) = 0$ does not imply the feasibility of the original problem. On the other hand, if in every solution of $\psi(X) = 0$ a certain relation holds (e. g. $x = 0$, $\bar{y} = 0$, $u\,\bar{v} = 0$, etc.), the same relation must also hold for every solution (if any!) of the original problem.

After having relaxed the Boolean function we work with, we shall also have to relax the consensus method, in order to avoid producing conjunctions of length exceeding ℓ. We can do this by simply changing step 2 of the consensus method: the consensus C_{ji} of C_j and C_i is entered into our list if and only if it is not absorbed by any member of the list, and it is of length $\leq \ell$. Each conjunction obtained at the end of this process will be called an ℓ-prime implicant (which is not necessarily a prime implicant, but is always an implicant).

The determination of the ℓ-resolvent of a system of linear inequalities and the determination of all its ℓ-prime implicants will be called the logical analysis of the constraints.

It is not difficult to see that the analysis usually incorporated into a branch-and-bound type method is actually the logical analysis, when $\ell = 1$. Of course, by taking $\ell = 1$, interactions of different constraints can not be detected.

We shall give the details of a particular algorithm (described in [11]) based on the above ideas of logical analysis; ℓ was chosen to be equal to 3. Because of the role played in it by order relations, this procedure was called APOSS (A Partial Order in the Solution Space). (The particular version of APOSS which was tested on a computer has actually $\ell = 2$).

APOSS consists of the following steps:

1. Logical analysis. If the relaxed consensus method produces an implicant of length 0 (the constant 1) then the given problem is infeasible. If it produces an implicant of length 1 then the corresponding varialbe has a fixed value in every solution (if any). If it produces two implicants of length 2 relative to the same two variables x_i and x_j, then in every solution of the original problem (if any), $x_i = x_j$ (or $x_i = \bar{x}_j$). If finally implicants of length 2 have been produced (but not in pairs) then a certain order relation must hold in the solution space.

2. Synthesis. If in step 1 we find that in every solution certain variables are equal to certain fix values, then introducting these values into the problem we produce a new one with less variables. If in step 1 we find that in every solution of our problem certain pairs of varialbes are equal $(x_i = x_j)$ or complementary $(x_i = \bar{x}_j)$, by introducing these relations into our constraints, we produce a new problem with less variables.

3. Linear programming. Solve the LP associated to our problem by replacing each condition $x_j \varepsilon \{0,1\}$ by $0 \leq x_j \leq 1$. If the optimal solution is integer, stop. Also stop if the optimal value is smaller then the value of the objective function in a feasible integer point (if such a point is known). Otherwise, add to the constraints the order relations (if any) obtained in step 1 and unused in step 2. These new relations $(x_i \leq x_j$, $x_h + x_k \leq 1$, etc.) will act as cuts. If the new LP-optimum is integer stop. Otherwise, if fo(X) is the objective function and ζ its optimal value, add $f_o(X) \leq [\zeta]$ to the constraints and go to

4. Branching or backtracking, which are carried out in the usual way. It is to be remarked that the knowledge of certain order relations may be useful in selecting the branching-variable.

Example 5. The following example has been discussed in some recent works: Maximize

$$f = 12\bar{x}_1 + 5\bar{x}_2 + 9x_3 + 5x_4 + 4\bar{x}_5 + 8\bar{x}_6 + 12x_7 + 3x_8 + 10x_9 + \bar{x}_{10} + 7x_{11} + 7x_{12} \quad (43)$$

subject to

$$12\bar{x}_3 + 7x_6 + 5\bar{x}_{11} + 3x_2 + 3x_{10} + \bar{x}_1 + \bar{x}_5 + \bar{x}_7 + \bar{x}_{12} \leq 15 \quad (44\text{-}1)$$

$$7\bar{x}_2 + 6x_5 + 3x_1 + \bar{x}_4 \leq 8 \quad (44\text{-}2)$$

$$11x_1 + 9x_{11} + 7\bar{x}_4 + 5\bar{x}_9 + 2x_7 + x_3 + \bar{x}_6 + x_8 \leq 9 \quad (44\text{-}3)$$

(44-4) $$12\bar{x}_5 + 8\bar{x}_{10} + 7x_6 + 6x_3 + 5x_2 + 5x_{12} + 3x_8 + x_9 \le 28$$

(44-5) $$8\bar{x}_6 + 7\bar{x}_1 + 7x_9 + 7\bar{x}_{12} + 5\bar{x}_3 + 3x_4 + 2\bar{x}_8 + \bar{x}_2 + x_5 + x_{10} \le 23$$

(44-6) $$5\bar{x}_8 + 4\bar{x}_4 + 3\bar{x}_7 + 2\bar{x}_1 + \bar{x}_9 + x_{11} + x_{12} \le 11$$

and

(45) $$x_j \varepsilon \{0,1\} \qquad (j = 1,\ldots, 12)$$

It can be seen that the 3-resolvent contains the conjunctions x_1 (from (44-3)), $\bar{x}_3\,\bar{x}_{11}$ (from (44-1)) and $x_3\,x_{11}$ (from 44-3)) showing that in every solution of our problem $x_1 = 0$ and $\bar{x}_3 = x_{11}$. Introducing $x_1 = 0$ into our constraints and replacing x_{11} by $\bar{x}_3$ we find

(46-1) $$7\bar{x}_3 + 7x_6 + 3x_2 + 3x_{10} + \bar{x}_5 + \bar{x}_7 + \bar{x}_{12} \le 9$$

(46-2) $$7\bar{x}_2 + 6x_5 + \bar{x}_4 \le 8$$

(46-3) $$8\bar{x}_3 + 7\bar{x}_4 + 5\bar{x}_9 + 2x_7 + \bar{x}_6 + x_8 \le 8$$

(46-4) $$12\bar{x}_5 + 8\bar{x}_{10} + 7x_6 + 6x_3 + 5x_2 + 5x_{12} + 3x_8 + x_9 \le 28$$

(46-5) $$8\bar{x}_6 + 7x_9 + 7\bar{x}_{12} + 5\bar{x}_3 + 3x_4 + 2\bar{x}_8 + \bar{x}_2 + x_5 + x_{10} \le 16$$

(46-6) $$5\bar{x}_8 + 4\bar{x}_4 + 3\bar{x}_7 + \bar{x}_9 + \bar{x}_3 + x_{12} \le 9$$

(and $x_1 = 0$, $x_{11} = \bar{x}_3$);

it can be seen that the 3-resolvent will contain the terms $\bar{x}_3\,x_6$ (from (46-1)) and $\bar{x}_3\,\bar{x}_6$ (from (46-3)) showing that $\bar{x}_3 = 0$, i.e. $x_3 = 1$ and $x_{11} = 0$. Introducing these values into the system, we get

(47-1) $$7x_6 + 3x_2 + 3x_{10} + \bar{x}_5 + \bar{x}_7 + \bar{x}_{12} \le 9$$

(47-2) $$7\bar{x}_2 + 6x_5 + \bar{x}_4 \le 8$$

(47-3) $$7\bar{x}_4 + 5\bar{x}_9 + 2x_7 + \bar{x}_6 + x_8 \le 8$$

(47-4) $$12\bar{x}_5 + 8\bar{x}_{10} + 7x_6 + 5x_2 + 5x_{12} + 3x_8 + x_9 \le 22$$

(47-5) $$8\bar{x}_6 + 7x_9 + 7\bar{x}_{12} + 3x_4 + 2\bar{x}_8 + \bar{x}_2 + x_5 + x_{10} \le 16$$

(47-6) $$5\bar{x}_8 + 4\bar{x}_4 + 3\bar{x}_7 + \bar{x}_9 + x_{12} \le 9$$

(and $x_1 = x_{11} = 0$, $x_3 = 1$).

Here for the first time we shall have to apply the modified consensus method to the new 3-resolvent:

$$\psi = x_2 x_6 \vee x_6 x_{10} \vee \bar{x}_2 x_5 \vee \bar{x}_4 \bar{x}_9 \vee \bar{x}_4 x_7 \vee \bar{x}_4 \bar{x}_6 x_8 \vee \bar{x}_5 x_6 \bar{x}_{10} \vee x_2 \bar{x}_5 \bar{x}_{10} \vee$$
$$\vee \bar{x}_5 \bar{x}_{10} x_{12} \vee \bar{x}_5 x_8 \bar{x}_{10} \vee x_2 \bar{x}_5 x_6 \vee \bar{x}_5 x_6 x_{12} \vee \bar{x}_6 x_9 \bar{x}_{12} \vee x_4 \bar{x}_6 x_9 \vee$$
$$\vee \bar{x}_6 \bar{x}_8 x_9 \vee x_4 \bar{x}_6 \bar{x}_{12} \vee \bar{x}_6 \bar{x}_8 \bar{x}_{12} \vee x_4 x_9 \bar{x}_{12} \vee \bar{x}_4 \bar{x}_7 \bar{x}_8 \vee \bar{x}_4 \bar{x}_8 \bar{x}_9 \vee$$
$$\vee \bar{x}_4 \bar{x}_8 x_{12} ,$$

showing that

$$\psi = \bar{x}_4 \vee x_6 \vee x_9 \vee \bar{x}_{12} \vee \bar{x}_2 x_5 \vee \bar{x}_2 \bar{x}_{10} \vee \bar{x}_5 \bar{x}_{10}$$

Hence, in every solution of our problem

$$x_4 = 1, \quad x_6 = 0, \quad x_9 = 0, \quad x_{12} = 1.$$

Introducing these values, we find the new system:

(48-1) $\quad 3x_2 + 3x_{10} + \bar{x}_5 + \bar{x}_7 \le 9 \quad \checkmark$

(48-2) $\quad 7\bar{x}_2 + 6x_5 \le 8$

(48-3) $\quad 2x_7 + x_8 \le 2$

(48-4) $\quad 12\bar{x}_5 + 8\bar{x}_{10} + 5x_2 + 3x_8 \le 17$

(48-5) $\quad 2\bar{x}_8 + \bar{x}_2 + x_5 + x_{10} \le 5 \quad \checkmark$

(48-6) $\quad 5\bar{x}_8 + 3\bar{x}_7 \le 7$

(and $x_1 = x_6 = x_9 = x_{11} = 0$, $x_3 = x_4 = x_{12} = 1$). The two obviously redundant constraints of this system have been marked by a ✓ and will be omitted from now on. From (48-3) and (48-6) we see that $x_7 x_8 = \bar{x}_7 \bar{x}_8 = 0$, and hence we can replace everywhere x_8 by $\bar{x}_7$. We get

(49-2) $\quad 7x_2 + 6x_5 \le 8$

(49-3) $\quad x_7 \le 1 \quad \checkmark$

(49-4) $\quad 12\bar{x}_5 + 8\bar{x}_{10} + 5x_2 + 3\bar{x}_7 \le 17$

(49-6) $\quad 2x_7 \le 4 \quad \checkmark$

$$(\text{and} \quad x_1 = x_6 = x_9 = x_{11} = 0, \quad x_3 = x_4 = x_{12} = 1, \quad x_8 = \bar{x}_7).$$

The new 3-resolvent is now

$$\psi' = x_2 x_5 \vee \bar{x}_5 \bar{x}_{10} \vee x_2 \bar{x}_5 \bar{x}_7$$

and the modified consensus method gives

$$\psi' = x_2 x_5 \vee x_2 \bar{x}_7 \vee x_2 \bar{x}_{10} \vee \bar{x}_5 \bar{x}_{10};$$

no direct information can be obtained from it.

The solution of the corresponding LP:

$$\text{maximize } -5x_2 - 4x_5 + 12x_7 - x_{10}$$

$$\text{s.t.} \quad 7x_2 + 6x_5 \le 8$$

$$5x_2 - 12x_5 - 3x_7 - 8x_{10} \le -6$$

$$0 \le x_j \le 1 \qquad (j = 2, 5, 7, 10)$$

is $x_2^* = x_5^* = 0$, $x_7^* = 1$, $x_{10}^* = 3/8$ and the objective function has the optimal value $11\frac{5}{8}$. Hence in the discrete optimum

$$-5x_2 - 4x_5 + 12x_7 - x_{10} \le 11,$$

which gives no further information. Introducing into the LP as a cut the violated constraint $\bar{x}_5 \bar{x}_{10} = 0$ (in its linear form, $x_5 + x_{10} \ge 1$), the new LP optimum is $x_2^{**} = x_5^{**} = 0$, $x_7^{**} = x_{10}^{**} = 1$. The LP optimum being integer, the solution is completed. The optimal solution of the original problem is hence

$$(0, 0, 1, 1, 0, 0, 1, 0, 0, 1, 0, 1)$$

IV. QUADRATIC UNCONSTRAINED MAXIMIZATION

Let us consider now a problem which has been given considerable attention in the last time; the problem is to maximize

$$(50) \qquad Q(X) = c_o + \sum_{k=1}^{n} c_k x_k + \sum_{\substack{i,j=1 \\ i\neq j}}^{n} d_{ij} x_i x_j ,$$

when $x_j \varepsilon B$ $(j = 1,\ldots,n)$. The order relations among literals play here an even more important role than in the linear case, because they allow a "linearization" of $Q(X)$. Indeed,

if $x_i x_j = 0$ then $d_{ij} x_i x_j = 0$

if $x_i \bar{x}_j = 0$ then $d_{ij} x_i x_j = d_{ij} x_i$

if $\bar{x}_i x_j = 0$ then $d_{ij} x_i x_j = d_{ij} x_j$

if $\bar{x}_i \bar{x}_j = 0$ then $d_{ij} x_i x_j = -d_{ij} + d_{ij} x_i + d_{ij} x_j$

Of course, after making use of the fact that

$$(51) \qquad x_{i_o}^{\alpha_{i_o}} \, x_{j_o}^{\alpha_{j_o}} = 0$$

the remaining problem can no longer be considered unrestricted. Let us put

$$m_{i_o j_o} = \begin{cases} M & \text{if } (2\alpha_{i_o} - 1)(2\alpha_{j_o} - 1) d_{i_o j_o} < 0 \\ d_{i_o j_o} & \text{otherwise} \end{cases}$$

where M is a "very large" positive number. It is easy to show that if in every maximizing point of $Q(X)$ the relation (51) holds, then the maximizing points of $Q(X)$ coincide with those of $Q'(X)$, where

$$(52) \qquad Q'(X) = Q(X) - m_{i_o j_o} x_{i_o} x_{j_o} - (1 - \alpha_{i_o})(1 - \alpha_{j_o}) d_{i_o j_o} +$$

$$+ (1 - \alpha_{j_o}) d_{i_o j_o} x_{i_o} + (1 - \alpha_{i_o}) d_{i_o j_o} x_{j_o} .$$

Hence, when $(2\alpha_{i_o} - 1)(2\alpha_{j_o} - 1)d_{i_o j_o} > 0$, (52) produces a quadratic function $Q'(X)$ having less quadratic terms than $Q(X)$.

We shall proceed as follows:

- if we discover variables having fixed values in the optimum, we fix them, and re-examine the new problem;

- if we discover two relations of type (51) regarding hhe same two variables x_{i_o} and x_{j_o}, then we can either fix one of them, or conclude that in the optimum they are equal, or complementary; in both cases we can eliminate a variable and re-examine the new problem;

- if we discover relations of type (51) with $(2\alpha_{i_o} - 1)(2\alpha_{j_o} - 1)d_{i_o j_o} > 0$, form $Q'(X)$ using (52) and re-examine the new problem;

- it might be worth making use of transformation (52) even for those relations (51) for which $(2\alpha_{i_o} - 1)(2\alpha_{j_o} - 1)d_{i_o j_o} < 0$.

If no further conclusion is available, we branch on one of the variables. In order to discover binary relations of the type (51) which have to hold in the optimum, three devices will be used.

1. Derivatives. Let us call

(53)
$$\Delta_j = c_j + \sum_{\substack{i=1 \\ i \neq j}}^{n} (d_{ij} + d_{ji})x_i \qquad (i = 1,\ldots,n)$$

the j-th derivative of $Q(X)$. It is easy to notice that for those values of $(x_1,\ldots,x_{j-1},x_{j+1},\ldots,x_n)$ for which Δ_j is positive, x_j must be equal to 1 in all maximizing points, while for those for which Δ_j is negative, x_j must be equal to 0. We shall use this remark in only three specific cases:

a) If Δ_j has a constant sign, we fix accordingly the value of x_j. E.g. If $\Delta_1 = -4 + x_2 + 2\bar{x}_3$, then Δ_1 is always negative, and hence x_1 will always be equal to 0 in the maximum.

b) If there is a variable x_i and a value ξ_i such that by putting $x_i = \xi_i$ the sign of Δ_j becomes constant, we derive from here a binary relation between x_i and x_j. E.g. If $\Delta_1 = -4 + 5x_2 + 4x_3$ then putting $x_2 = 1$, Δ_1 becomes positive, hence x_1 becomes 1 ; we deduce $\bar{x}_1 x_2 = 0$. Similarly, if $\Delta_1 = -4 + 3\bar{x}_2 + 2x_3 + x_4$, then $x_2 = 1$ implies $\Delta_1 < 0$, we deduce $x_1 x_2 = 0$.

c) If there are two variables x_i and x_k and values ξ_i and ξ_k asuch that $x_i = \zeta_i$ and $x_k = \xi_k$ fixes the sign of Δ_j, then we deduce a ternary relation between x_i, x_k and x_j. E.g. If again $\Delta_1 = -4 + 3\bar{x}_2 + 2x_3 + x_4$ then $\bar{x}_2 = x_3 = 1$ implies $\Delta_1 > 0$ and hence $x_1 = 1$; we deduce $\bar{x}_1 \bar{x}_2 x_3 = 0$.

Another similar device allows to discover more relationships between literals.

2. <u>Second order derivatives</u>. Let us call

$$\Delta_{ij} = \Delta_i - \Delta_j + (d_{ij} + d_{ji})(x_i - x_j) \quad (i \neq j; i,j = 1,\dots,n) \tag{54}$$

the <u>second order derivatives</u> of $Q(X)$. It is easy to notice that for any maximizing point

$$\Delta_{ij}(X) > 0 \Rightarrow x_i \geq x_j$$

$$\Delta_{ij}(X) < 0 \Rightarrow x_i \leq x_j \quad .$$

We shall use these implications only in two special cases:

a) If Δ_{ij} has a constant sign, we deduce a binary relation between x_i and x_j. E.g. If $\Delta_{12} = -3 + x_5 + \bar{x}_6$ then $\Delta_{12} < 0$ for any X, and hence in the maximum $x_1\bar{x}_2 = 0$.

b) If there is a variable x_k and a value ξ_k such that by putting $x_k = \xi_k$, the sign of Δ_{ij} becomes constant, then we deduce a relationship between $x_i x_j$ and x_k. E.g. if $\Delta_{12} = -3 + x_3 + 4\bar{x}_4$, then $\bar{x}_4 = 1 \Rightarrow \Delta_{12} > 0 \Rightarrow$ $\Rightarrow \bar{x}_1 x_2 = 0$ (i.e. $\bar{x}_1 x_2 \bar{x}_4 = 0$), and also $x_4 = 1 \Rightarrow \Delta_{12} < 0 \Rightarrow x_1\bar{x}_2 = 0$ (i.e. $x_1\bar{x}_2 x_4 = 0$).

3. Penalty-Relaxation Inequalities. P. Hansen's additive penalities [13], have been used in [7] for associating a linear inequality to $Q(X)$, such that every maximizing point of $Q(X)$ satifies it.

Let p be the sum of c_o and of all positive coefficients c_j and d_{ik} in (50); let us put

$$p_j^1 = \max(0, -c_j)$$

$$p_j^o = -\min(0, -c_j) - \tfrac{1}{2}\left(\sum_{k\neq j} \min(0, -d_{jk}) + \sum_{k\neq j} \min(0, -d_{kj})\right)$$

$$p_j = \min(p_j^o, p_j^1)$$

$$\pi_j = |p_j^o - p_j^1|$$

$$\alpha_j = \begin{cases} 1 \text{ if } p_j^1 \geq p_j^o \\ 0 \text{ otherwise} \end{cases}$$

and

$$L(X) = -p + \sum_{j=1}^{n} p_j + \sum_{j=1}^{n} \pi_j x_j^{\alpha_j}$$

$L(X)$ will be called the <u>penalty-relaxation form (PRF)</u> of $Q(X)$, and it can easily be shown that for any bivalent vector X,

(55) $$L(X) \leq -Q(X)$$

Hence, if X^* is an arbitrary bivalent vector, then every maximizing point of $Q(X)$ will satisfy the linear inequality

(56) $$L(X) \leq -Q(X^*)$$

(the "penalty-relaxation inequality" (PRI) of (50)). New linary relations can be obtained from (56) using the techniques of the previous chapters. A considerably stronger linear lower bound $\Lambda(X)$ $(L(X) \leq \Lambda(X) \leq Q(X))$ will be given in [5].

A reasonable way of finding a good starting point X^* is to put $X^1 = (x_1^1,\ldots,x_n^1)$, where $x_j^1 = \begin{cases} 1 \text{ if } c_j > 0 \\ 0 \text{ if } c_j \leq 0 \end{cases}$. Let $X^{k+1} = (x_1^{k+1},\ldots,x_n^{k+1})$ be defined by

$$x_j^{k+1} = \begin{cases} x_j^k & \text{if } (2x_j^k - 1)\,\Delta_j(X^k) \geq 0 \\ \overline{x_j^k} & \text{if } (2x_j^k - 1)\,\Delta_j(X^k) < 0 \end{cases}.$$

For the first k such that $X^k = X^{k+1}$ we put $X^* = X^k$. Obviously k is finite.

<u>Example 6</u>. Let

$K(X) = -3x_1 - x_4 + 3x_6 - 2x_7 - 2x_1x_4 + 2x_4x_7 + 3x_1x_7 + x_6x_7$ and let us assume that $d_{kj} = 0$ for $k > j$.

Here $p = 9$, and p_j^1, p_j^o, p_j, π_j and α_j are given below:

j	1	4	6	7
p_j^1	3	1	0	2
p_j^o	$\frac{3}{2}$	1	$\frac{7}{2}$	3
p_j	$\frac{3}{2}$	1	0	2
π_j	$\frac{3}{2}$	1	$\frac{7}{2}$	1
α_j	1	0	0	0

Hence,

$$L(X) = -9 + \frac{3}{2} + 1 + 2 + \frac{3}{2}x_1 + \frac{7}{2}\bar{x}_6 + \bar{x}_7 \quad .$$

If we chose as X^* the point with $x_1^* = x_4^* = x_7^* = 0$, $x_6^* = 1$, then the PRI of $K(X)$ is

$$\frac{3}{2}x_1 + \frac{7}{2}\bar{x}_6 + \bar{x}_7 \leq \frac{3}{2} \quad .$$

(implying $\bar{x}_6 = x_1\bar{x}_7 = 0$).

Using the three techniques outlined above, a set of elementary conjunctions $C_r (r = 1,\ldots,R)$ is derived. Considering now their union

$$\Phi = C_1 \vee \ldots \vee C_R ,$$

it is obvious that $\phi(X) = 0$ is a necessary condition for a 0-1 point X to be an optimum. $\phi(X)$ will be called - by analogy to the case in Chapter III - the <u>3-resolvent</u> of $Q(X)$. Applying the modified consensus method to Φ we find that

$$\Phi = \Phi_1 \vee \Phi_2 \vee \Phi_3$$

where $\Phi_i (i = 1,2,3)$ is a union of 3-prime implicants of length i. We conclude that

(i) those literals which appear in Φ_1 have fixed values - **and fix** them accordingly;

(ii) those pairs of literals which appear in two conjunctions of Φ_2 are equal or complementary (if Φ_2 contains $x\bar{y}$ and $\bar{x}y$ then $x = y$; if Φ_2 contains xy and $\bar{x}\,\bar{y}$ then $x = \bar{y}$) - and we substitute everywhere x by y or by $\bar{y}$;

(iii) those pairs of literals which appear only in one conjunction in Φ_2 are linked by a binary relation - and we may linearize the corresponding quadratic term .

After simplifying in this way our original problem, we derive new conclusions regarding $Q'(X)$, etc. If no conclusions can be obtained by the above methods, we branch.

It is to be assumed that good bounds can be obtained by using a continuous quadratic programming procedure (it has been remarked in [9] that the matrix (d_{ij}) can always be made positive or negative semidefinite, by using the fact that $x = x^2$ in B).

Example 7. Let us maximize

(57) $$Q = -7x_2 - 6x_3 - 8x_5 + x_6 - 2x_7 - 3x_8 - 2x_1x_2 - 3x_1x_3 + 5x_1x_5 - 2x_1x_6 - 2x_1x_8 + 4x_2x_3 + 2x_2x_4 - 3x_2x_6 - x_2x_7 - x_2x_8 - 6x_3x_4 + 4x_3x_5 - 3x_3x_7 + 3x_3x_8 - 2x_4x_5 - x_4x_6 + 2x_4x_7 + 4x_5x_6 + 3x_5x_7 + 5x_6x_8 + x_7x_8 ,$$

where

$$x_j \varepsilon B \qquad (j = 1,\ldots,8) .$$

We see that $\Delta_2 = -7 -2x_1 +4x_3 +2x_4 -3x_6 -x_7 -x_8$ is always negative, and hence in the optimum

(58) $$x_2 = 0$$

Substituting (58) into (57) we get

$$Q^I = -6x_3 - 8x_5 + x_6 - 2x_7 - 3x_8 - 3x_1x_3 + 5x_1x_5 - 2x_1x_6$$
$$-2x_1x_8 - 6x_3x_4 + 4x_3x_5 - 3x_3x_7 + 3x_3x_8 - 2x_4x_5 - x_4x_6$$
$$+2x_4x_7 + 4x_5x_6 + 3x_5x_7 + 5x_6x_8 + x_7x_8 \,.$$

Here,

$$\Delta_1 = -2x_2 - 3x_3 + 5x_5 - 2x_6 - 2x_8 = -7 + 3\bar{x}_3 + 5x_5 + 2\bar{x}_6 + 2\bar{x}_8$$
$$\Delta_3 = -18 + 3\bar{x}_1 + 4x_2 + 6\bar{x}_4 + 4x_5 + 3\bar{x}_7 + 3x_8$$
$$\Delta_4 = -9 + 6\bar{x}_3 + 2\bar{x}_5 + \bar{x}_6 + 2x_7$$
$$\Delta_5 = -10 + 5x_1 + 4x_3 + 2\bar{x}_4 + 4x_6 + 3x_7$$
$$\Delta_6 = -3 + 2\bar{x}_1 + \bar{x}_4 + 4x_5 + 5x_8$$
$$\Delta_7 = -5 + 3\bar{x}_3 + 2x_4 + 3x_5 + x_8$$
$$\Delta_8 = -5 + 2\bar{x}_1 + 3x_3 + 5x_6 + x_7$$

From Δ_1 we deduce that in the optimum $\phi_1 = 0$, where

$$\phi_1 = x_1x_3\bar{x}_5 \vee x_1\bar{x}_5x_6 \vee x_1\bar{x}_5x_8 \,.$$

Similarly we see that $\phi_j = 0$ $(j = 3,\dots,8)$, where

$$\phi_3 = x_1x_3 \vee x_3x_4 \vee x_3\bar{x}_5 \vee x_3x_7$$
$$\phi_4 = x_3x_4 \vee x_4x_5x_6 \vee x_4x_5\bar{x}_7 \vee x_4x_6\bar{x}_7$$
$$\phi_5 = \bar{x}_1\bar{x}_3x_5 \vee \bar{x}_1x_5\bar{x}_6$$
$$\phi_6 = x_5\bar{x}_6 \vee \bar{x}_6x_8$$
$$\phi_7 = \bar{x}_3x_5\bar{x}_7 \vee x_3\bar{x}_5x_7$$
$$\phi_8 = \bar{x}_1x_6\bar{x}_8 \vee x_3x_6\bar{x}_8 \vee x_6x_7\bar{x}_8 \vee x_1\bar{x}_6x_8 \vee \bar{x}_3\bar{x}_6x_8$$

We conclude that $\Phi = \phi_1 \vee \phi_3 \vee \phi_4 \vee \phi_5 \vee \phi_6 \vee \phi_7 \vee \phi_8 = 0$; the modified consensus method gives

$$\Phi = x_1x_3 \vee x_3x_4 \vee x_3\bar{x}_5 \vee x_3x_7 \vee x_5\bar{x}_6 \vee \bar{x}_6x_8 \vee x_3\bar{x}_6 \vee x_4x_5 \vee x_3\bar{x}_8 \vee x_5\bar{x}_8 \vee x_6\bar{x}_8 \vee x_1\bar{x}_5x_6 \vee x_1\bar{x}_5x_8$$
$$\vee x_4x_6\bar{x}_7 \vee \bar{x}_1\bar{x}_3x_5 \vee \bar{x}_3x_5\bar{x}_7 \vee x_1x_5\bar{x}_7 \vee \bar{x}_1x_5x_7 \vee x_4\bar{x}_7x_8 \vee x_1x_4x_6 \vee x_1x_4x_8 \vee x_1x_6\bar{x}_7 \vee x_1\bar{x}_7x_8$$

From $x_6\bar{x}_8 = \bar{x}_6 x_8 = 0$ we conclude

(59) $x_6 = x_8$

Substituting into Q^I we get

$$Q^{II} = -6x_3 - 8x_5 + 3x_6 - 2x_7 - 3x_1x_3 + 5x_1x_5 - 4x_1x_6 - 6x_3x_4 + 4x_3x_5 + 3x_3x_6 -3x_3x_7 - 2x_4x_5 - x_4x_6 + 2x_4x_7 + 4x_5x_6 + 3x_5x_7 + x_6x_7$$

From $\Delta_{16} = -3 - 6x_3 + x_4 + x_5 - x_7$ which is always negative, we obtain

$$x_1\bar{x}_6 = 0$$

From $\Delta_{37} = -4 - 3x_1 - 8x_4 + x_5 + 2x_6$, which is always negative, we obtain

$$x_3\bar{x}_7 = 0$$

From $\Delta_{46} = -3 + 4x_1 - 9x_3 - 6x_5 + x_7$, we obtain

$$\bar{x}_1 x_4 \bar{x}_6 = 0 \quad .$$

The other second order derivatives give no new information. Applying the modified consensus method to

$$\Phi' = \underline{\Phi} \vee x_1\bar{x}_6 \vee x_3\bar{x}_7 \vee \bar{x}_1 x_4 \bar{x}_6 \quad ,$$

where $\underline{\Phi}$ is the expression of Φ obtained after substituting into it $x_8 = x_6$, we get

$$\Phi' = x_3 \vee x_5\bar{x}_6 \vee x_4x_5 \vee x_1\bar{x}_5 \vee x_1\bar{x}_7 \vee x_1x_4 \vee x_1\bar{x}_6 \vee \bar{x}_1x_5 \vee x_5\bar{x}_7 \vee x_4\bar{x}_6 \vee x_4\bar{x}_7 \quad .$$

Hence in every optimal solution

(60) $x_3 = 0$

(61) $x_5 = x_1$

Substituting, we get

$$Q^{III} = -3x_1 + 3x_6 - 2x_7 - 2x_1x_4 - x_4x_6 + 2x_4x_7 + 3x_1x_7 + x_6x_7 \quad .$$

Further, $-x_4x_6 = -x_4 + x_4\bar{x}_6$, and applying (52) we see that Q^{III} becomes

$$Q^{IV} = -3x_1 - x_4 + 3x_6 - 2x_7 - 2x_1x_4 + 2x_4x_7 + 3x_1x_7 + x_6x_7 \quad .$$

Here $\Delta_6 = 3 + x_6$ shows that

(62) $x_6 = 1$

in every optimal solution. Also, $\Delta_{14} = -2 + x_7$ shows that

$$x_1\bar{x}_4 = 0$$

in every optimal solution.

Hence, $\Phi'' = \underline{\Phi}' \vee \bar{x}_6 \vee x_1 \bar{x}_4$, where $\underline{\Phi}'$ is the expression of Φ' obtained by susbtituting into it $x_3 = 0, x_6 = 1$, $x_5 = x_1$; i.e.

$$\Phi'' = x_1 \vee x_4\bar{x}_7 \quad .$$

Introducing

(63) $$x_1 = 0$$

into Q^{1V} we get

$$Q^V = 3 - x_4 - x_7 + 2x_4x_7 \quad .$$

Here $\Delta_4 = -1 + 2x_7$, hence

$$x_4\bar{x}_7 = \bar{x}_4x_7 = 0$$

It follows that

(64) $$x_4 = x_7 ,$$

and Q^V becomes

$$Q^{V1} = 3 \quad .$$

Hence, the maximum of Q is 3 and we see from (64), (63), (62), (61), (60), (59) and (58) that the two maximizing points of Q are (0, 0, 0, t, 0, 1, t, 1) where $t \varepsilon B$ is a free parameter.

V. QUADRATIC 0-1 PROGRAMMING

If our problem is to maximize a quadratic function in 0-1 variables subject to linear and quadratic constraints, we shall proceed as follows:

a) associate to each quadratic constraint

$$Q(X) \geq 0$$

a linear relaxation

(65) $$L(X) \leq 0 \ ,$$

where $L(X)$ is the penalty-relaxation form of $Q(X)$;

b) form the 3-resolvent Φ of the system of given linear constraints and penalty-relaxation constraints (65) of the given quadratic constraints;

c) apply the modified consensus method to Φ ; if it fixes variables or if it forces certaine variables to be equal or complementary, then introduce the corresponding conclusions into the original problem and re-examine the new simplified problem; if binary relations

(66) $$x_{i_o}^{\alpha_{i_o}} x_{j_o}^{\alpha_{j_o}} = 0$$

are deduced, linearize in every quadratic form appearing in the objective function and in the constraints, the corresponding terms, using formula

(67) $$Q'(X) = Q(X) - d_{i_o j_o} x_{i_o} x_{j_o} - (1 - \alpha_{i_o})(1 - \alpha_{j_o}) d_{i_o j_o} + (1 - \alpha_{j_o}) d_{i_o j_o} x_{i_o} + (1 - \alpha_{i_o}) d_{i_o j_o} x_{j_o} \ ,$$

adding

(68) $$x_{i_o}^{\bar{\alpha}_{i_o}} + x_{j_o}^{\bar{\alpha}_{j_o}} \geq 1$$

to the system of constraints, and re-examine the new simplified problem;

d) if no further simplifications can be deduced, branch;

e) obviously, after a first feasible solution was found, a new constraint is introduced, bounding the objective function by its value in the last feasible solution.

The techniques are extremely similar to those examined in the previous chapters and will be illustrated by an example.

Example 8. Let us determine the largest value not exceeding -3 of the quadratic function

$$Q(X) = 4x_1 - 5x_2 + 3x_3 + 4x_4 + 2x_5 + 6x_6 + 3x_1x_2 + 5x_1x_3 - 7x_1x_5 - 2x_1x_6$$
$$+ 6x_2x_3 - 6x_2x_4 + x_2x_5 - 4x_2x_6 + 2x_3x_4 - 3x_4x_5 - 5x_4x_6 - 5x_5x_6 \quad ,$$

whose 0-1 variables are subject to the first three linear constraints (37) - (39) of Example 2.

The 3-resolvent of (37) - (39) is

$$\psi = \phi_1 \vee \phi_2 \vee \phi_3 \quad ,$$

where ϕ_1, ϕ_2, ϕ_3 are given by (23), (24), (25) . Applying the modified consensus method we find

$$\psi = \psi_2 \vee \psi_3 \quad ,$$

where

$$\psi_2 = x_1x_3 \vee \bar{x}_2x_6 \vee x_3\bar{x}_4 \vee x_3x_6 \vee \bar{x}_4x_6$$
$$\psi_3 = x_1x_2\bar{x}_4 \vee x_1x_2\bar{x}_5 \vee x_1x_2\bar{x}_6 \vee x_1\bar{x}_5x_6 \quad .$$

From the fact that $\psi_2 = 0$, we see that $Q(X)$ becomes

$$Q'(X) = 4x_1 - 5x_2 + 3x_3 + 4x_4 + 2x_5 + 6x_6 + 3x_1x_2 - 7x_1x_5 - 2x_1x_6 + 6x_2x_3$$
$$-6x_2x_4 + x_2x_5 - 4x_6 + 2x_3 - 3x_4x_5 - 5x_6 - 5x_5x_6 \quad ,$$

or,

$$Q'(X) = 4x_1 - 5x_2 + 5x_3 + 4x_4 + 2x_5 - 3x_6 + 3x_1x_2 - 7x_1x_5 - 2x_1x_6 + 6x_2x_3$$
$$-6x_2x_4 + x_2x_5 - 3x_4x_5 - 5x_5x_6 \quad .$$

Here the PRF of $-Q'(X)$ is

$$L(X) = -21 + \tfrac{1}{2}\bar{x}_1 + 8\bar{x}_2 + 5x_3 + \tfrac{1}{2}\bar{x}_4 + \frac{11}{2}\bar{x}_5 + \frac{13}{2}\bar{x}_6 \quad ,$$

and the PRI of $-Q'(X) \geq 3$ is $L(X) \leq -3$, showing that

$$\bar{x}_2x_3\bar{x}_5 \vee \bar{x}_2x_3\bar{x}_6 \vee \bar{x}_2\bar{x}_5\bar{x}_6 = 0 \quad .$$

Adding these conjunctions to ψ , this becomes

$$\psi' = \psi_2' \vee \psi_3' \quad ,$$

where

$$\psi_2' = x_1x_3 \vee x_1\bar{x}_5 \vee \bar{x}_2x_3 \vee \bar{x}_2\bar{x}_5 \vee x_2x_6 \vee x_3\bar{x}_4 \vee x_3x_6 \vee \bar{x}_4x_6$$
$$\psi_3' = x_1x_2\bar{x}_4 \vee x_1x_2\bar{x}_6 \quad .$$

Taking into account $\psi_2' = 0$, Q' becomes

$$Q''(X) = -1 - 3x_1 - 4x_2 + 11x_3 + 4x_4 + 3x_5 - 3x_6 + 3x_1x_2 - 2x_1x_6 - 6x_2x_4$$
$$- 3x_4x_5 - 5x_5x_6 \quad ;$$

the new $L'(X)$ is

$$L'(X) = -21 + 4\bar{x}_1 + 11x_3 + \tfrac{1}{2}\bar{x}_4 + \bar{x}_5 + \frac{13}{2}\bar{x}_6 \quad ;$$

from $L'(X) \leq -3$ we conclude that

$$\bar{x}_1\bar{x}_2x_3 \vee \bar{x}_1x_3\bar{x}_6 \vee \bar{x}_2x_3\bar{x}_4 \vee \bar{x}_2x_3\bar{x}_5 \vee \bar{x}_2x_3\bar{x}_6 \vee x_3\bar{x}_5\bar{x}_6 = 0 \quad .$$

Adding these terms to the 3-resolvent, this becomes

$$\psi'' = \psi_1'' \vee \psi_2'' \vee \psi_3'' \quad ,$$

where

$$\psi_1'' = x_3$$
$$\psi_2'' = x_1\bar{x}_5 \vee \bar{x}_2\bar{x}_5 \vee \bar{x}_2x_6 \vee \bar{x}_4x_6$$
$$\psi_3'' = x_1x_2\bar{x}_4 \vee x_1x_2\bar{x}_6 \quad .$$

Hence, $x_3 = 0$. No further information seems to result from the linear constraints, or from $-Q''(X) \geq 3$. Hence, we have to branch. As $\Delta_2 = -4 + 3x_1 - 6x_4$ is always negative, we first consider the branch $x_2 = 0$. From $\psi_2'' = 0$ it will follow that $x_6 = 0$ and $x_5 = 1$. For these values Q'' becomes $2 - 3x_1 + x_4$ and it is obvious that it can not take values below -3. Hence, no solutions can be found on this branch and we have to turn to the branch $x_2 = 1$.

Now

$$\psi^{III} = \bar{x}_2 \vee x_3 \vee x_1\bar{x}_4 \vee x_1\bar{x}_5 \vee x_1\bar{x}_6 \vee \bar{x}_4 x_6 ,$$

and hence

$$Q^{III}(X) = -5 - 2x_1 - 2x_4 + 3x_5 - 3x_6 - 3x_4x_5 - 5x_5x_6 .$$

No information being obtainable from the constraints we shall have to branch. Because $\Delta_5 > 0$, we shall first consider the branch $x_5 = 1$. The inequality $Q \leq -3$ becomes $8\bar{x}_6 + 5\bar{x}_4 + 2\bar{x}_1 \leq 14$ and implies $\bar{x}_1\bar{x}_4\bar{x}_6 = 0$; the new 3-resolvent is

$$\psi^{IV} = \bar{x}_2 \vee x_3 \vee \bar{x}_4 \vee \bar{x}_5 \vee x_1\bar{x}_6$$

Hence $x_4 = 1$ and Q^{III} becomes $-7 - 2x_1 - 8x_6$. The only constraint being now $x_1\bar{x}_6 = 0$, we can easily see that the optimum along this branch is reached in the point (0, 1, 0, 1, 1, 0) and Q has the value -7 in this point. Turning now to the branch $x_5 = 0$, we find that the optimum is on this branch, its value is -5 and it is reached in (0, 1, 0, 0, 0, 0). Hence, under the given constraints, $Q_{max} = Q(0, 1, 0, 0, 0, 0) = -5$.

VI. UNCONSTRAINED POLYNOMIAL OPTIMIZATION.

It is known (see [10]) that every pseudo-Boolean function $f(x_1,\dots,x_n)$ has a polynomial expression

$$\sum_{i=1}^{n} c_i \prod_{j\in K_i} x_j \, , \qquad (K_i \subseteq \{1,\dots,n\}) \quad . \tag{69}$$

A method for maximizing such an unconstrained pseudo-Boolean function based on successive elimination of variables is given in Chapter VI.§1 of [10] ; methods based on branching according to the conjunctions $\prod_{j\in K_i} x_j$ (rather than according to individual variables) were developed in [1], [4], [8], [12] and [14] . The method of [8] has been successfully run on computers and the results are given in Chapter VII.

The method we are going to describe briefly here is based on a combination of the branching of [8], of the idea of 3-resolvent and of PRF. It will contain the following steps.

1. If in any stage of the solution process we arrive to a "saturated" function, i.e. to one where all c_i's (with the possible exception of the coefficient c_o corresponding to $K_o = \emptyset$) have the same sign, then its optimum can be obtained directly by

(i) putting all $x_j = 1$ (if all $c_i \geq 0 (i \neq 0)$)

(ii) putting all $\prod_{j\in K_i} x_j = 0$ (if all $c_i \leq 0 (i \neq 0)$) .

2. The k-th derivative of f will be defined as

$$\Delta_k = \sum_{i=1}^{m}{}' c_i \prod_{j\in H_i} x_j \, ,$$

where $\sum'$ means that the summation involves only those indices i for which $j \varepsilon K_i$, and where $H_i = K_i \quad \{k\}$ (with the usual convention that product over the empty set is considered equal to 1). As in the quadratic case, $\Delta_k > 0$ implies $x_k = 1$ in the optimum, and $\Delta_k < 0$ implies $x_k = 0$. These implications will be taken into account only if by fixing at most two of the variables the sign of Δ_i becomes constant.

3. The knowledge of a "good" point $X^* = (x_1^*, \ldots, x_n^*)$ can be used as in [1] to derive supplementary information. Indeed, let $C = \sum_{i=1}^{m} \max(0, c_i)$. Berman has remarked that in the optimum

(i) $c_i > 0$ and $c_i > C - f(X^*)$ implies $x_j = 1$ for all $j \varepsilon K_i$;

(ii) $c_i < 0$ and $-c_i > C - f(X^*)$ implies $\prod_{j \varepsilon K_i} x_j = 0$ for all $j \varepsilon K_i$.

In order to get a "good" point, we might start by defining $X^1 = (x_1^1, \ldots, x_n^1)$ such that

$$x_j^1 = \begin{cases} 1 \text{ if } c_{i(j)} \geq 0 \\ 0 \text{ otherwise,} \end{cases}$$

where $i(j)$ is that index i for which $K_i = \{j\}$. Examining now the implications $\Delta_k(X^1) > 0 \Rightarrow x_k^1 = 1$, and $\Delta_k(X^1) < 0 \Rightarrow x_k^1 = 0$, we select an index j_1 for which the implication does not hold (if any) and define X^2 by putting

$$x_j^2 = \begin{cases} x_j^1 & \text{for } j \neq j_1 \\ \overline{x_{j_1}^1} & \text{for } j = j_1 . \end{cases}$$

Obviously we arrive in a finite number of steps to a point X^t where all the implications hold, and put $X^* = X^t$ (X^* will be a "local optimum").

4. Similarly to the quadratic case we can define a PRI in the following way. Let p be the sum of c_o and all the positive coefficients $c_j (j \neq 0)$ in (69); let us put

$$p_j^1 = \max(0, -c_{i(j)})$$

$$p_j = -\min(0, -c_{i(j)}) - \sum_{\substack{i=1\\ i\neq i(j)}}^{m} \frac{1}{|K_i|} \min(0,-c_i)$$

$$p_j = \min(p_j^o, p_j^1)$$

$$\pi_j = |p_j^o - p_j^1|$$

$$\alpha_j = \begin{cases} 1 \text{ if } p_j^1 \geq p_j^o \\ 0 \text{ otherwise} \end{cases}$$

and

$$L(X) = -p + \sum_{j=1}^{n} p_j + \sum_{j=1}^{n} \pi_j x_j^{\alpha_j}$$

$L(X)$ will be again called the PRF of f, and the inequality

$$L(X) \leq -f(X^*) \tag{70}$$

where (X^*) is the point defined at step 3 will be called the PRI of f.

We form the 3-resolvent using the implications given in step 2, the conclusions derived in step 3 and those derived from the PRI (70); we apply the modified consensus method to this 3-resolvent. If they allow us to fix the value of some variables, we introduce this value into (69) and re-examine the new function. If a conclusion of the form $x_i = x_j$ or $x_i = 1 - x_j$ can be obtained from the 3-resolvent, we eliminate one of the variables and re-examine the new problem. Similarly to the quadratic case we can apply simplifications of the form "if $x_h \bar{x}_j$, is a 3-prime implicant of the 3-resolvent, if $h, j \in K_i$ and $c_i < 0$ then replace the term $c_i \prod_{j \in K_i} x_j$

by $c_i \prod_{j \in K_i'} x_j$, where $K_i' = K_i \setminus \{j\}$". Obviously similar simplifications might be obtained from every 3-prime implicant. Finally, if no simplifications can be detected, then we select the index i_o for which $|c_{i_o}| = \max_i |c_i|$ and branch. The first branch to be examined will be $\prod_{j \in K_{i_o}} x_j = 1$ if $c_{i_o} > 0$, and $\prod_{j \in K_{i_o}} x_j = 0$ if $c_{i_o} < 0$.

Example 9. Maximize the pseudo-Boolean function

$$
\begin{aligned}
(71) \qquad f = {} & 6x_1 - 3x_2 - x_3 + 8x_4 - 2x_5 - 2x_6 + 3x_7 - 4x_8 + 4x_1x_2 + 2x_1x_3 \\
& -8x_1x_4 + 3x_2x_3 + x_2x_5 + 3x_2x_6 - x_3x_4 + x_3x_5 + x_4x_6 - 5x_5x_7 \\
& -x_5x_8 + 2x_1x_2x_5 + 3x_1x_2x_6 - 4x_1x_4x_5 - 9x_2x_3x_4 + 3x_2x_5x_6 \\
& +x_2x_7x_8 - 2x_4x_5x_6 - 2x_6x_7x_8 + 4x_5x_6x_7x_8 .
\end{aligned}
$$

Here,

$$\Delta_1 = 6 + 4x_2 + 2x_3 - 8x_4 + 2x_2x_5 + 3x_2x_6 - 4x_4x_5 ,$$

$$\Delta_2 = -3 + 4x_1 + 3x_3 + x_5 + 3x_6 + 2x_1x_5 + 3x_1x_6 - 9x_3x_4 + 3x_5x_6 + x_7x_8 ,$$

$$\Delta_3 = -1 + 2x_1 + 3x_2 - x_4 + x_5 - 9x_2x_4 ,$$

$$\Delta_4 = 8 - 8x_1 - x_3 + x_6 - 4x_1x_5 - 9x_2x_3 - 2x_5x_6 ,$$

$$\Delta_5 = -2 + x_2 + x_3 - 5x_7 - x_8 + 2x_1x_2 - 4x_1x_4 + 3x_2x_6 - 2x_4x_6 + 4x_6x_7x_8 ,$$

$$\Delta_6 = -2 + 3x_2 + x_4 + 3x_1x_2 + 3x_2x_5 - 2x_4x_5 - 2x_7x_8 + 4x_5x_7x_8 ,$$

$$\Delta_7 = 3 - 5x_5 + x_2x_8 - 2x_6x_8 + 4x_5x_6x_8 ,$$

$$\Delta_8 = -4 - x_5 + x_2x_7 - 2x_6x_7 + 4x_5x_6x_7 .$$

From Δ_1 we find $\bar{x}_1\bar{x}_4 = \bar{x}_1 x_2 x_6 = 0$, from Δ_3 (and also from Δ_4) we find $x_2x_3x_4 = 0$, from Δ_7 we find $\bar{x}_5\bar{x}_7 = 0$, and from Δ_8 we find $x_6x_7x_8 = 0$.

Let us form the PRI. The "good print" is $X^* = (1,1,0,1,0,1,1,0)$, $f(X^*) = 15$, and the determination of X^* is illustrated in the following table.

j	X^1	Δ^1	X^2	Δ^2	X^*	Δ^*
1	1	-2	1	2	1	5
2	[0]	1	1	1	1	7
3	0	0	0	-6	0	-6
4	1	0	1	0	1	1
5	0	-11	0	-8	0	-7
6	0	-1	[0]	5	1	5
7	1	3	1	3	1	3
8	0	-4	0	-3	0	-5

The penalties are given by the following table.

	1	2	3	4	5	6	7	8
p_j^1	0	3	1	0	2	2	0	4
p_j^o	$\frac{32}{3}$	$\frac{17}{2}$	3	$\frac{17}{2}$	$\frac{11}{3}$	5	$\frac{13}{3}$	$\frac{4}{3}$
p_j	0	3	1	0	2	2	0	$\frac{4}{3}$
π_j	$\frac{32}{2}$	$\frac{11}{2}$	2	$\frac{17}{2}$	$\frac{5}{3}$	3	$\frac{13}{3}$	$\frac{8}{3}$
α_j	0	0	0	0	0	0	0	1

Hence the PRI (70) will be

$$(72) \quad 64\bar{x}_1 + 33\bar{x}_2 + 12\bar{x}_3 + 51\bar{x}_4 + 10\bar{x}_5 + 18\bar{x}_6 + 26\bar{x}_7 + 16x_8 \leq 88 \quad .$$

From (72) we see that $\bar{x}_1\bar{x}_2 = \bar{x}_1\bar{x}_4 = \bar{x}_1\bar{x}_7 = \bar{x}_1\bar{x}_3\bar{x}_6 = \bar{x}_1\bar{x}_3x_8 = \bar{x}_1\bar{x}_5\bar{x}_6 = \bar{x}_1\bar{x}_5x_8$ $= \bar{x}_1\bar{x}_6x_8 = \bar{x}_2\bar{x}_3\bar{x}_4 = \bar{x}_2\bar{x}_4\bar{x}_5 = \bar{x}_2\bar{x}_4\bar{x}_6 = \bar{x}_2\bar{x}_4\bar{x}_7 = \bar{x}_2\bar{x}_4x_8 = \bar{x}_3\bar{x}_4\bar{x}_7 = \bar{x}_4\bar{x}_6\bar{x}_7$ $= \bar{x}_4\bar{x}_7x_8 = 0$. Forming now the 3-resolvent obtained from the conclusions derived from the Δ_j's and from the PRI, and applying the modified consensus method to it, we find:

$$\phi_1 = \bar{x}_1 \vee \bar{x}_5\bar{x}_7 \vee x_2x_3x_4 \vee \bar{x}_2\bar{x}_3\bar{x}_4 \vee \bar{x}_2\bar{x}_4\bar{x}_5 \vee \bar{x}_2\bar{x}_4\bar{x}_6 \vee \bar{x}_2\bar{x}_4\bar{x}_7 \vee \bar{x}_2\bar{x}_4x_8 \vee \bar{x}_3\bar{x}_4\bar{x}_7 \vee \bar{x}_4\bar{x}_6\bar{x}_7$$
$$\vee \bar{x}_4x_6x_8 \vee \bar{x}_4\bar{x}_7x_8 \vee \bar{x}_5x_6x_8 \vee x_6x_7x_8 \quad .$$

The following conclusions can be derived from here regarding every optimal solution:

1) $x_1 = 1$

2) The term $4x_5x_6x_7x_8$ can be dropped (for,its coefficient is positive,

and in every optimal solution $x_6x_7x_8 = 0$). Hence, f becomes

$$f^I = 6 + x_2 + x_3 - 2x_5 - 2x_6 + 3x_7 - 4x_8 + 3x_2x_3 + 3x_2x_5 + 6x_2x_6 - x_3x_4 + x_3x_5 - 4x_4x_5 + x_4x_6 - 5x_5x_7 - x_5x_8 - 9x_2x_3x_4 + 3x_2x_5x_6 + x_2x_7x_8 - 2x_4x_5x_6 - 2x_6x_7x_8 \ .$$

From $\Delta_8^I = -4 - x_5 + x_2x_7 - 2x_6x_7 < 0$ we get $x_8 = 0$. From $\Delta_7^I = 3 - 5x_5 + x_2x_8 - 2x_6x_8$ we see that $x_5x_7 = \bar{x}_5\bar{x}_7 = 0$, i.e. $x_7 = \bar{x}_5$. Replacing everywhere in f^I x_7 by $1-x_5$, and putting $x_8 = 0$, we get

$$f^{II} = 9 + x_2 + x_3 - 5x_5 - 2x_6 - 4x_8 + 3x_2x_3 + 3x_2x_5 + 6x_2x_6 - x_3x_4 + x_3x_5 - 4x_4x_5 + x_4x_6 - 9x_2x_3x_4 + 3x_2x_5x_6 - 2x_4x_5x_6 \ .$$

From $\Delta_6^{II} = -2 + 6x_2 + x_4 + 3x_2x_5 - 2x_4x_5$, we see that $x_2\bar{x}_6 = \bar{x}_2x_6 = 0$, i.e. $x_2 = x_6$. Substituting everywhere in f^{II} x_6 by x_2 we get:

$$f^{III} = 9 + 5x_2 + x_3 - 5x_5 + 3x_2x_3 + x_2x_4 + 6x_2x_5 - x_3x_4 + x_3x_5 - 4x_4x_5 - 9x_2x_3x_4 - 2x_2x_4x_5 \ .$$

From Δ_2^{III} we find $\bar{x}_2\bar{x}_4 = 0$. From Δ_5^{III} we find $x_4x_5 = 0$. From Δ_5^{III} we find $x_2\bar{x}_4\bar{x}_5 = 0$. It follows that $x_4x_5 = \bar{x}_4\bar{x}_5 = 0$ i.e. $x_5 = \bar{x}_4$. Substituting we find

$$f^{IV} = 4 + 11x_2 + 2x_3 + 5x_4 + 3x_2x_3 - 5x_2x_4 - 2x_2x_4 - 2x_3x_4 - 9x_2x_3x_4 \ .$$

Applying step 3 of the algorithm, we see that here $C = 25$, $f^{IV}(X^*) = 15$, and we conclude that $x_2 = 1$. Now f^{IV} becomes

$$f^V = 15 + 5x_3 - 11x_3x_4 \ .$$

From $\Delta_3^V = 5 - 11x_4$ we see that $x_3x_4 = \bar{x}_3\bar{x}_4 = 0$, hence $x_4 = \bar{x}_3$.

Substituting we get

$$f^{VI} = 15 + 5x_3 \quad .$$

f^{VI} is saturated, hence $x_3 = 1$. We find

$$f_{max} = 20$$

and the optimal point determined from $x_3 = 1$, $x_4 = \bar{x}_3 = 0$, $x_2 = 1$, $x_5 = \bar{x}_4 = 1$, $x_6 = x_2 = 1$, $x_8 = 0$, $x_7 = \bar{x}_5 = 0$, $x_1 = 1$ is (1,1,1,0,1,1,0,0).

VII. COMPUTATIONAL EXPERIENCE

Computational experience is available at present only for a variant of the algorithm described in Chapter III for linear 0-1 programs, and for a rudimentary variant of the algorithm of Chapter VI for unconstrained polynomial optimization.

1. A variant of the algorithm of Chapter III for linear 0-1 programs was coded in FORTRAN and test problems were run on the CDC-6600 of the University of Montreal.

The major characteristics of the coded variant are the following:

a) instead of using the 3-resolvent it only works with the 2-resolvent;

b) equality or complementarity of variables although discovered by the process are not exploited;

c) the number of order relations introduced as cuts in the LP is never more than 5.

In spite of the obvious limitations of the coded variant, it performed much above expectations; test problems were run involving up to a few hundred 0-1 variables, and in most of the cases the computational times were better than in other methods, while the number of iterations was substantially reduced. Some results of these experiements are reported in the following Tableau:

Number of variables	Duration (secs)	Number of iterations
30	.35	20
40	.90	45
50	7.86	343
60	5.59	198
70	10.70	472
80	14.14	486
90	10.44	370
100	20.19	653
110	5.00	212
120	32.71	1088
130	17.96	513
140	85.41	2261
150	21.82	542
160	15.87	406
170	65.21	1578
180	50.21	968
190	23.65	632

The reported results refer to average times and average numbers of iterations for groups of 5 randomly generated test problems involving always 15 constraints. For more details see [11] .

2. A very rudimentary variant of the algorithm of Chapter VI for unconstrained polynomial optimization was coded in FORTRAN and tested on the IBM 360/50 of the Technion. In the coded variant, the 3-resolvent is not used at all. In spite of this, the results, sumarized in the following Tableau are very encouraging. For more details see [8].

Problem No.	No. of variables	No. terms in the polynomial	No. iterations	Execution time (sec). including input/output
1	10	5	6	0.48
2	10	10	8	2.02
3	10	11	6	0.90
4	10	20	16	2.31
5	10	29	33	7.35
6	10	48	34	11.40
7	20	10	12	2.02
8	20	19	17	4.99
9	20	29	28	9.36
10	20	34	97	37.96
11	20	40	118	65.44
12	20	50	148	110.95
13	30	30	50	24.06
14	30	38	120	106.17
15	30	50	212	239.03
16	30	70	>77	>116.22
17	40	54	>84	>117.57
18	40	74	>154	>238.38
19	40	143	>6	>72.45
20	50	80	>9	>57.44
21	50	98	>7	>51.62

REFERENCES

1. BERMAN, G.: A Branch and Bound Method for the Maximization of a Pseudo-Boolean Function. *TECH. REPORT. FACULTY OF MATHEMATICS, UNIVERSITY OF WATERLOO*, Waterloo, Ontario, Canada.

2. GRANOT, F. and P. L. HAMMER: On the Use of Boolean Functions in 0-1 Programming. *METHODS OF OPERATIONS RESEARCH*, 12 (1972), pp. 154-184.

3. GRANOT, F. and P. L. HAMMER: On the Role of Generalized Covering Problems. *PUBL. CENTRE DE RECHERCHES MATHEMATIQUES, UNIVERSITE DE MONTREAL* No. 220, September 1972.

4. HAMMER, P. L.: BBB Methods. I. Maximization of a Pseudo-Boolean Function. *PUBL. CENTRE DE RECHERCHE MATHEMATIQUES, UNIVERSITE DE MONTREAL*, No. 220, September 1972.

5. HAMMER, P. L.: Linear Bounds on Quadratic Pseduo-Boolean Functions. *TECHNICAL REPORT. DEPT. OF COMBINATORICS AND OPTIMIZATION. UNIVERSITY OF WATERLOO* (forthcoming).

6. HAMMER, P. L.: Monotonous Bivalent Programs. *TECHNICAL REPORT. DEPT. OF COMBINATORICS AND OPTIMIZATION. UNIVERSITY OF WATERLOO* (forthcoming).

7. HAMMER, P. L. and P. HANSEN: Quadratic 0-1 Programming. *CORE DISCUSSION PAPER* 7219, April 1972.

8. HAMMER, P. L. and U. N. PELED: On the Maximization of a Pseudo-Boolean Function. *JOURNAL OF THE ASSOCIATION FOR COMPUTING MACHINERY* 19, 1972, 165-282.

9. HAMMER, P. L. and A. RUBIN: Some Remarks on Quadratic Programming with 0-1 Variables. *REVUE FRANCAISE D'INFORMATIQUE ET DE RECHERCHE OPERATIONELLE*, 4, 1970, V-3 pp.67-79.

10. HAMMER, P. L. and S. RUDEANU: *BOOLEAN METHODS IN OPERATIONS RESEARCH AND RELATED AREAS*. Springer-Verlag, Berlin-Heidelberg-New York, 1968.

11. HAMMER, P. L. and SANG NGUYEN: APOSS - A Partial Order in the Solution Space of Bivalent Programs. *PUBL. CENTRE DE RECHERCHES MATHEMATIQUES, UNIVERSITE DE MONTREAL*. No. 163, April 1972.

12. HANSEN, P.: Un Algorithme SEP pour les programmes pseudo-Booléens non linéaires. *CAHIERS DU CENTRE D'ETUDES DE RECH. OPERATIONNELLE*. 11, 1969, 26-44.

13. HANSEN, P.: Pénalités additives pour les programmes en variables zéro-un. *COMPTES RENDUS DE L'ACADEMIE DES SCIENCES DE PARIS*. 273 1971, 175-177.

14. TAHA, H. A.: A Balasian - Based Algorithm for Zero-One Polynomial Programming. *MANAGEMENT SCIENCE*. 18, 1972, pp. 328-343.

Mathematical Programming in Theory and Practice,
P.L. Hammer and G. Zoutendijk, (Eds.)

ON POLAROID INTERSECTIONS

by

Claude-Alain Burdet

Management Sciences Research Group
Graduate School of Industrial Administration
Carnegie-Mellon University
Pittsburgh, Pennsylvania 15213

ABSTRACT

Polaroid sets and functions have been introduced as a new tool for non-convex problems in non-linear programming, particularly for quasi-concave and integer optimization problems over a linearly constrained set of feasible solutions.

The name polar programming applies to a general class of non-linear mathematical programming problems which can be solved by the polaroid approach. In integer programming polaroids yield non-trivial extensions of the intersection cut approach.

This paper builds on the properties of polaroid sets (particularly complete convex polaroids) and focuses on the following intersection problem:

Given a point $\bar{x}$ belonging to the polaroid set P^*, find the intersection point u^* of a one-dimensional ray u with the boundary of P^* : $u^* \in (\text{bd } P^* \cap u)$

We also present a theoretical comparison of the relative merits of several currently proposed cutting planes for integer and/or concave programming.

This report was prepared as part of the activities of the Management Sciences Research Group, Carnegie-Mellon University, under Contract N00014-67-A-0314-0007 NR 047-048 with the U. S. Office of Naval Research.

0) Introduction

0.1: Polaroid sets and functions have been introduced in [9], where we mentioned areas of application in non-linear programming, particularly in quasi-concave and integer optimization problems over a linearly constrained set of feasible solutions.

Polaroids can be used to improve the efficiency of the algorithms outlined by Hoang Tuy in [16] for the maximization (minimization) of a quasi-convex (concave) objective function over a polyhedral set. They also improve in a similar way the modified version of this approach, presented in [12] by Glover and Klingman. Other algorithms can also be extended by using polaroids and, in [10], the name polar programming was given to a general class of non-linear mathematical programming problems which can be solved by the polaroid approach in a variety of ways.

In integer programming polaroids yield non-trivial extensions of the intersection cut approach [1, 6] as indicated in [9]; in [7], a special type of polaroid is presented and we show how it brings under one roof the enumerative approach of [6] and the extension method of Balas [3]; it also illustrates a connection between concave and integer programming, of a nature different from that of Raghavachari [14].

0.2: Consider the following general principle for solving the (arbitrary) mathematical programming problem:

maximixe $g(x)$, subject to $x \epsilon P$

where x is an n vector, P a subset of $\mathbb{R}^n$ and

g a real valued function of $x \epsilon \mathbb{R}^n$.

Method: Construct a (finite) collection of n-dimensional simplices (see Fig. 1) $S_j \subset \mathbb{R}^n$, $j \epsilon J$ such that

a) $P \subset S = \bigcup_{j \epsilon J} S_j$

b) suppose that $\forall j \epsilon j$ one can establish that

$$\max_{x \epsilon (S_j \cap P)} g(x) \leq \Delta_j$$

where Δ_j is a known lower bound:

i.e. $\Delta_j \leq g(x^j)$, for some $x^j \epsilon P$

Then, one easily shows that the optimal solution $\tilde{x}$ of the original problem is delivered by (finite) search: $\tilde{x} \epsilon \{x^j, j \epsilon J\}$ such that $g(\tilde{x}) \geq g(x^j)$, $\forall j \epsilon J$.

Particular examples of this general method are described in [10]; they can be classified as follows:

- cutting plane algorithms (see, for instance, [16], [1], [6], [7])
- enveloping algorithms (see [10])
- partitioning algorithms (see [10], [16], [4])

0.3: One of the basic ingredients in the above general method lies in the construction of the simplex S_j, $\forall j \epsilon J$, that is, in the determination of its (n+1) vertices.

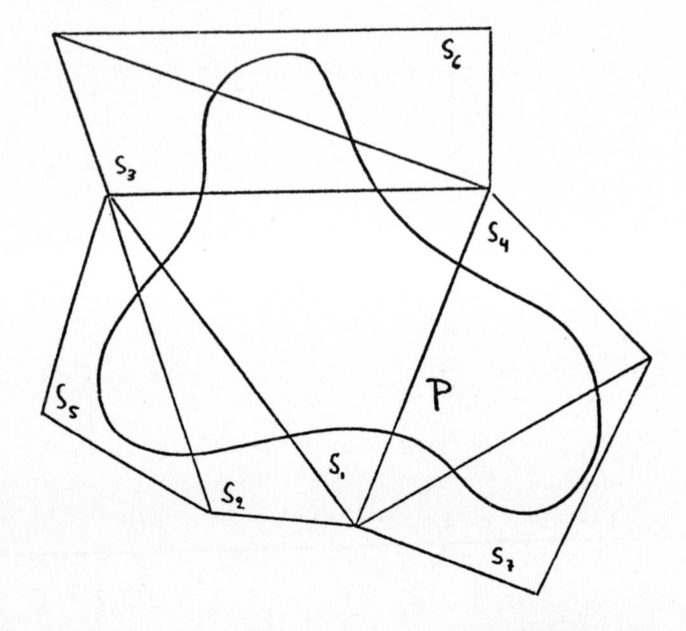

Figure 1: A simplicial covering of the feasible set P, illustrating the general method based on polaroids P*(k). (See section 1.2)

This paper builds on the properties of polaroid sets (particularly complete convex polaroids, as defined in [9]) to derive a construction of the simplices S_j with the desired properties. Thus we focus our attention here on the determination of the vertices of S_j ; these vertices can be obtained by solving an intersection problem of the following type:

> Given a point $\bar{x}$ belonging to the polaroid set P^* (for a definition see section 1 below), find the intersection point u^* of a one-dimensional ray u, with the boundary of P^* : $u^* \epsilon$ (bd $P^* \cap u$).

First we outline the basic methodology which leads to polaroid intersection algorithms; a second part contains some typical examples of polaroid functions and sets.

The interested reader is referred to [3,4,7,12,] for application areas and to [9], [10] and [21] for further aspects of the theory of polaroids.

1) Definitions

1.1: In order to make this report self contained, we briefly reproduce here some definitions from [9].

Let the polaroid function $f = f(x,y)$ be real valued with two arguments x and y, both n-vectors; let P denote a closed set in $\mathbb{R}^n$.

Definition 1: The polaroid set $P^*(k)$ defined by the polaroid function f with respect to the set P and the parameter $k \epsilon f(P, R^n)$ is

$$P^*(k) = \{y \mid f(x,y) \leq k \quad , \quad \forall x \epsilon P\}$$

One has the

<u>Theorem 1</u>: The <u>polaroid set</u> $P^*(k)$ is <u>convex</u> $\forall k$

if the <u>polaroid function</u> $f = f(x,y)$ is

<u>quasi-convex</u> in y, for all $x \in P$.

<u>Proof</u>: When $f(x,y)$ is quasi-convex in y, the polaroid $P^*(k)$ can be viewed as the intersection of convex sets R:

$$P^*(k) = \bigcap_{x \in P} R(x)$$

with, for $x \in P$, $R(x) = \text{lev}_k\ f(x,y)$.

(For an algebraic proof see [9]) QED

Although non-convex polaroid sets P^* may also be of occasional interest (see [9]), we are mainly concerned here with convex polaroids. An interesting special case of this type occurs when the function $f(x,y)$ is <u>linear in y</u>; the corresponding convex polaroid (theorem 1) is called <u>simple</u>.

<u>Definition 2</u>: Consider the point $\bar{x} \in P^*(k)$ and a vector $a \in \mathbb{R}^n$; we define the ray u, as the following half-line originating at $\bar{x}$ in the direction $(-a)$:

$$u = \{u(\lambda) \mid u = \bar{x} - \lambda a \quad , \quad \lambda \geq 0\}$$

Denote by F_a the real valued function

$$F_a(\lambda) = \max_{x \in P} \{f(x, \bar{x} - \lambda a)\}$$

<u>Proposition 1</u>: Let λ^* satisfy

1) $\lambda^* \geq 0$

2) $F_a(\lambda^*) = k$

then $u^* = u(\lambda^*)$ lies on the "boundary" of $P^*(k)$, i.e.

$$u^* \in \text{"bd"} P^*(k) = \{y \in P^*(k) \mid f(x,y) = k \text{ , for some } x \in P\}$$

The term "boundary" has been adopted here because this definition corresponds to the intuitive concept for the important special case where $P^*(k)$ is convex; for an arbitrary function f however, P^* need not be convex and "bd" P^* no longer need correspond to the topological concept.

Proof: From the hypothesis 2), $\exists\, x^*$ such that

$$f(x^*, \bar{x} - \lambda^* a) = f(x^*, u^*) = k \; . \qquad \text{QED}$$

1.2: We now briefly indicate below (a more detailed study can be found in [9]) how polaroids find their justification and application in non-linear optimization problems: the following results make it possible to solve (in principle) a very broad class of problems using polaroids and the general method outlined in section 0.2 .

Definition 3: The polaroid $P^*(k)$ is said complete if $P \subset P^*(k)$.

Theorem 2: Let $\Delta = g(\tilde{x}) = \max_{x \in P} g(x)$, with $g(x) = f(x,x)$. If $P^*(\Delta)$ is complete then every optimal solution $\tilde{x}$ of the problem

$$\text{maximize } g(x) \quad , \quad \text{subject to } x \in P$$

lies on the "boundary" of $P^*(\Delta)$.

Proof: Immediate since $f(\tilde{x},\tilde{x}) = \Delta$ and thus, $\tilde{x} \in P \cap P^*(\Delta)$. QED

Proposition 2: Assume that $P^*(\Delta)$ is a complete convex polaroid; then any collection of simplices S_j, $j \in J$ such that $S_j \subset P^*(\Delta)\ \forall j \in J$ and $P \subset S = \bigcup_{j \in J} S_j$ yields a sufficient optimality condition.

Proof: A condition for convexity of $P^*(\Delta)$ is given in Theorem 1; for completeness, see Corollary 1.1, for instance. Optimality of $\tilde{x}$

now follows from the observation that $f(x,y) \leq \Delta = g(\tilde{x})$, $\forall x \in P$, $\forall y \in S_j$, $\forall j \in J$ since, by assumption, $P \subset S \subset P^*(\Delta)$ QED

The general method described in the introduction now becomes operational when the simplices S_j are constructed as subsets of the polaroids $P^*(k)$, because, for $k \leq \Delta$, one has $S_j \subset P^*(k) \subset P^*(\Delta)$ (a proof is given in [9]); in practice, the value of k represents the current value of the maximum and it is gradually increased (step-wise) by the algorithm, until the optimal value Δ is reached. Thus the only problem remaining in the design of an algorithm is the practical construction of the simplices S_j: This is where the intersection problem (with complete convex polaroids) can be used to characterize the vertices of S_j as intersection points u^*.

<u>1.3</u>: Finally, to conclude this introductory exposition of polaroids, let us mention a result which will be found useful in estimating the relative depth of polaroid cuts in non-convex programming.

<u>Corollary 1.1</u>: Let the polaroid function $f(x,y)$ satisfy the hypothesis

$$f(x,y) \leq \max\{g(x), g(y)\} \quad , \quad \forall x,y$$

where $g(z) = f(z,z)$;

then one has

$$P^*(\Delta) \supset \text{lev}_\Delta\, g \supset P \text{ , when } \Delta \geq \max_{x \in P} g(x)$$

i.e., $P^*(\Delta)$ is <u>complete</u>.

<u>Proof</u>: The first inclusion follows from the definition of polaroids: $P^*(\Delta) = \{y \mid f(x,y) \leq \Delta, \forall x \in P\}$. Indeed one has $\forall x \in P$, $\forall y \in \text{lev}_\Delta g$ $\max\{g(x) , g(y)\} \leq \Delta$; hence $f(x,y) \leq \max\{g(x) , g(y)\} \leq \Delta$

shows that $y \in P^*(\Delta)$; and $P^*(\Delta) \supset lev_\Delta g$ is proved. The second inclusion $lev_\Delta g \supset P$ is a mere consequence of the choice of Δ.

Q.E.D.

Examples:

1.3.1: $$f(x,y) = \sum_{i=1}^{n} x_i y_i \leq \sqrt{\sum_{i=1}^{n} x_i^2}\sqrt{\sum_{i=1}^{n} y_i^2} \leq \max\left\{\sum_{i=1}^{n} x_i^2 , \sum_{i=1}^{n} y_i^2\right\}$$
(Cauchy-Schwartz)

1.3.2: Corollary 1.1 can be applied to the maximization problem of a convex function g over the set P; the corollary 1.3 of section 3 shows how this result relates to Hoang Tuy's extended function approach.

2) The Intersection Method

2.1: Proposition 1 indicates that the determination of intersection points can be obtained by a parametric optimization process of the following type:

Find $\bar{\lambda} > 0$ such that $F_a(\bar{\lambda}) = \max_{x \in P} f(x, \bar{x} - \bar{\lambda} a) \leq k$;

(i) if $F_a(\bar{\lambda}) < k$ then $\bar{u} = u(\bar{\lambda})$ can be seen to lie in the "interior" of P^*:

"Int" $P^*(k) = \{y \mid f(x,y) < k \ , \ \forall x \in P\}$;

(ii) if $F_a(\bar{\lambda}) = k$, then u lies on the "boundary" set "bd" $P^*(k)$ defined in section 1 and it is therefore a solution u^* to the intersection problem.

This polaroid intersection problem can be seen to contain the following two parts:

P1) A Mathematical programming part:

maximize $z(x) = f(x, \bar{x} - \bar{\lambda} a)$, $\bar{\lambda}$ given

subject to $x \in P$

P2) A parametric search problem:

$$\text{Find } \lambda^* = \max_{\lambda > 0} \{\lambda \mid F_a(\lambda) \le k\}$$

In general neither of these optimizations can be executed exactly and one often prefers a good approximate solution which is obtained in relatively few computations. This approximation, however, must be such that the approximate intersection point $\tilde{u}$ belongs to the set $P^*(k)$; for the problem P1) this means that one needs an upper bounding solution $\tilde{x}$ such that

$$\min_{x \in P} z(x) \le \tilde{z} = z(\tilde{x}) = F_a(\tilde{\lambda}) \le k .$$

and for the problem P2) this implies that $\tilde{\lambda} \le \lambda^*$ must hold true.

2.2: In principle the method of section 0.2 is quite general; practically, however, one is easily convinced that the condition b) may be difficult to satisfy except for special problems which possess additional properties.

We therefore restrict our attention, in this section, to intersections with convex polaroids because, in this case, intersection points can be used directly to generate simplices S_j which (automatically) satisfy this condition b).

Corollary 1.2: If the polaroid function $f(x,y)$ is quasi-convex in y, $\forall x \in P$, then $\forall a \in \mathbb{R}^n$ the intersection function $F_a(\lambda)$ is quasi-convex in λ.

Proof: By definition one has $\forall a \in \mathbb{R}^n$:

$$F_a(\lambda) \geq f(x, y = \bar{x} - \lambda a) \quad , \quad \forall x \in P$$

Let $y^1 = \bar{x} - \lambda^1 a$, $y^2 = \bar{x} - \lambda^2 a$, $\lambda^3 = \mu\lambda^1 + (1-\mu)\lambda^2$ with $\mu \in [0,1]$:

$$y^3 = \bar{x} - [\mu\lambda^1 + (1-\mu)\lambda^2]a = \mu y^1 + (1-\mu)y^2 \ ;$$

for $i=1,2,3$ one has by hypothesis $F_a(\lambda^i) \geq f(x,y^i)$, $\forall x \in P$, with

$$F_a(\lambda^i) = f(x^i, y^i) \ , \text{ for some } x^i \in P \ ;$$

hence $F_a(\lambda^3) = f(x^3,y^3) = f(x^3, \mu y^1 + (1-\mu)y^2) \leq \max\{f(x^3,y^1), f(x^3,y^2)\}$

$$\leq \max\{f(x^1,y^1), f(x^2,y^2)\} = \max\{F_a(\lambda^1), F_a(\lambda^2)\}$$

Q.E.D.

Theorem 3: Let $f(x,y)$ be quasi-convex in y , $\forall x \in P$; and

let $\lambda^* = \max_{\lambda > 0} \{\lambda \mid F_a(\lambda) \leq k\}$

then $F_a(\lambda) \leq F_a(\lambda^*) = k$, $\forall \lambda \in [0, \lambda^*]$

Proof: By hypothesis $\bar{x} \in P^*(k)$, implying $f(x,\bar{x}) \leq k$, $\forall x \in P$ thus $F_a(0) \leq k$; the quasi-convexity of F_a established in the corollary 1.1 completes the proof.

Q.E.D.

The above theorem corroborates a well-known property of convex sets with respect to their intersection by a one-dimensional line (ray); the situation is complicated here by the fact that the function $F_a(\lambda)$ is not explicitly known but merely defined in terms of the optimization problem P1; naturally the determination of an intersection point could also be formulated as one single optimization problem in the variables x and λ ; however, the independence of the constraints in x and λ variables respectively as well as the separate properties of f (with respect to x and y respectively) motivates a separate treatment of the optimization in x and in λ ; the result of theorem 3 indicates that the parametric search problem P2 can be solved by increasing stepwise the value of λ until λ^* (or $\tilde{\lambda} \leq \lambda^*$) is reached; each step consists in a (post-) optimization of the mathematical programming problem P1. The resulting algorithm becomes a <u>non-linear descent</u> algorithm of a particular type.

<u>Algorithm</u>: (f is assumed quasi-convex in y)

<u>Step 0</u>: Let $\lambda_0 = 0$, $i = 1$; choose $\lambda_1 > 0$ such that

$$f(x,\bar{x} - \lambda_1 a) \leq k \quad , \quad \forall x \epsilon P \ , \quad \text{i.e.} \quad \bar{x} - \lambda_1 a = y_1 \ \epsilon \ P^*(k)$$

<u>Step 1</u>: Solve the mathematical programming problem P1 :

$$\max z(x) = f(x,\bar{x} - \lambda_i a) \quad , \quad \text{s.t.} \quad x \epsilon P$$

<u>Step 2</u>:

2.1: If $\bar{z} = \max_{x \epsilon P} z(x) = k$ then STOP

2.2: If $\bar{z} > k$, then choose $\lambda_{i+1} \ \epsilon \ (\lambda_{i-1}, \lambda_i)$

2.3: If $\bar{z} < k$, then choose $\lambda_{i+1} > \lambda_i$

set $i = i + 1$ and go to Step 1.

<u>Remarks</u>:

1) The above algorithm merely represents an unpolished set of guidelines; depending on the particular problem at hand it can be refined in order to increase computational efficiency (see section 3).

2) Clearly this algorithm need not be finite; practically, however, one should remember that there usually is no need for an accurate value λ^*;

thus the exit criterion of step 2 normally will be made to contain a (often large) "fuzz" factor $\varepsilon > 0$ and reads:

2.1: If $\bar{z} \in [k-\varepsilon\ ,\ k]$ then STOP

2.2: (Unchanged)

2.3: If $\bar{z} < k-\varepsilon$ then choose $\lambda_{i+1} > \lambda_i$

The algorithm then delivers an approximate intersection point $\tilde{u} \in P^*(k)$ in a finite number of iterations (this number clearly depends on the quantity ε).

3) Some Examples of Polaroid Intersections

This section presents below a list of particular polaroids, ranked by order of increasing complexity of the corresponding intersection problems.

3.1) Bilinear polaroids: $f(x,y) = Ax + By + y^T Cx$

In this case the polaroid $P^*(k)$ is convex and the determination of an intersection point u^* is obtained by solving a parametric linear program when P is polyhedral:

$$\max z(x) = B(\bar{x}-\lambda a) + [A + (\bar{x}-\lambda a)^T C]\ x$$

$$\text{s.t. } x \in P$$

Problem P1 is an ordinary L.P. and the increment of λ in step 2 of the algorithm can be determined by the previous L.P. optimal solution $\bar{x}$ of step 1. Because $F_a(\lambda)$ is known to be piece-wise linear, the optimal value λ^* can be obtained by linear algebra in a finite number of iterations.

3.1.1.: For the case where C is a symmetric positive definite matrix, Balas has first recognized in [3] the possibility to use sets he calls outer-polars to generate valid cutting planes in integer programming with the help of parametric linear programming (the outer-polar is a (generalized) polar set (see [15]) which may also be viewed as a polaroid set [9]). The same technique is used by Balas and Burdet in [4] to solve quadratic concave programs.

3.1.2: For indefinite as well as semi-definite quadratic programs, polaroid cuts have been implemented successfully in a facial decomposition schema for solving the general quadratic programming problem [21,22].

3.1.3: Another application for valid cuts in integer programming is given in [7]; there polaroid cuts are constructed from a homogeneous (A=B=0) bilinear polaroid function which is centered at the origin (i.e.at $\bar{x}$); this case presents some computational advantages because the optimization problem P1 and P2 can be combined in a single linear program (not parametric); indeed, one can verify by inspection that the problem P1,

$$\text{i.e.,} \quad \max_{x \in P} -(\lambda a^T Cx) \leq k ,$$

directly delivers the optimal value

$$(\lambda^*)^{-1} = \frac{1}{k}\left[\max_{x \in P} \ (-a^T Cx)\right]$$

of problem P2.

3.2) The previous bilinear polaroids are a particular case of the following family (with parameter α):

$$f(x,y) = g(x) + \alpha(y-x)^T F(x) , \quad \text{where} \quad g(x) = x^T F(x) ,$$

(F is a vector valued function).

An example of this type is obtained by choosing $F(x) = \nabla g(x)$, the gradient of g ; as an illustration, consider the (quadratic) case where $g(x) = \frac{1}{2}x^T Cx$; $\nabla g(x) = F(x) = Cx$ (C = n by n symmetric positive definite matrix), one has

$$f(x,y) = (1/2-\alpha)\, x^T Cx + \alpha\, y^T Cx.$$

3.2.1: $\alpha = 1/2$: $f(x,y) = y^T Cx$, bilinear polaroid (see 3.1 above)

3.2.2: $\alpha = 1$: $f(x,y) = -1/2\, x^T Cx + y^T Cx$.

This is a quadratic illustration of the general polaroid function $f(x,y) = g(x) + (y-x)^T \nabla g(x)$ which generates a family of hyperplanes supporting (at the point $(g(x),x)$) the graph $\{(x_o,x) | x_o \geq g(x),\ x \in P\} \subset \mathbb{R}^{n+1}$. The corresponding polaroid set $P^*(k)$ represents an analytical characterization of the level set of the maximal convex (concave) extension of the convex (concave) function $g(x)$ with respect to the set P, which has been characterized, in geometrical terms, by Hoang Tuy in [16] (see proposition 3, below).

3.2.3: $\alpha > 1/2$: In this case (which contains, in particular, the previous one 3.2.2), the mathematical programming problem P1 is a convex quadratic programming problem over the feasible set P and it can be solved by the corresponding classical algorithms.

3.2.4: $\alpha < 1/2$: Here problem P1 is a concave program (i.e. max(min)imization of a convex (concave) objective function).

When $0 < \alpha \leq 1$, an illustration of the unifying insight provided by the polaroid approach in the fields of concave and/or integer programming can be found in [7].

We describe here a more general situation than in [7]:

assume that $\Delta = g(\tilde{x}) = \max_{x \in P} x^T Cx$; then, for a small enough value $\bar{\alpha} > 0$, the polaroid $P^*_{\bar{\alpha}}(\Delta)$ will be the (intersecti of the) halfspace(s) determined by the hyperplane(s) supporting P at the optimal solution(s) $\tilde{x}$. This is the "largest" polaroid P^* in the family, with the desirable property that $\tilde{x} \notin \text{Int } P^*$, as shown in the proposition 3 below. Larger values $\alpha > \bar{\alpha}$ yield a lower bound for the intersection parameter λ, which can be used as an approximation; the intersection algorithm could thus conceivably refined by changing the parameter α (starting with, say, the L.P. value $\alpha = 1$) and post-optimizing the (new) problem P2 to find a better value for λ ; ultimately this process determines the intersection parameter $\bar{\lambda}$ with the largest polaroid $P^*_{\bar{\alpha}}(\Delta)$.

In integer programming, the outer-domain theory (see, for instance, [6] and [7]) indicates that $P^*_{\bar{\alpha}}(\Delta)$ is the intersection of a collection of halfspaces, each belonging to a feasible vertex of the unit cube; decreasing the parameter α (down from $\alpha = 1$) corresponds to introducing additional integrality requirements into the linear program obtained in 3.1 (for more details, see [7]).

3.2.5: $\alpha = 0$: In this case $f(x,y) = g(x)$ and is independent of y ; hence P^* is the whole space whenever $k \in f(P, \mathbb{R}^n)$, and $P^* = \emptyset$ otherwise.

Proposition 3:[1/] Define $P^*_\alpha(k) = \{y \mid f_\alpha(x,y) = g(x) + \alpha(y-x)^T F(x) \leq k, \forall x \in P\}$ where g and F are real resp. vector valued functions. Assume $\alpha_1, \alpha_2 > 0$ and $P \subset \text{lev}_k\, g(x)$, i.e.: $g(x) \leq k$, $\forall x \in P$.

Then one has

(i) $\alpha_1 > \alpha_2$ if $P^*_{\alpha_1}(k) \subset P^*_{\alpha_2}(k)$ (in the strict sense)

(ii) $P^*_{\alpha_1}(k) \subseteq P^*_{\alpha_2}(k)$ if $\alpha_1 \geq \alpha_2$

Proof to (i): Take $y_1 \in$ "bd" $P^*_{\alpha_1}(k)$ and choose $x_1 \in P$ such that

$$\text{then } f_{\alpha_1}(x_1,y_1) = g(x_1) + \alpha_1(y_1-x_1)^T F(x_1) = k \; ;$$

it follows that

$$\alpha_1(y_1-x_1)^T F(x_1) = k - g(x_1) \geq 0 \quad (\text{since } x_1 \in P \text{ and } P \subset \text{lev}_k\, g)$$

and finally $(y_1-x_1)^T F(x_1) \geq 0$.

[1/] For the case where g is a positive definite quadratic form (with $F(x) = \nabla g(x)$), E. Balas was first to make the interesting observation that the cuts derived from Hoang Tuy's maximal extension are uniformly dominated by the outer-polar cuts. Proposition 3 corroborates this fact for a larger class of polaroid cuts: here the Hoang Tuy cuts correspond to the particular case $\alpha = 1$ and $F(x) = \nabla g(k)$.

We now show that $y_1 \varepsilon P^*_{\alpha_2}(k)$ implies $\alpha_2 < \alpha_1$:

$$g(x_1) + \alpha_2(y_1-x_1)^T F(x_1) =$$

$$= g(x_1) + \alpha_1(y_1-x_1)^T F(x_1) + (\alpha_2-\alpha_1)(y_1-x_1)^T F(x_1) \leq k \ ;$$

hence

$$(\alpha_2-\alpha_1)(y_1-x_1)^T F(x_1) \leq 0$$

holds true. Now either $\exists$ a pair $(y_1 \ \varepsilon \ \text{bd} \ P^*_{\alpha_1}(k), \ x_1 \ \varepsilon \ P)$ such that $(y_1-x_1)^T F(x_1) \neq 0$, then we obtain the desired result $(\alpha_2-\alpha_1) < 0$ (since $\alpha_1 = \alpha_2$ is contradictory to the strict inclusion $P^*_{\alpha_1} \subset P^*_{\alpha_2}$), or else we have $\forall \alpha \geq 0$:

$f_\alpha(x_1,y_1) = k$, $\forall y_1 \ \varepsilon \ \text{bd} \ P^*_{\alpha_1}(k)$ and $x_1 \ \varepsilon \ P$; thus, the boundary sets of the convex sets $P^*_\alpha(k)$ (and therefore also the sets themselves) are identical, which contradicts the strict inclusion $P^*_{\alpha_1}(k) \subset P^*_{\alpha_2}(k)$.

<u>To (ii)</u>: The converse statement is also true:

$$\alpha_2 \leq \alpha_1 \quad \text{implies} \quad y_1 \ \varepsilon \ P^*_{\alpha_2}(k) \ .$$

The proof is by contradiction: suppose $y_1 \notin P^*_{\alpha_2}(k)$; then there exists a point $x_2 \ \varepsilon \ P$ such that $g(x_2) + \alpha_2(y_1-x_2)^T F(x_2) = k' > k$;

it follows that $\alpha_2(y_1-x_2)^T F(x_2) = k'-g(x_2) > 0$ and hence $(y_1-x_2)^T F(x_2) > 0$.

Furthermore, since $y_1 \ \varepsilon \ P^*_{\alpha_1}(k)$, one has

$g(x_2) + \alpha_2(y_1-x_2)^T F(x_2) + (\alpha_1-\alpha_2)(y_1-x_2)^T F(x_2) \leq k$ hence

$(\alpha_1-\alpha_2)(y_1-x_2)^T F(x_2) \leq k-k' < 0$ which implies $(\alpha_1-\alpha_2) < 0$, a contradiction to the hypothesis $\alpha_2 \leq \alpha_1$; thus y_1 must belong to $P^*_{\alpha_2}(k)$.

In summary, we have shown that the boundary of $P^*_{\alpha_1}(k)$ lies (entirely) in $P^*_{\alpha_2}(k)$ if $\alpha_1 \geq \alpha_2$; but both sets $P^*_{\alpha_1}(k)$ and $P^*_{\alpha_2}(k)$ are closed convex; hence proposition (ii) holds true. Q.E.D.

Corollary 1.3: Let $g(x)$ be convex and differentiable (differentiability is assumed here for convenience of the exposition but the result is easily extended to the non-differentiable case); define a polaroid function $f_\alpha(x,y) = g(x) + \alpha(y-x)^T \nabla g(x)$; then $\forall \alpha \in (0,1]$ the polaroid $P^*_\alpha(\Delta)$ is complete when

$$\Delta \geq \max_{x \in P} g(x) = \Delta_o.$$

Proof: Take $\alpha = 1$; convexity of g implies that

$$f_1(x,y) \leq g(y) \leq \max \{g(x), g(y)\}, \ \forall x,y \text{ because } g(y) \geq g(x) + (y-x)^T \nabla g(x).$$

Hence the assertion (for $\alpha=1$) follows from corollary 1.1; furthermore one has

$$P^*_1(\Delta_o) \supset \text{lev}_{\Delta_o} g \text{ by hypothesis;}$$

thus Proposition 3 can now be applied $\forall \alpha \in (0,1]$; finally the proof is complete upon noticing that $P^*_\alpha(\Delta) \supset P^*_\alpha(\Delta_o), \ \forall \Delta \geq \Delta_o$. Q.E.D.

In paragraph 3.2.2 we mentioned that Hoang Tuy's extended function corresponds to the case $\alpha = 1$; Proposition 3 and Corollary 1.3 now indicate a whole class of polaroids $P^*_\alpha(\Delta)$ which can be used in a similar manner, and which uniformly dominate Tuy's extended level set.

One may conclude from proposition 3 that there is a reward for solving increasingly difficult problems; for small α, however, the intersection problem is often of the order of difficulty of the original problem, one is trying to solve with the help of polaroids. Note that parametric linear programs (section 3.2.1), however, hold a position which seems to make them computationally quite attractive in that respect.

3.3 Simple polaroids: This class contains all the previous cases and is characterized by functions $f(x,y)$ which are linear in y, $\forall x \varepsilon P$; the convexity of the corresponding polaroid sets $P^*(k)$ is readily verified (see theorem 1).

Of course this class may contain a variety of different types of optimization problems P1: convex, concave, discrete, etc.. . . but linearity in y makes it easier to predict adequate increment for λ in the algorithm. One should also note that the convexity of P^* makes a cutting plane approach possible for problems which do not possess convex level sets of the objective function $g(x)$. An example of this type can be found in [21] and a general discussion of polar programs is presented in [10].

3.4 A more general class yet is defined by polaroid functions $f(x,y)$ which are merely required to satisfy the hypothesis of theorem 1 in order to yield convex polaroid sets. Here, except for very special functions, there are no artifices available to render the problem P2 more easily tractable, and one has to resort to the stepwise incremental method described in the algorithm to solve the intersection problem (depending on the actual function f it may become more or less difficult to determine adequate increments $(\lambda_{i+1} - \lambda_i)$).

3.5 One may also use polaroid sets which are not assumed in advance to be convex; after solving the intersection problems (for a cutting plane approach, for instance), one has yet to test the validity of the cut; this task is essentially different from checking a quasi-convex property of $f(x,y)$; non-convex polaroids may be useful in the following situations:

- when one can relatively easily prove (a posteriori) that the constructed cut is indeed valid
- when non-convex polaroids allow for the construction of deeper intersection cuts than other convex outer-domains.

The above comment (of course) merely indicates a direction for possible research, based on the evidence that for some simple examples, the convexity requirement can be advantageously relaxed in the intersection cut theory.

References

[1] Balas, E.: "Intersection Cuts - A New Type of Cutting Planes for Integer Programming." Operations Research, Vol. 19, No. 1, Jan.-Feb. 1971, pp. 19-39.

[2] ________: "Intersection Cuts from the Maximal Extensions of the Ball and the Octahedron." Management Science Report #214, Graduate School of Industrial Administration, Carnegie-Mellon University, 1970.

[3] ________: "Integer Programming and Convex Analysis." Management Science Report #246, Graduate School of Industrial Administration, Carnegie-Mellon University, 1971.

[4] Balas, E. and Burdet C.-A.: "Quadratic Concave Programming." In preparation.

[5] Burdet, C.-A.: "A Class of Cuts and Related Algorithms in Integer Programming." Management Science Report #220, Graduate School of Industrial Administration, Carnegie-Mellon University, 1970.

[6] ____________: "Enumerative Cuts I." Submitted to Operations Research, 1971.

[7] ____________: "Enumerative Cuts II." Submitted to Operations Research, 1972.

[8] ____________: "Integer and Concave Programming." Lecture notes for a Conference at the Thomas J. Watson IBM Research Center, 1971.

[9] ____________: "Polaroids: a new tool in non-convex and in integer programming." W.P. 78-72-1, Graduate School of Industrial Administration, Carnegie-Mellon University, to appear in Naval Research Logistics Quarterly, 1973.

[10] ____________: "Polar Programming." W.P. 98-71-2, Graduate School of Industrial Administration, Carnegie-Mellon University, 1972.

[11] Glover, F.: "Cut-Search Methods in Integer Programming." University of Colorado, September, 1970.

[12] Glover, F. and Klingman, D.: "Concave Programming Applied to a Special Class of 0-1 Integer Programs." University of Texas at Austin, 1971.

[13] Minkowski, H.: "Theorie der Konvexen Körper insbesondere Begründung ihres Oberflächenbegriffes." Gesammelte Abhandlungen, Vol. 2, Leipzig, 1911.

[14] Raghavachari, M.: "On the Zero-One Integer Programming Problem." Operations Research, 17, 1969, pp. 680-684.

[15] Rockafellar, R. T.: Convex Analysis. Princeton University Press, 1970.

[16] Tuy, Hoang: "Concave Programming Under Linear Constraints," (Russian). Doklady Akademii Nauk SSSR, 1964. English translation in Soviet Mathematics, 1964, pp. 1437-1440.

[17] ________: "Vê môt lop qui hoach phi-tuyên." Toán Lý, No. 1, 1963. Publications de la Section de Physique et Mathématiques du comité d'Etat des Sciences, République Démocratique du Vietnam.

[19] Burdet, C.-A. and Mikhail, O. I.: "Centralization of Computer Systems under Economies of Scale." W.P. 99-71-2, Graduate School of Industrial Administration, Carnegie-Mellon University, 1972.

[20] Burdet, C.-A.: "Some Enumerative Inequalities and Their Application in Integer Programming Algorithms." Published in Mathematical Programming, Vol. 2 No. 1, (1972) pp. 32-64.

[21] ________: "Simple Polaroids for Non-convex Quadratic Programming." W.P. 101-71-2, Graduate School of Industrial Administration, Carnegie-Mellon University, 1972.

[22] ________: "General Quadratic Programming." Management Science Report #272, Graduate School of Industrial Administration, Carnegie-Mellon University.

[23] ________: "Generating All the Faces of a Polyhedron." Management Science Report #271, Graduate School of Industrial Administration, Carnegie-Mellon University. Submitted to SIAM Journal on Applied Mathematics, 1972.

[24] ________: "The Facial Decomposition Method." W.P. 104-71-2, Graduate School of Industrial Administration, Carnegie-Mellon University, 1972.

Mathematical Programming in Theory and Practice,
P.L. Hammer and G. Zoutendijk, (Eds.)

"THE ROUTING OF A MINIMAL CIRCULAR FLOW"

M. S. Rosa
The University of Birmingham, England

The following paper is concerned with the question of the minimum circular flow which insures that each arc of the network contains a minimum flow given. The solution requires the construction of a new chain decomposition, which is described in detail.

Introduction

Consider a network $G(N;A)$. Associated with every arc (i,j) in A there is a non-negative integer number c_{ij}, the minimum flow required through that arc. In this study we shall be interested in determining a minimum total circular flow, given the minimum required on each arc of the network. By a circular flow is meant a flow both starting and ending at the same node. We shall restrict ourselves in this note to a network of directed arcs and assume that feasible flows exist. This can be done without loss of generality.

Formulation

This problem can be formulated as a minimum-cost flow problem. For this purpose, let the cost of shipping one unit of flow from node i to node j be a_{ij}. The problem can be formulated as follows:

$$\text{Minimize} \sum_i \sum_j a_{ij} \; x_{ij} \tag{1}$$

subject to

$$\sum_j x_{ij} = \sum_k x_{ki} \qquad \text{for all } i \in N \tag{2}$$

$$c_{ij} \leqslant x_{ij} < \infty \qquad \text{for all } ij \in A \tag{3}$$

where:

$$a_{ij} = \begin{cases} 1 \text{ if } j = S \\ 0 \text{ otherwise} \end{cases}$$

S is the "source" and the "sink" of the flow

x_{ij} is the flow from node i to node j

c_{ij} is the minimum flow required on arc ij.

Actually, according to the values set for a_{ij}, (1) is equivalent to

$\text{Minimize} \sum_i x_{iS} = V$, where V is the value of the total flow.

The value of the flow can be found for instance by the "OUT-OF-KILTER" algorithm, adapted to the case when there are no upper limits to the capacity of the arcs. Following D. R. Fulkerson, define a set of integer variables $\{\pi(i)\}$ called the node numbers. The initial values of these variables may be chosen arbitrarily and are adjusted by the algorithm. Define a set of arc numbers $\bar{a}_{ij}=a_{ij}+\pi_i-\pi_j$: Each arc is then in one of the following states:

(α)	$\bar{a}_{ij} > 0,$	$x_{ij} = c_{ij}$
(β)	$\bar{a}_{ij} = 0,$	$x_{ij} \geqslant c_{ij}$
(α_1)	$\bar{a}_{ij} > 0,$	$x_{ij} < c_{ij}$
(α_2)	$\bar{a}_{ij} > 0,$	$x_{ij} > c_{ij}$
(β_1)	$\bar{a}_{ij} = 0,$	$x_{ij} < c_{ij}$

Arcs of states (α) and (β) are "in-kilter" and the corresponding kilter numbers are zero. Arcs of states (α_1), (α_2) and (β_1) are "out-of-kilter" and have positive kilter numbers as listed below:

(α_1)	$c_{ij}-x_{ij}$
(α_2)	$\bar{a}_{ij}(x_{ij}-c_{ij})$
(β_1)	$c_{ij}-x_{ij}$

The algorithm starts with any circulation $X = \{x_{ij}\}$ and any price vector $\Pi = \pi$ (i) and then uses a slightly modified labeling procedure to adjust an arc of the network that fails to satisfy the optimality conditions. That is to say, the algorithm concentrates on a particular out-of-kilter arc, and gradually puts it in kilter, in such a way that all in-kilter arcs stay in-kilter, whereas any other out-of-kilter arc either improves, its kilter number being reduced, or stays the same.

The subsequent routing of the flow must be specified by a new chain decomposition algorithm.

Example

We illustrate the method by using the following example. Suppose that we have a road network of an urban area which must be cleaned periodically by a number of road sweepers, given the smallest number of sweeps necessary in each link to perform the task satisfactorily.

All sweepers start at the same point S, and return to it eventually. They may sweep either on the way out, or on the way back. Some of them might have to pass through a cycle, or part of a cycle, more than once. However, no sweeper starts again once he has returned to S.

Let us consider the following network in Figure 1:

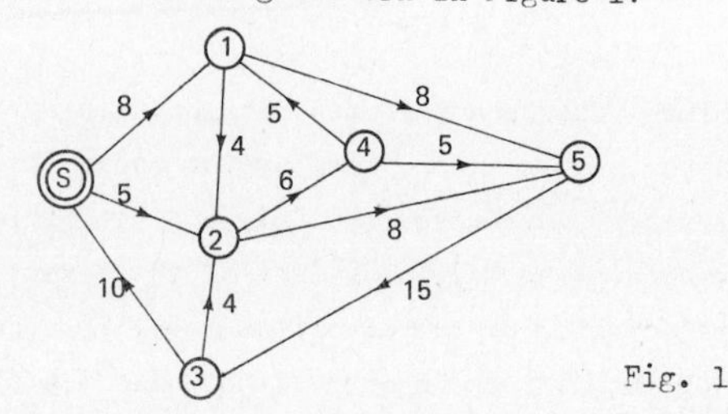

Fig. 1

It will be noticed that this network contains a number of cycles, e.g. 1-2-4-1 and 2-4-5-3-2. The numbers attached to an arc ij represent the value of c_{ij}.

The out-of-kilter algorithm produces the following solution, with V=13, as in Figure 2:

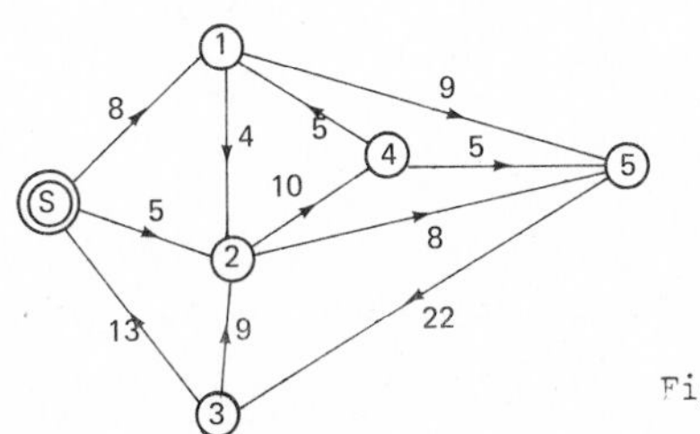

Fig. 2

We imagine now that every individual sweeper must be given a feasible route through the network, starting at S. Of course, a group of sweepers might have identical instructions. The routing to be followed is not immediately obvious from the out-of-kilter flow.

The programme of routes can be obtained by the following "chain decomposition algorithm". A chain is a sequence of links, taken in their directed sense starting at S and returning to S, without excluding the possibility of cycles.

Chain decomposition algorithm

Step 0 : Call the solution obtained for the minimal flow, the "original" arc flows x_{ij}. Go to Step 1.

Step 1 : Choose an initial arc (S,i_1) where $x_{S,i_1} = \max_i (x_{S,i})$.

Let $F = x_{S,i_1}$. Go to Step 2.

Step 2 : Repeat this procedure at node i_1, and continue choosing successively the arcs (i_p, i_{p+1}) of a chain using the criterion:

$$x_{i_p, i_{p+1}} = \max_{i \neq S} (x_{i_p,i}) \neq 0$$

— If no such i_{p+1} exists, let $i_{p+1} = S$ and go to Step 5.

— If i_{p+1} is a node that we have encountered earlier (that is to say, $i_{p+1} = i_j$ where $j < p$), go to Step 3.

— Otherwise, repeat Step 2, with p+1 replacing p.

Step 3 : A cycle $C = \{i_j, i_{j+1}, \ldots, i_p, i_{p+1} = i_j\}$ has been found.

Call $L = \{i_1, i_2, \ldots, i_j, i_{j+1}, \ldots i_p, i_j\}$ and record the nodes of L immediately after those (if any) already recorded in previous applications of Step 2. Go to Step 4.

Step 4 : Let $F_1 = \min_{ij \in L} (F, X_{ij})$. F_1 is the smallest arc flow in L).

Change the value of F : $F = F_1$.

Change the arc flows:

$$\text{new value of } x_{ij} = \begin{cases} \text{old } x_{ij} - F & \text{if } ij \in C \\ \text{old } x_{ij} & \text{if } ij \notin C \end{cases}$$

Let node i_j be new i_1, and return to Step 2.

Step 5 : Record the set of node pairs $K = \{i_1, i_2, \ldots, i_p, i_{p+1}\}$ (note $i_{p+1} = S$) after those previously recorded in Step 3 (if any).

Let $F_2 = \min \{F, \min_{ij \in K} (x_{ij})\}$

Let the new value of $F = F_2$.

Note that at this point a chain (which may include cycles) bearing a flow $F \leqslant V$ has been found, and recorded in Steps 2 through 5. Go to Step 6.

Step 6 : Let n_{ij} be the number of times arc ij is traversed in the chain just found. For each arc of this chain, let:

The new flow x_{ij} = (original x_{ij}) - $n_{ij}F$.

The chain and the value of F are recorded. A new flow is produced with flow $V - F \geqslant 0$, redefining the "original x_{ij}". If $V-F = 0$, terminate, Otherwise, re-enter Step 0.

When this procedure terminates the minimal flow will have been decomposed into chain flows that may include cycles.

In our example we obtain first, $F = x_{S,1} = 8$ (Step 1).

The first application of Steps 2 and 3 gives,

$C = L = \{(1,5),(5,3),(3,2),(2,4),(4,1)\}$.

By Step 4, subtract $\min_{ij \in L} (8, x_{ij}) = 5$ from old x_{ij} for $ij \in C$

Next application of Steps 2 and 3 gives,

$C = \{(5,3),(3,2),(2,5)\}$ and $L = \{(\underline{1,5}),(5,3),(3,2),(2,5)\}$.

Again by Step 4, subtract $\min_{ij \in L} (5, x_{ij}) = 4$ from old x_{ij} for $ij \in C$

The remaining flow is shown in Figure 3 :

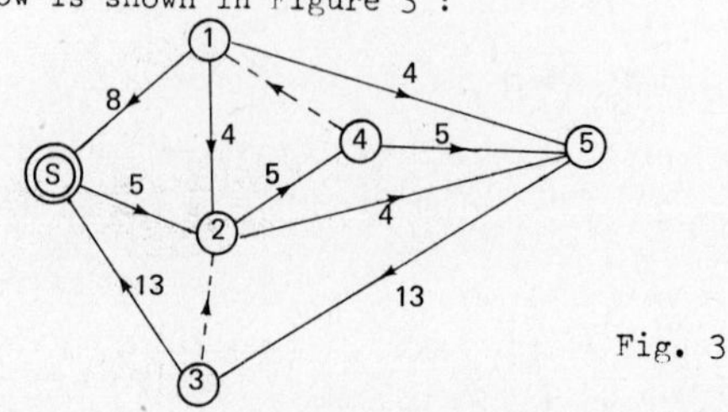

Fig. 3

Another application of Steps 2 and 5 gives: $5 \to 3 \to S$, i.e.

$K = \{(5,3),(3,S)\}$. Compute $\min_{ij \in K} (4, x_{ij}) = 4$.

Thus we have found a first chain, viz,

$\{(S,1),(1,5),(5,3),(3,2),(2,4),(4,1),(1,5),(5,3),(3,2),(2,5),(5,3),(3,S)\}$

of value 4, just computed, represented in Figure 4:

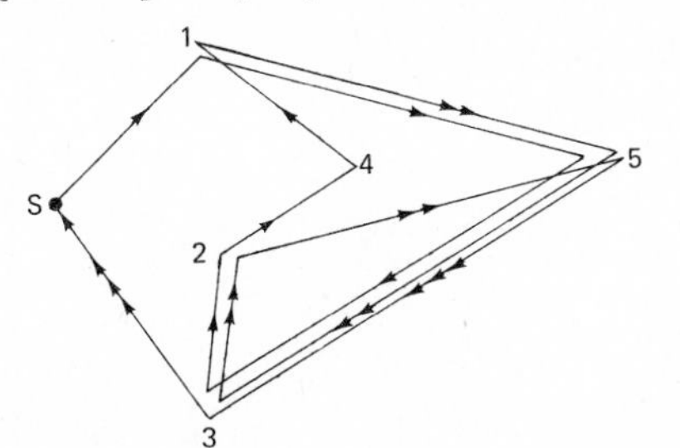

Fig. 4

By Step 6, the new flow x_{ij} is produced, redefining the "original x_{ij}" to re-enter Step 0. After two further applications of the routine, we obtain the following two chains:

$\{(S,2),(2,4),(4,5),(5,3),(3,2),(2,4),(4,5),(5,3),(3,S)\}$ of value 1, and

$\{(S,1),(1,2),(2,5),(5,3),(3,S)\}$ of value 4. (There are no cycles in this and subsequent chains).

The "original x_{ij}" are now shown in Figure 5:

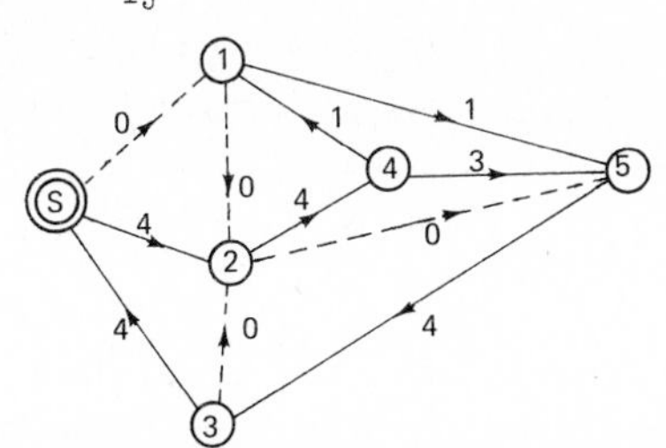

Fig. 5

Carrying on, we obtain the last two chains:

$\{(S,2),(2,4),(4,5),(5,3),(3,S)\}$ of value 3, and

$\{(S,2),(2,4),(4,1),(1,5),(5,3),(3,S)\}$ of value 1 (Figure 6).

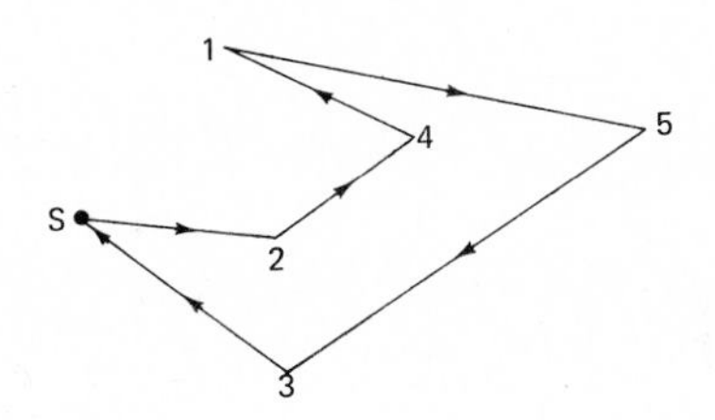

Fig. 6

Termination of the algorithm

It is easily seen that each application of Steps 2-4 removes one or more cycles from a chain, before going on to Step 5. Moreover, due to the manner in which the new values of x_{ij} are computed in Step 4, the flow conservation equations (2) are satisfied each time Step 2 is going to be applied. So, because the number of cycles is assumed to be finite as well as the values of the "original x_{ij}", the number of Steps 2 through 4 must be a finite number. This proves that a chain is found after a finite number of applications of Steps 2 through 5.

After application of Step 6, the redefined "original x_{ij}" to re-enter Step 0, must still satisfy the flow conservation equations as a consequence of what was said above.

Finally, because the initial V is finite, the number of chains produced by the entire routine is finite and the routine itself will terminate when V-F=0 and consequently when the new "original x_{ij}" to re-enter at Step 0 are all equal to zero.

References

D. R. Fulkerson, "An Out-of-Kilter method for minimal cost-flow problems" SIAM Journal, vol. 9 pp 18-27 (1961)

Ford, Jr., L. R. and Fulkerson, D. R. "Flows in Networks", Princeton (1962)

Mathematical Programming in Theory and Practice,
P.L. Hammer and G. Zoutendijk, (Eds.)

GOAL PROGRAMMING
AND A MANPOWER PROBLEM

Dr. W. L. Price
Defence Research Analysis Establishment
Canada

1. The Standard Linear Programming Model

Let us consider the familiar linear programming model, which deals with the optimization of a linear objective function subject to linear constraints. In the usual notation this can be expressed as:

$$\text{minimise } (cx') \tag{1}$$

$$\text{subject to } Ax' \leqslant b' \tag{2}$$

where
- A is the structural matrix of the constraint;
- b' is the column vector of the right-hand sides of the constraints;
- x' is the column vector of main variables;
- c is the row vector of objective function coefficients.

For solution of the problem, a column vector of slack variables, r, will be added to the problem so that the constraints now have the form:

$$Ax' + r' = b' \tag{3}$$

Geometrically, one can interpret both the original constraints, and the slack variables. The original constraints can be interpreted as hyperplanes which delimit a region of the problem space including all feasible solutions to the problem. The slack variable associated with a constraint is a measure of how "far" a given basic feasible solution (vertex) is from a given constraint (hyperplane).

The value of the objective function is a concrete measure of the goodness of a given solution. Typical objective functions are of the cost minimization or profit maximization type.

2. The Question of Ill-Defined Objectives and Constraints

The standard linear programming model is quite suitable for the many cases where the problem specification is clear, the constraints consistent, and the objective function is well defined.

However, as Eilon[1] has pointed out in a recent article, such is not always the case. Often, constraints are not absolute, but can be "bent" somewhat if necessary. Worse still, if some constraints are not bent, no feasible solution will exist, because the constraints conflict. Worst of all, there is no simple objective function. The decision makers will voice their objectives in terms of "meeting all goals, in the measure that it is possible" rather than in terms of finding the absolute optimum for some single objective function.

A number of authors have done work in the field of multi-criterion optimization in recent years. One thinks immediately of the work of Roy[2] on the ELECTRE system and the work of Charnes and Cooper[3, 4] in applications of goal programming. It is this latter technique that we shall now examine.

3. The Goal Programming Model

Let us consider the case where the constraints as defined in (2) are conflicting, so that there is no feasible solution space, no matter what the objective function may be. In such a case the decision maker will often be interested in a solution which "comes closest" to satisfying all of his constraints. (Later, we will discuss what is meant by "comes closest".)

We will introduce, at this point, a column vector d' to the conflicting constraints of (2). The elements of d' are unrestricted in sign so that the constraints which started out as strict inequalities can now be looked upon as expressions of the <u>goals</u> of the decision maker.

$$Ax' + d' = b' \tag{4}$$

Since the vector d' can have positive or negative elements, the goals can be either overshot or undershot, always bearing in mind that we wish to remain as close to the right-hand side goals as possible.

This problem can be cast in the normal linear programming format as follows: divide the vector d' into the difference between two non-negative vectors r' and s', and minimise the sum of deviations from the goals. In fact, since some goals will be more important than others, we will amost certainly wish to introduce vectors of weights, c_1 and c_2, to reflect this. The problem now becomes:

$$\text{minimise } (c_1 r' + c_2 s') \tag{5}$$
$$\text{subject to } Ax' + r' - s' = b' \tag{6}$$

Note that for any given vector x', there will be an infinite number of pairs of vectors r' and s' which will solve (6), however, since in (5) we are minimising, this number will be reduced to one pair in which either the element of r' or the corresponding element of s will be zero.

Geometrically we can make the following interpretation of the deviation vectors r' and s': since there is no feasible region in x-space defined by (2), we have added further dimensions to form (x, r, s) - space as in (6). There are an infinite number of solutions in this space, however, we wish to select the one that is, roughly speaking, "closest" to being contained in x-space.

4. A Problem in Manpower Planning

Let us examine a manpower system such as that represented in Figure 1. This system can be characterized as a multi-channel hierarchy. People are drawn in to the hierarchy at the lowest level, and within a given occupational classification may be promoted through a number of ranks until finally they join a senior rank that is unclassified. It is also possible that they will not be promoted, or will be lost through attrition. Note that with modifications of detail, one can consider intake at all rank levels, cross-classification transfers and so on.

A common problem of manpower planning within such a system can be described as follows: Given budgetary limitations that restrict the total number of people who can be trained and employed, given operational requirements that fix maximum and minimum numbers that can be employed in a given classification at a given rank, given personnel policies on promotion rates, and given those variables, such as attrition rates, that are not directly under control, what should be the manning plan for the coming three to five years. By manning plan, we refer to the intake quotas, manning levels, and promotion quotas in a given year.

This problem has been formulated as a goal-programming problem by more than one group of authors. Charnes, Cooper, Niehaus, and Sholtz[4] have examined the problem of planning for the civilian workers of the U.S. Navy. Clough, Dudding and Price[6] used a similar approach in a proposal in the manpower planning for pilots of the Canadian Forces. Price and Piskor[7] have reported on the construction of a goal-programming model for officers' manpower planning within the Canadian Forces. It is within the context of this latter problem that we will consider the question of the actual construction of an objective function for a goal-programming model.

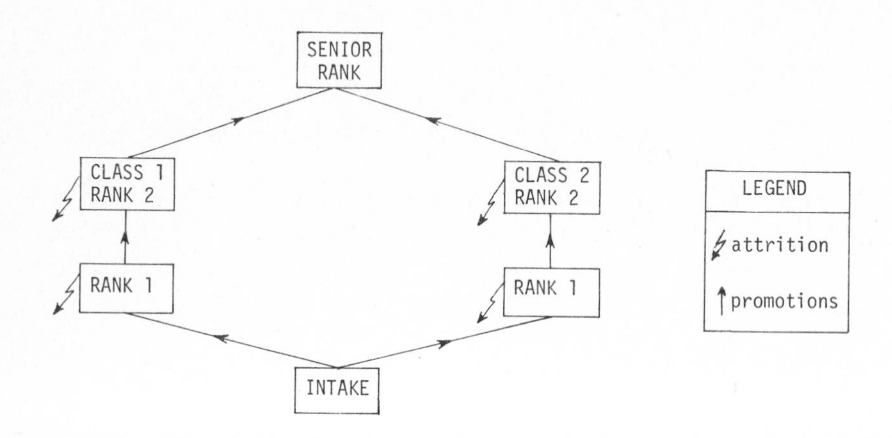

Figure 1: A Simplified Manpower System

CONSTRUCTION OF AN OBJECTIVE FUNCTION

1. Nature of the Organization. Commissioned officers of the Canadian Forces are divided into eight rank levels (referred to as R1 to R8 in this paper) and into thirty-one classifications (referred to as C1 to C31). The first two of these classifications are for senior ranks (R5 to R8) and refer to general service officers and specialist officers such as doctors, lawyers, etc. The next twenty-eight classifications are for lower ranks (R1 through R4) and refer to occupational classifications. The final classification is for officers of ranks R1 through R4 who may not have been allocated to an occupational classification. (For example commissioned officers undergoing training). Classifications C3 through C31 may feed into C1 or C2 but not both. Specialist officers may not hold ranks higher than R6. Recruits are accepted only in R1. A diagram of this manpower system is shown in Figure 2.

2. The Problem of Manpower Planning and Planning Objectives. In this context the problem of manpower planning can be reduced to one of fixing promotion quotas and recruiting quotas for the eight rank levels and thirty-one classifications for a given planning horizon. In our case, a planning horizon of three years, divided into three one-year periods, is used. Notice that by fixing promotion and recruiting quotas, one is also determining the manning levels within each rank-classification state, the total number of officers, the total number of promo-

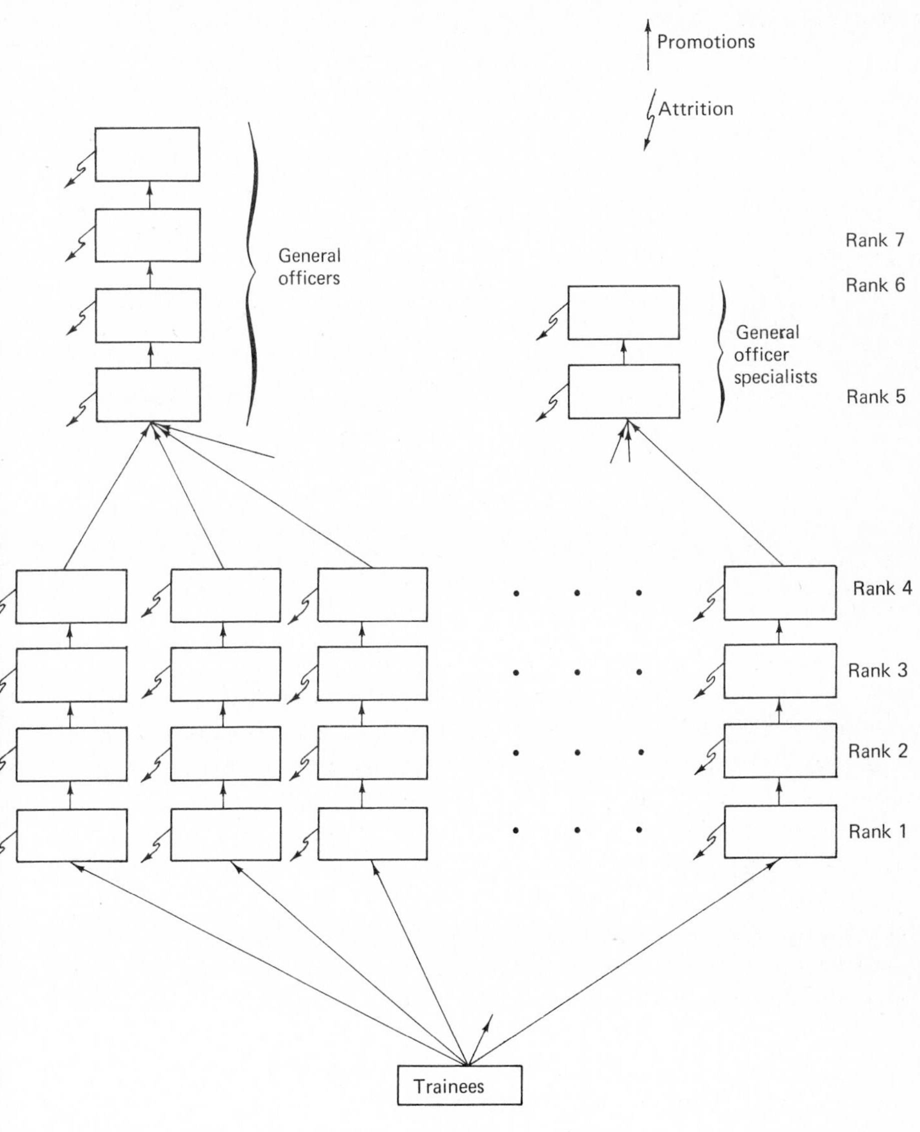

Figure 2

tions, and so on.

It is obvious that such promotion and recruiting quotas are not derived in a vacuum, but are determined taking into consideration a number of overall objectives. These objectives can be summarized as follows:

Budgetary: The budget available for salaries is limited, and therefore so is the total number of officers at each rank level. Given these total end-strengths and an estimate of attrition, the total numbers to be promoted from one rank to the next one can be calculated.

Military Effectiveness: Because of the nature of the work done, it is possible to establish, for each rank and classification, a minimum number of officers below which military effectiveness is impaired. It is also possible to fix a "preferred" manning level for each rank/classification state.

Promotion Equitability: It has been established that in ranks R1 through R4 it is desirable to have as small a variation in promotion rate among classifications as it is possible to attain.

Stability of Rank Structure: It is highly desirable not to have large and rapid fluctuations from year to year in promotion rates and end-strengths, both within rank/classification states and overall, because of the effect on morale and therefore upon effectiveness.

As pointed out earlier, these objectives conflict with each other in some measure. For example, if all classifications are accorded the same promotion rate, at some rank level, then there is an excellent chance of violating the total end strength and total promotion objectives as well as the end strength objectives within certain classifications. Again, if the total promotion objectives are respected, promotion objectives within individual classifications may be violated. Many other such conflicts could be cited.

3. Expression of the Constraints. We will now consider the constraints on the movements of people through this system. A verbal description of these constraints is given below, and the mathematical formulation is summarized in Tables 1 and 2.

(1) Financial Constraints. Budgetary limitations obviously fix the numbers who can be paid, and to provide input to the model are translated into a permitted total strength for each of the eight rank levels. Given these strengths and estimates of attrition, limits on the total numbers of promotions from one rank level to the next are calculated. This procedure is used, rather than one which would use the budget figures within the goal programming model, because no direct trade-off is desired between strengths and promotions of one rank level and those of another. This is a requirement if the organizational structure and the responsibi-

lity and authority position of each rank is to be maintained. Direct trade-offs are permitted at the same rank level between occupational classifications.

(2) Manning Constraints. While the total number of people employed at a given rank level is fixed through financial limitations, the number of people at that rank in a given occupational classification is permitted to vary somewhat. There is a minimum number below which operational effectiveness is affected, and a "preferred" number (treated in practice as a maximum) which will facilitate the maintenance of a career structure within the occupational classification. The strength of a given rank/ classification state should always lie within these bounds.

(3) Promotion Constraints. In order to provide a measure of equitability to personnel, it is desirable to maintain approximately equal promotion rates in all classifications. This is not always possible because attrition rates vary from classification to classification, however,minima and maxima are placed (as goals) on the numbers who may be promoted from one rank of classification to the next higher rank. This has the effect of narrowing the range of variation in the promotion rates.

(4) Manpower Accounting Constraints. For each possible transition, a manpower accounting constraint is inserted to ensure that people are not "created" or "destroyed" at the boundary of two states.

(5) General Officers. Constraints of all the types noted above exist for officers of general rank, however the detailed structure of these constraints varies. In the interest of brevity, and without loss of generality, the description of these constraints is omitted.

4. Expression of the Objective Function.

a. Ranking of the Objectives. It is necessary to rank the policy objectives which have been established because, as has been pointed out, some of these will be in conflict with others (were there no conflicts, the problem of manpower planning would be simple). The model must have a means of deciding which policy will be respected or to what degree it will be violated in cases of conflict. Since policies are represented as constraints within the model, each constraint is assigned a numerical value representing its importance. These weights are determined by three factors:

TABLE 1

DEFINITION OF VARIABLES

SYMBOL	MEANING
i	• rank level ($i = 1, 2, \ldots\ldots 8$)
j	• occupational classification ($j = 1, 2, \ldots\ldots 28$)
t	• time period
$x_{ij}(t)$	• strength of rank i, classification j, at the end of period t
$E_i(t)$	• the total strength permitted for rank i at the end of period t
$y_{ij}(t)$	• the number promoted out of rank i (to rank i + 1), classification j during period t
$P_i(t)$	• the total number of promotions out of rank i which are permitted during period t
φ_{ij}	• a lower bound on $x_{ij}(t)$
Φ_{ij}	• an upper bound on $x_{ij}(t)$
γ_{ij}	• a lower bound on $y_{ij}(t)$
Γ_{ij}	• an upper bound on $y_{ij}(t)$
$a_{ij}(t)$	• attrition from rank i, classification j, during period t
$g_{ij}(t)$	• gains to rank i, classification j, during period t

TABLE 2

EXPRESSION OF CONSTRAINTS

TYPE	NUMBER	FORM
Financial-Total Strengths	4	$\sum_{j} x_{ij}(t) = E_i(t)$
Financial-Total Promotions	5	$\sum_{j} y_{ij}(t) = P_i(t)$
Manning Levels	232*	$\varphi_{ij} \leq x_{ij}(t) \leq \Phi_{ij}$
Promotion Levels	232*	$\gamma_{ij} \leq y_{ij}(t) \leq \Gamma_{ij}$
Manpower Accounting Constraints	116*	$x_{ij}(t-1) - a_{ij}(t) + g_{ij}(t) + y_{i-1,j}(t) - y_{ij}(t) = x_{ij}(t)$
General Officers' Constraints (all types)	33	
TOTAL	622	

Note that an "unallocated" occupational classification has been included in the model to allow for future expansion, and that two "goal variables" are added to each constraint, to complete the formulation.

- type of constraint (policy)
- rank associated with constraint
- classification associated with constraint

(1) Type of Constraint. First a numerical range is assigned to each constraint-type indicating the relative importance of policies. The relations among the ranges used in the current version of the model are shown in Figure 3.

Figure 3.

ORDERING OF POLICIES AS EXPRESSED BY WEIGHTS

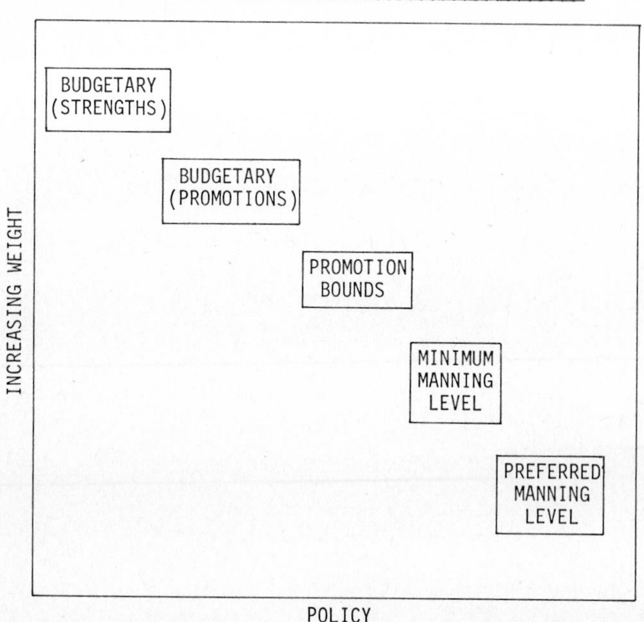

The budgetary constraints are assigned a range of weights much higher than any other policy constraints, thus ensuring that they will be respected. The promotion constraints have the next highest weights (because the promotion bound values for a given rank/classification are used as controls over the solution). The manning level policy constraints follow, with it being judged more important to exceed (or equal) the minimum manning level than to reach the preferred manning level.

(2) Rank Associated with Constraints. Each of the "policy blocks" is divided into four major divisions, one for each of the ranks. For example, the weights associated with minimum manning level policy constraints may range, let us say, from 1600 to 1800, and within this block we obtain R4 minimum manning level constraints between 1700-1800, R3 constraints, from 1700-1750, R2 constraints from 1650-1700 and R1 constraints 1600-1650. This is illustrated in Figure 4.

Figure 4.

MINIMUM MANNING LEVEL POLICY CONSTRAINTS

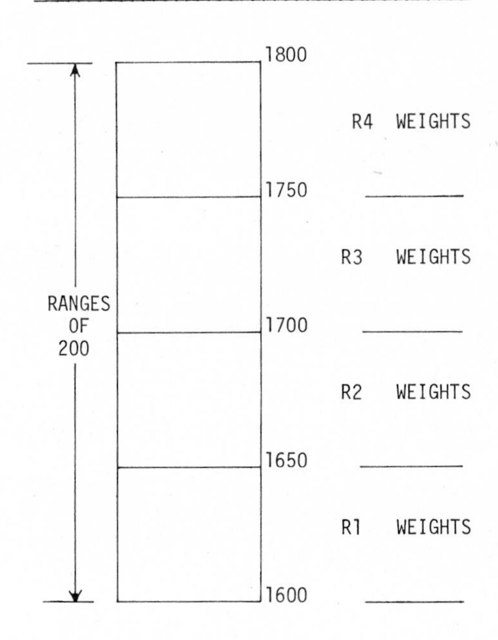

(3) Priority of Classifications. There is no classification that is not necessary to Canadian Forces operations, and policies on promotion equitability ensure that there are no "second-class" classifications. However, for computational reasons certain classifications must be considered before others in the calculation of promotion quotas and end strengths. For example, to a classification with a very low end strength, a change in promotion quota of one man may be critical, whereas

for a classification that is overborne in a given rank, it will not materially affect operational effectiveness of the classification or of the Forces. It is therefore necessary to establish an ordering in which classifications will be considered. This ordering is an input to the program and can be varied at will. All runs up to the present have used an ordering placing classifications which are "lean" towards the top of the list. When ordering the policy constraints, within the policy-type block, and within the rank division, a weight is assigned to a constraint according to the classification ranking. Table 3 shows some examples of the results.

The ranking of classification is an integral part of the model, however in testing, we have not found that minor variations in the ordering significantly affected results. The values assigned to the right hand sides of the constraints control the solution to a much higher degree.

TABLE 3

THE EFFECT OF CLASSIFICATION ON CONSTRAINT WEIGHTINGS

POLICY CONSTRAINT TYPE	RANK	CLASSIFICATION	WEIGHT
Minimum Manning Level	R4	C1	777.4
		C2	775.8
		C3	774.2
Maximum Promotions	R4	C1	2388.7
		C2	887.9
		C3	1887.1
Preferred Manning Level	R1	C1	341.1
		C2	338.7
		C3	336.3
Maximum Promotions	R1	C1	813.7
		C2	812.9
		C3	812.1

b. Exceptions to the Initial Ranking. The initial ranking produced by the procedure described in the previous section is modified in a number of ways to take into account particular circumstances. For example, in a given rank/classification state the maximum and minimum bounds on promotions are equal. This is taken as an indication that it is highly desirable for promotions to equal this figure in the eventual solution, even at the expense of violating manning level constraints or promotion bounds in other classifications. This condition therefore is recognized by the program generating the weights, and the weights for violating these equal bounds are reset to higher levels. The same holds true for cases where the bounds on the manning levels are made equal.

c. Control Over the Solution. To OR practicioners it is a fairly common experience to produce a solution which is "almost" acceptable - where all variables with the exception of a few are in ranges which are acceptable to the decision-maker. Since the objective function, whatever it may be, is only a model, an imperfect representation, of the goals of the organization, it is unwise in many cases for the analyst to insist that this solution be accepted. It is often more reasonable to adjust the values of the offending variables, while disturbing those remaining (and therefore the value of the objective function) as little as possible.

To this end the authors have provided to the user a form of "manual control" over solutions. It has already been noted in sections 5a and 5b that the weights assigned to violation of the promotion bounds are fairly high and are increased if the bounds are made equal. The increase is such that a request for a specific number of promotions will almost certainly be met unless the budgetary constraints would be violated.

Using this facility, a number of critical promotion rates and manning levels can be set at desired values, and the model will automatically adjust the remaining variables in accordance with the objective function. The number of specific requests made should be kept as small as possible, however, so as not to use up all the degrees of freedom available to the model.

COMPUTATIONAL ASPECTS

1. A Numerical Example

This example deals with a manpower system having two occupational classifications, two junior ranks and one senior rank into which people from the two classifications may be promoted. A flow diagram of this system is shown in Figure 5.

Figure 5.

A Three-Rank, Two-Classification System

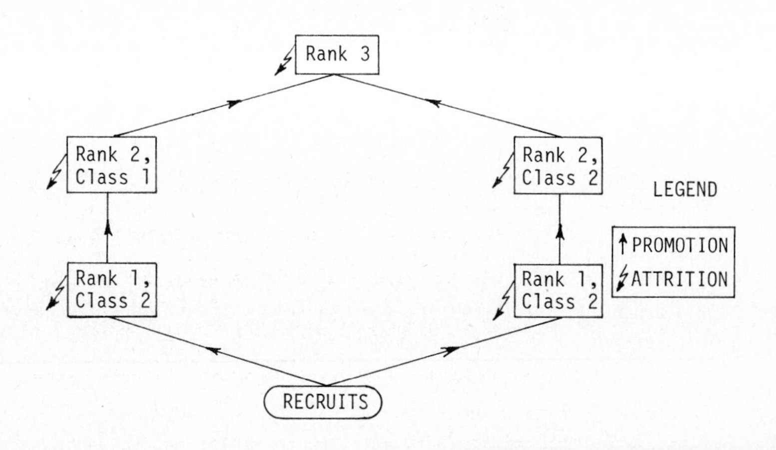

The constraints that govern this system are those which are shown in Figure 6, which contains the following information:

VARIABLES: X_{11} to X_{TOP} - the manning levels in the various ranks and classifications;

Y_{01} to Y_{22} - the promotion quotas (including the intake quotas);

P1 to P26, M1 to M26 - the goal variables;

CONSTRAINTS:

No 1 - 5: accounting constraints;

No 6 -13: constraints giving the desired upper and lower bounds to manning levels;

No 14 -25: constraints giving the desired upper and lower bounds to promotion and intake quotas;

FIGURE 6. PROBLEM CONSTRAINTS

	x_{11}	x_{12}	x_{21}	x_{22}	x_{TOP}	y_{10}	y_{11}	y_{12}	y_{20}	y_{21}	y_{22}		GOALS	OBJECTIVE FUNCTION WEIGHTS: NOT REACHING	OBJECTIVE FUNCTION WEIGHTS: EXCEEDING
Continuity Constraints	x_{11}					$-y_{10}$	$+y_{11}$						82		
		x_{12}					$-y_{11}$	$+y_{12}$					27		
			x_{21}						$-y_{20}$	$+y_{21}$			113		
				x_{22}						$-y_{21}$	$+y_{22}$		41		
					x_{TOP}			$-y_{12}$			$-y_{22}$		5		
Endstrength MAX and MIN	x_{11}											$+p_1 - m_1$	= 100	0	271
	x_{11}											$+p_2 - m_2$	= 90	593	0
		x_{12}										$+p_3 - m_3$	= 35	0	370
		x_{12}										$+p_4 - m_4$	= 30	642	0
			x_{21}									$+p_5 - m_5$	= 125	0	274
			x_{21}									$+p_6 - m_6$	= 115	589	0
				x_{22}								$+p_7 - m_7$	= 50	0	368
				x_{22}								$+p_8 - m_8$	= 42	638	0
Promotions MAX and MIN						y_{10}						$+p_9 - m_9$	= 30	39	700
						y_{10}						$+p_{10} - m_{10}$	= 20	790	0
							y_{11}					$+p_{11} - m_{11}$	= 10	49	815
							y_{11}					$+p_{12} - m_{12}$	= 2	815	0
								y_{12}				$+p_{13} - m_{13}$	= 2	59	840
								y_{12}				$+p_{14} - m_{14}$	= 0	840	0
									y_{20}			$+p_{15} - m_{15}$	= 30	39	798
									y_{20}			$+p_{16} - m_{16}$	= 20	798	0
										y_{21}		$+p_{17} - m_{17}$	= 15	49	823
										y_{21}		$+p_{18} - m_{18}$	= 3	823	0
											y_{22}	$+p_{19} - m_{19}$	= 2	59	335
											y_{22}	$+p_{20} - m_{20}$	= 1	848	0
Endstrenth Totals	x_{11}		x_{21}									$+p_{21} - m_{21}$	= 220	8000	8000
		x_{12}		x_{22}								$+p_{22} - m_{22}$	= 83	9000	9000
					x_{TOP}							$+p_{26} - m_{26}$	= 8	9900	9900
Promotion Totals						y_{10}			y_{20}			$+p_{23} - m_{23}$	= 48	25	25
							y_{11}			y_{21}		$+p_{24} - m_{24}$	= 18	9500	9500
								y_{12}			y_{22}	$+p_{25} - m_{25}$	= 3	9900	9900

No 26-28: financial constraints giving the ceilings on rank manning totals;

No 29-31 financial constraints giving the ceilings on promotion and intake quotas.

The objective function weights are shown at the right of Figure 6.

Figure 7 shows the objective function weights that were determined according to the methods set out in the paper. Figure 8 shows the manning totals and promotion quota totals that must be adhered to (these are the right-hand sides of constraints 26-31). In Figure 9 for each rank and classification, we find the following information:

RESIDUAL STRENGTH: The right-hand sides of constraints 1 - 5. Referring to the manpower accounting constraints of Table 2, it is the quantity.

$$x_{ij}\ (t-1) - a_{ij}\ (t) + g_{ij}\ (t)$$

MANNING LEVELS: The right-hand sides of constraints 6 - 13.

PROMOTION LEVELS: The right-hand sides of constraints 14 - 25.

MANNING PLAN: The calculated values for manning levels, promotions quotas and intake quotas, resulting from the input goals and objective function.

The MPSX 'PICTURE' output for this problem is given in Figure 10.

OBJECTIVE FUNCTION WEIGHTS

CONSTRAINT TYPE	CLASSIFICATION	RANK	PENALTIES FOR NOT REACHING	EXCEEDING
ACCOUNTING	ALL	ALL	N/A	N/A
MANNING LEVELS				
MAX	1	1	0	271
MIN	1	1	593	0
MAX	1	2	0	370
MIN	1	2	640	0
MAX	2	1	0	274
MIN	2	1	589	0
MAX	2	2	0	368
MIN	2	2	638	0
PROMOTION LEVELS				
MAX	1	0	39	790
MIN	1	0	790	0
MAX	1	1	49	815
MIN	1	1	815	0
MAX	1	2	59	840
MIN	1	2	840	0
MAX	2	0	39	798
MIN	2	0	798	0
MAX	2	1	49	823
MIN	2	1	823	0
MAX	2	2	59	335
MIN	2	2	848	0
MANNING TOTALS				
		1	8000	8000
		2	9000	9000
		3	9900	9900
PROMOTION TOTALS				
		0	5000	5000
		1	9500	9500
		2	9900	9900

NOTE: Rank of 0 refers to intake quota

Figure 7.

RANK	MANNING TOTALS	PROMOTION TOTALS
INTAKE	N/A	48
1	220	18
2	83	3
3	8	N/A

Figure 8.

CLASSIFICATION	RANK	RESIDUAL STRENGTH	MANNING LEVELS MAX	MIN	PROMOT. LEVELS MAX	MIN	PLAN STRENGTH	PROMOTIONS
1	0	——	——	——	30	20	——	23
1	1	82	100	90	10	2	96	9
1	2	27	35	30	2	0	35	1
2	0	——	——	——	30	20	——	20
2	1	113	125	115	15	3	124	9
2	2	41	50	42	2	1	48	2
——	3	5	——	——	——	——	8	——

Figure 9. NOTE: Rank of 0 refers to intake quota

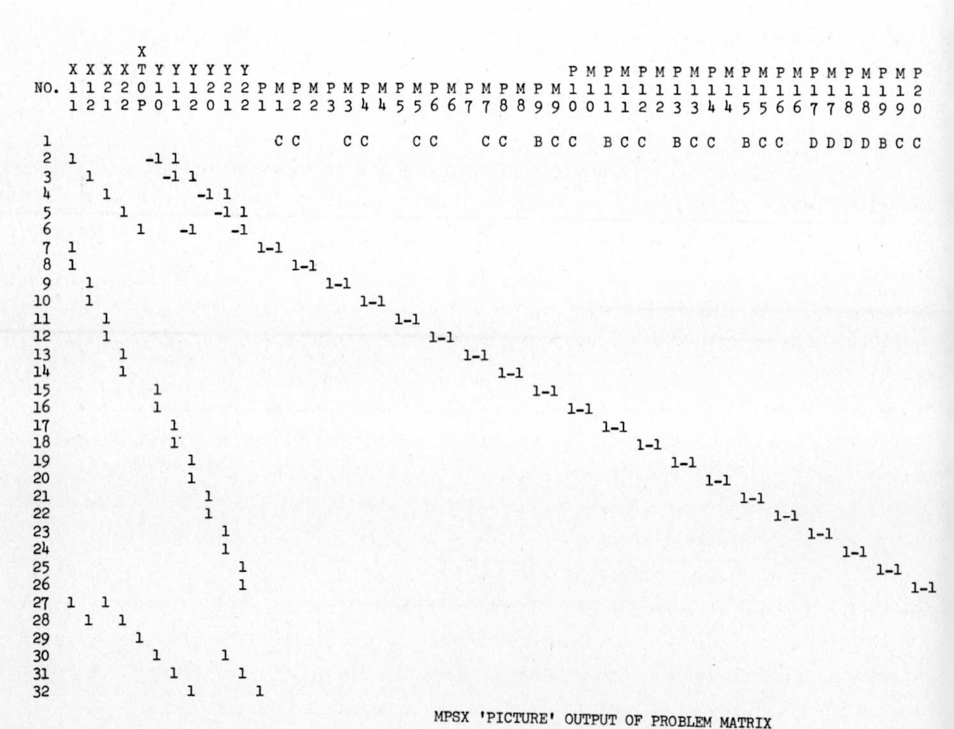

MPSX 'PICTURE' OUTPUT OF PROBLEM MATRIX

Figure 10.

2. SOLVING A LARGE PROBLEM

a. Setting up the Constraint Matrix

In even moderate size problems, entering the problem matrix via punched cards in the MPSX format would be a time-consuming error-prone process. Fortunately, there is an alternative for many problems, including the goal-programming problem described in these lectures. It can be seen that many of the constraints resemble each other, with only different variable subscripts to distinguish them. In this case, it is then possible to write a computer program that will generate the constraint matrix and store it in MPSX format, on a magnetic device. Not only is this process quicker than the hand punching of cards, but error detection is often easier because many data elements will be affected, making the error more visible.

b. The Design of a Software System

Once the constraint matrix has been set up it will most likely be changed but rarely. The objective function and right-hand sides are another matter however. The decision-maker will probably often want to change his goals and the priorities assigned to them. It is not, therefore, practical to enter RHS and objective function data on punched cards or even through a "once off" matrix generator. In addition, these data will probably be of continual interest to the decision maker, who will wish to see them presented in a form more familiar, and more easily read, than the MPSX input format.

One solution to this problem is to construct, on a magnetic device, a separate data base to store input data. The data can be formatted as required for MPSX by a computer program, and the initial problem modified using the MPSX "REVISE" instruction.

When the problem has been solved, the solution values can be written unto the same storage unit, and output reports, in a form acceptable to decision maker, can be printed.

A functional flowchart of the system used in solving the problem described in Lecture 2 is given in Figure 11.The UPDATE program is used to maintain the manpower data base, which contains both input data and solution values. It checks the data format syntax, performs a number of checks (for example, it verifies that lower bounds are in fact smaller than upper bounds), issues any required diagnostic messages, and modifies the data base. Once the data base is satisfactorily amended, the data are translated to the MPSX format, the linear programming problem is solved, and the solution values are transferred back to the data base. Various characteristic reports are printed so that the planners can detect any unexpected effects before proceeding. When an acceptable solution has been obtained, reports are printed for management.

Figure 11

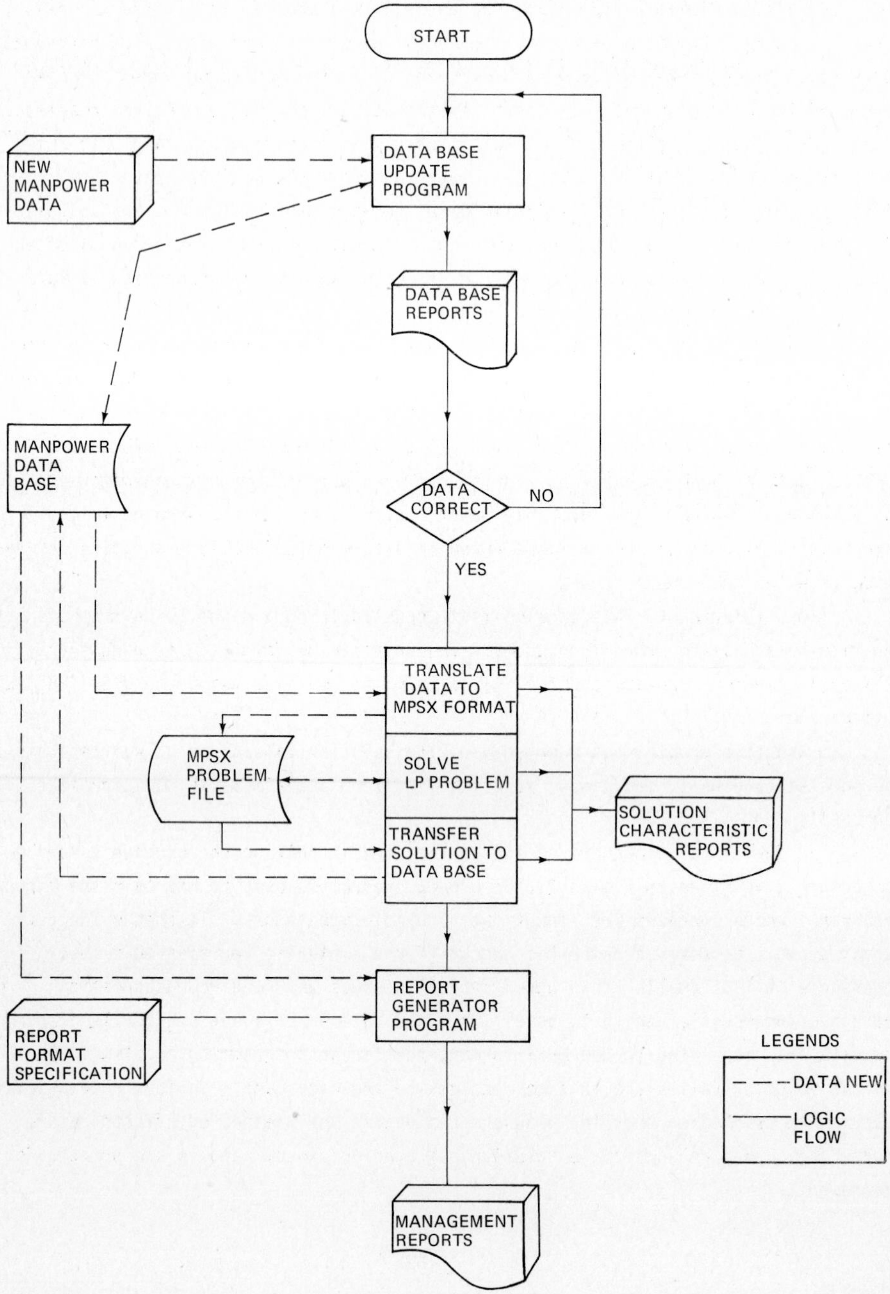

START

NEW MANPOWER DATA

DATA BASE UPDATE PROGRAM

DATA BASE REPORTS

MANPOWER DATA BASE

DATA CORRECT

NO

YES

TRANSLATE DATA TO MPSX FORMAT

MPSX PROBLEM FILE

SOLVE LP PROBLEM

TRANSFER SOLUTION TO DATA BASE

SOLUTION CHARACTERISTIC REPORTS

REPORT GENERATOR PROGRAM

REPORT FORMAT SPECIFICATION

LEGENDS

– – – DATA NEW

——— LOGIC FLOW

MANAGEMENT REPORTS

BIBLIOGRAPHY

(1) Eilon, S; "Goals and Constraints in Decision Making"; ORQ, vol 23, no. 1; March 1972

(2) Roy B; "Décisions avec critères multiples: problèmes et méthodes"; Revue METRA, vol XI, no. 1, 1972

(3) Charnes, A. and W. W. Cooper; "Management Models and Industrial Applications of Linear Programming"; (New York; John Wiley and Sons Inc, 1961)

(4) Charnes, A., W. W. Cooper and R. J. Niehaus- "A Goal Programming Model for Manpower Planning" Management Science in Planning and Control; (Technical Association of the Pulp and Paper Industry, New York 17, New York; 1968).

(5) Charnes, A., W. W. Cooper and R. J. Niehaus, and D. Sholtz; "A Model for Civilian Manpower Management in the U.S. Navy"; Models of Manpower Systems; (A. R. Smith, Editor; English Universities Press, 1970).

(6) Clough, D. J. and W. P. McReynolds: "State transition model of an educational system incorporating a constraint theory of supply and demand"; Ontario Journal of Educational Research; vol 9, no. 1 (Autum, 1966).

(7) Price, W. L. and Piskor W. G.; "The Design of an Objective Function for a Goal-Programming Manpower Model"; Annual Conference of Canadian OR Society, Toronto, Canada; June 5-7, 1972

Mathematical Programming in Theory and Practice,
P.L. Hammer and G. Zoutendijk, (Eds.)

The DOAE Reinforcement and Redeployment Study: A Case Study in Mathematical Programming

by

E.M.L. Beale[1], G.C. Beare[2] and P. Bryan Tatham[1]

Abstract

This paper illustrates the technical aspects of formulation and running a large mathematical programming model. The model is concerned with minimizing transportation costs in a defence context, but the methods used are equally relevant to similar problems in industry and commerce.

The early history of the project is described by Beare and Mc.Indoe (1970). This is reviewed. The main emphasis of the paper is on the development of a revised matrix generator program in 1971 that followed the input data to be expressed in user (i.e. military logistic) terms rather than in algebraic terms. This was done by building up the algebraic formulation and the specification of the input data formats in parallel. It is typical of this type of work that major changes in the algebraic formulation only stopped when the client had completed the task of expressing a real problem on our data sheets.

One technical problem of formulation concerns the representation of continuous time in a discrete time-period model. This inevitably requires approximations. But it is shown how the use of a sophisticated special-purpose matrix generator enables one to use better approximations than would otherwise be convenient.

1) Scientific Control Systems Limited
2) Defence Operational Analysis Establishment

---

1. INTRODUCTION

There is a mass of literature on the techniques of operational research in general, and mathematical programming in particular. There is also a fair amount of literature on applications of these techniques, at least at the level of outlining the type of problem studied and the type of mathematical model used. Some books and papers include detailed formulations and results for small-scale models. This paper is an attempt to indicate what is involved in setting up a large-scale mathematical programming model. This is done with reference to a particular project on which we have been working since the middle of 1971. The project is concerned with minimizing transport costs in a defence context, but the methods used are equally relevant to similar problems in industry and commerce. The project seems particularly suitable as an illustration because it has already been discussed is the open literature, and the recent developments on which we concentrate do not present any security problems.

As with many applications of mathematical programming, the problem required a few special, and possibly original, pieces of formulation technique. But we hope that the main interest of the paper will be in the unoriginal parts, since we believe that they represent a fairly typical large-scale practical mathematical programming application. The idea that an application is only worth publishing if it is atypical does not help to disseminate an understanding of practical applications work.

Section 2 of this paper discusses the general methodology of formulating a class of mathematical programming models in a form suitable for the development of an effective matrix generator. We then turn to the DOAE Reinforcement and Redeployment Study. Section 3 discusses the background to the project. Section 4 discusses the development of the current matrix generator in the light of the principles of Section 2. Section 5 discusses some specific formulation problems arising in this work: these are largely concerned with the problem of making a reasonably accurate approximations to effects that really depend on continuous time in a model working with a small number (here 2) of discrete time steps. Finally Section 6 discusses some of the computer programming aspects of the development of the matrix generator and output analyser.

In this presentation we have tried to steer a middle course between vapid generality and a mass of technical detail. For the sake of simplicity we have omitted a few self-contained features of the practical model that seem of little general interest. But the tables at the end of the paper include otherwise complete lists of the subscripts, variables, constraints and data sheets used in the model. These include some details not discussed elsewhere, in the hope that this will convey more of the flavour of the project than would otherwise be possible in a paper of reasonable length.

2. GENERAL METHODOLOGY OF MATRIX GENERATION

Almost everyone who knows what mathematical programming is knows that one normally uses a computer to solve practical mathematical programming problems. Most of these people also know that one normally uses the computer to translate the numerical data for a particular application into a suitable form for input to a general computer program for linear (or nonlinear or integer) programming. Such a general computer program is known as a Mathematical Programming System, and the translation process is known as Matrix Generation. The computer can also be used to translate the output from the mathematical programming system into a form that is particularly convenient for the application. This process is known as Output Analysis.

The possibility of writing computer programs for matrix generation and output analysis arises essentially from the compactness of an algebraic formulation that uses subscripts and summation signs. By including summation over one or more subscripts, one can represent in a single line of algebra an equation connecting a large number of variables. Furthermore, by including other subscripts not covered by summation signs, one can extend the same single line of algebra to represent several similar equations. The mathematical programming system requires each nonzero numerical coefficient of each variable in each equation to be supplied individually. But, given the algebraic formulation, a computer program can do this very easily using FORTRAN DO loops, or similar facilities in other high or low level computer programming languages.

It is therefore natural to start the development of a computer system to solve any class of mathematical programming problems with an algebraic formulation. In practice if the system is modified and extended, all the documentation may fall behind the computer program developments, but this habit is not to be encouraged and is in any case hardly feasible for an initial formulation. But it is quite hard to write an algebraic formulation of a complex mathematical programming model in a way that someone other than the original formulator can understand it. Our recommendations on this matter are summarized in Table 1. It will be seen that we suggest listing the subscripts, sets, constants, variables and constraints in that order.

This may seem a perverse approach, since with a very simple model one can go straight to the heart of the matter, which is the constraints, and then explain them by defining the constants and variables. The subscriptions are then defined implicitly in the course of the other definitions.

Our approach is recommended for the following reasons:

a) In a model of any complexity, the constraints are hard to appreciate until the terms in them have been defined.

b) An explicit definition of the subscripts at the outset conveys a great deal of information about the scope of the model. Once the subscripts have been defined, they need not be repeated in the definitions of constants and variables using them. These definitions are then much abbreviated. Furthermore, this approach strongly discourages the habit of using two different letters as symbols for the same type of subscript to different constants or variables, and the equally confusing habit of using the same letter as a symbol for two different types of subscript.

c) Sets may not be needed at all in simple models, and will generally be developed at a relatively late stage in the formulation. But they often provide the only compact way of giving a precise definition of domains of summation, or conditions under which a constraint, a variable, or a constant, is defined. It is expedient to list them immediately after the subscripts, with which they are closely associated.

d) The constants, variables and constraints are normally developed in parallel. But it is expedient to list the constants first, since they implicitly define the model for the practical user who is willing to trust the competence and integrity of his mathematical collaborators. Such a user need only study the definitions of the constants, together with the subscripts and sets used in these definitions, to understand the scope of the model.

e) Given the constants, it is natural to proceed to the decision variables. The constraints are then comprehensible in terms of what has gone before.

f) It seems very desirable to keep a sharp distinction between the constants (or assumed quantities) and the decision variables to be determined by the model. We therefore use capital letters consistently for constants and lower case letters for decision variables (and subscripts).

g) When it is hard to find enough mnemonic letters to represent all the constants and variables of a problem, it is natural to use combinations of two or more letters. But it seems best to maintain the mathematical convention that multiplication is implied when two alphanumeric symbols are written side by side and not as subscripts. When a combination of letters is used to represent a single quantity we therefore write all letters after the first as subscripts. These "literal" subscripts are written as capital letters to distinguish them from subscripts that can take numerical values, which are written as lower case letters.

One other more detailed point of general methodology deserves a brief discussion. In a general algebraic formulation, it often happens that a set of constraints are of basically the same form as each other except that one or two members of the set include a few terms that are missing from the other members, or alternatively omit a few terms that are present in the others. For example the first member of a set of material balance constraints in a multi-time-period model may include a constant term representing the initial stock and may correspondingly omit a variable representing the stock at the end of the previous time period. In these circumstances it may then seem easiest to write down the exceptional cases as other new types of constraint and not to regard them as members of the set. But this is not recommended. The terms that are common to all members of the set have to be written twice, and the corresponding piece of program to produce these terms in the matrix generator must be duplicated - unless the matrix generator is written independently of the algebraic formulation which is probably even more undesirable.

A convenient notation to cope with these situations accurately is the following:

Define E_{Sij} as a constant equal to 1 if j = i and to 0 otherwise.

Define E_{Dijk} as a constant equal to 1 if k = i or k = j, and to 0 otherwise

Define $E_{Tijk\ell}$ as a constant equal to 1 if $\ell = i$ or if $\ell = j$ or if $\ell = k$, and to 0 otherwise.

One can then write for example $(1-E_{Dijk})$ to represent a coefficient equal to 1 if $k \neq i$ and $k \neq j$ and equal to 0 otherwise.

Note that E_{Sij} could also be written as a Kroneker delta. But this is not such a convenient notation in mathematical programming work, where δ is often used as a symbol for a zero-one variable. And there is no simple standard extension of the Kroneker delta notation to represent E_{Dijk}.

3. BACKGROUND TO THE REINFORCEMENT AND REDEPLOYMENT STUDY

We now turn to the main topic of this paper, which is the DOAE Reinforcement and Redeployment study. This section outlines the work on this study up to the middle of 1971. The main features of this work are described by Beard and McIndoe (1970).

The Study started from a decision in 1966 by Her Majesty's Secretary of State for Defence to set up a working party to study the most economical combination of aircraft, ships and stockpiled material which will provide adequate strategic mobility for British Forces in the future. The Defence Operational Analysis Establishment (DOAE) were commissioned to carry out analysis for the working party - on which they were represented.

It is clear in general terms that this type of problem in amenable to linear programming. This will be developed more fully in Section 4. But in fact DOAE decided that they should use Integer Programming, because some of the decisions variables could not be represented very naturally as continuous variables. Specifically the integer variables were

a) Indicator variables, representing the presence or absence of stockpiles at particular sites, since each stockpile has a significant fixed cost if it is used at any non-zero level.

b) The numbers of ships of each type to be purchased. Some of these were small integers.

c) Indicator variables to decide which of two alternative combinations of forces should be used in situations where either would be acceptable but a weighted average would not. The example quoted by Beard and McIndoe is the choice between an assault by paratroops or by Royal Marine Commandos.

When this model was first formulated, CEIR Ltd., as SCICON were then called, were probably the only organization with a significant mixed integer programming capability. By 1972 standards our LP/90/94 system, which ran on the IBM 7094 Computer, was very primitive. But it was capable of tackling DOAE's problems, which had 850 constraints, about 2000 continuous variable and 15 integer variables. The computational aspects of this work are described by Shaw (1970).

The original studies considered possible deployments to many different parts of the world. Subsequently, policy switched to a greater emphasis on the European role. Fortunately the same mathematical methods were appropriate to studies of the problems of deploying British Forces from the United Kingdom to support NATO on the continent of Europe in the event of an emergency.

Studies continued for several months. Their essential purpose was to indicate the total cost of providing the transport capabilities needed to meet any commitment HM Government might make to NATO, as well as suggesting how these commitments could best be met.

It is perhaps of interest that later work was done using a purely linear programming model. It had been found that, particularly with the European scenarios, similar results were obtained by solving the models in linear programming terms as when they were solved as integer programming problems. Linear programming has two important advantages over integer programming:

1) It is quicker

2) It allows a much more comprehensive range of post-optimal analysis.

In particular, paramatric programming can be used to see how the total cost varies as one changes such things as the permitted foreign exchange component of the total cost.

There was then a lull in the computations of over a year. But in the Spring of 1971 DOAE returned to Scicon for more computing on the same model but with updated numerical values for several parameters. These runs were ultimately successful, but they caused considerable difficulties. Such difficulties often arise when a complex computing project has to be re-activated for an urgent study after lying dormant for a long time. However the main reason was the fact that the only matrix generator for the model was the one described by Shaw (1970), which required input data in the form of rows of coefficients of named variabled in named constraints, rather than in the form of basic data user (i.e. military logistic) terms. Once this study was completed, we therefore turned to the task of developing a matrix generator and output analyser for Reinforcement and Redeployment studies. This task is the topic of the rest of this paper.

4. THE DEVELOPMENT OF THE CURRENT MATRIX GENERATOR

The development of the matrix generator for the Reinforcement and Redeployment study was based on the general principles described in Section 2.

The main features to be represented in the model were as follows:

1) The requirement to supply specified amounts of material of different types to specified concentration areas within specified times. A problem might involve more than one option, each defined by such a set of requirements. The problem is then to find the cheapest mix of facilities that could meet any one option, without necessarily being able to meet more than one option simultaneously.
2) The standard method of meeting any option is to supply the material from the United Kingdom. It can be sent to the continent of Europe by either sea or air.
3) Some ports or airfields of destination may be in concentration areas. But others may not, and consideration has to be given to the problem of sending on material by rail or road from a port or airfield of

destination to the appropriate concentration area.

4) The movement schedule must be consistent with capacity limitations at

    Ports or Airfields of Embarkation
    Ports or Airfields of Destination
    Railway stations of destination
    Road bottlenecks.

   They must also be consistent with the availability and capacities of ships, aircraft, trains, tank transporters (for the road transport of tanks), lorries (for the road transport of freight and ammunition) and buses (for the road transport of personnel). The program is free to chose to buy more military aircraft or ship. The other transport facilities are civilian and can be used only to the extent that they are assumed to be available.

5) The other way to reduce the transport requirements is to stockpile material. Stockpile sites on the Continent of Europe are assumed to be associated with either a concentration area, or an airfield or port of destination. This is because if Armoured Personnel Carriers (APCs) or lorries are stockpiled and are to be transported by road to their concentration areas, then drivers must be supplied from the UK. These drivers will then contribute to the required build-up of personnel in the concentration areas to which they drive.

6) We had to give careful thought to the facilities needed in the general formulation to allow consideration of possible overseas stockpile sites not on the Continent of Europe. For example one might want to assess the desirability of stockpiling material in an allied country close to the concentration area. It is often hard to decide just how much generality to provide in the model in such circumstances. Our final compromise between technical convenience and possible requirements was to say that such a stockpile site may have an associated Non UK airfield of embarkation, or an associated Non UK port of embarkation, or both. At most one stockpile site may be associated with any such airfield or port. But we have not restricted the pattern of air or sea routes from these airfields or ports.

7) The permissible flows of material or personnel are defined by specifying.

   - Air routes from airfields of embarkation to airfields of destination.
   - Sea routes from ports of embarkation to ports of destination.

Rail and road routes from airfields of destination, ports of destination or stockpile sites to concentration areas.

(The possibility that one might need to consider a rail or road link from a stockpile site in one concentration area to another concentration area was not considered in our original formulation. It was a requirement that emerged from our first set of realistic test data for the system, and forced us to define additional types of variable).

Armed with this general understanding of the problem, we could already write down a complete list of subscripts. This in longer than in any other mathematical programming model on which any of us has worked,and we have had to use rather unmnenonic symbols for some subscripts. Otherwise there is nothing very remarkable about the list,which is reproduced in Table 2.

It is worth noting that a study of the list of subscripts, together with a very general understanding of the nature of the problem and some general knowledge of linear programming formulations, implicity defines the scope of the model in fair detail.

The next stage of the work was to make this detail more precise. We therefore developed the lists of sets, constants, variables, and constraints as suggested in Section 2.

At the same time we concentrated on the definition of the input data quantities and formats. There is an important methodological lesson here. In theory, the logical approach to the definition of a model is to develop its mathematical form first, in order to identify the quantities whose numerical values are required from the input data. We can then consider how these data can be found, or derived from more fundamental data. But this approach often leads to the definition of large multi-dimensional arrays. For example we could define an array of constants A_{adfgkt} representing the number of kilotons of Load Type d that can be carried in Aircraft Type a from Airfield f to Airfield g in Time Period t under Option k. But we must then consider how such data could be captured. It is clear that they must depend on

a) the payload of the aircraft for the load type

b) the aircraft speed

c) the distance between the airfields

d) the time the aircraft spends on the ground per round trip

e) the length of the time period under this option.
These are the basic data from which any element in our 6 dimensional array can be computed.

But when we turn to the basic data we find two other factors which are not revealed by this approach. One is that the payload of the aircraft may be reduced over certain routes because they involve long non-stop flights needing extra fuel. This does not affect the form of the model. It is just an extra term in the calculation. But the other factor is more interesting and is discussed below.

When aircrafts are loaded with vehicles, there may be some additional space that cannot be used for more vehicles but which can be used to carry freight or ammunition. And when the primary load of the aircraft is either vehicles, freight or ammunition, there may be some additional space for personnel.

To allow for the personnel carried, the LP variable (decision variable) representing the number of flights must be given coefficients in the material balance row for personnel at the airfield of destination (or concentration area if the airfield is in a concentration area and has no onward rail or road link). We do not need a material balance row at the airfield of embarkation since we are not studying transport within the UK to the ports or airfields of embarkation.

But to allow for the freight or ammunition carried we have to define an additional row at each airfield of destination in each time period, representing the fact that the total amount of freight flown from the UK plus the total amount of ammunition flown from the UK is less than or equal to the total capacity for carrying freight or ammunition provided by flights from the UK by aircraft of all types.

This incident illustrates the fact that the precise definition of the input data quantities and formats provides the user with an even more complete description of the mathematical formulation. It is typical of this type of work that major changes in the algebraic formulation continued until DOAE had completed the task of expressing a real problem on our data sheets.

We close this section with one other illustration of how an apparently innocent extension to the input data made a significant addition to the algebraic formulation. The requirements of the system were originally defined by the quantities D_{Ccdkt} representing the number of kilotons of Load Type d that must be supplied to Concentration Area c in Time Period t under Option k. But further quantities $D_{AMINcdkt}$ were then introduced representing the numbers of kilotons that must be delivered directly by air from the United Kingdom. At first sight, these mandatory air loads represent just one more constraint for each load type in each concentration area for each time period under each option. But this is an oversimplified view because the airfield need not necessarily be in the concentration area. We therefore define a maximum permitted distance of the airfield from the concentration area, and include as contributions to the mandatory air loads material sent by rail or road from airfields that are near enough. This in turn causes possible confusion with stockpiled material held near the airfield of destination or (theoretically) with stockpiled material flown from an overseas stockpile site to the airfield of destination. To avoid this confusion we define separate material balance rows for stockpiled material at airfields of destination, and introduce separate variables for rail or road transport of stockpiled material.

Space does not permit a complete description of the model. But the list of variables is given in Table 3, and lists of types of constraint and types of input data card are given in Tables 4 and 5 respectively. Some more mathematical aspects of the formulation are discussed in the next section.

5. SOME FORMULATION PROBLEMS

Most of the formulation of the Reinforcement and Redeployment Study is straightforward once the factors to be represented have been completely identified. It follows from the principles described in Chapter 8 of Beale (1968). Most of the constraints can be identified as Capacity Rows or Material Balance Rows. But one specific and one more general problem derserve further discussion.

The specific problem concerns methods of representing the permissible combinations of loads on a ship. Ships of type v have a capacity of V_{KTSdv} kilotons for Load Type d, but each kiloton of capacity for Load Type d can be converted into space for V_{CONdv} kilotons of Load Type (d+1) for d = 1,2,3,4,5, or 6.

There are two alternative natural approaches to this problem. One is to define modes of loading a ship, representing all extreme points of the set of feasible loads. We then have variables $x_{Pkmpqtv}$ representing the number of voyages made by ships of Type v loaded in Mode m from Port p to Port q in Time period t under Option k. The other is to omit the subscript m from these variables but to introduce variables w_{Qdkqtv} representing the amount of Load Type d carried on ships of Type v from the UK to Part q in Time Period t under Option k and variables w_{Tdkqtv} representing the number of kilotons of load type d that could have been shipped from the UK omitted to provide space for load type (d+1). We must then introduce the constraints

$$\sum_p V_{KTSdv}\, x_{Pk\,pqtv} + (1-E_{S1d})\, V_{CONd-1,\,v}\, w_{Td-1,kqtv} - (1-E_{S7d})\, w_{Tdkqtv} - w_{Qdkqtv} \geqslant 0.$$

The disadvantage of the first of these approaches is that it introduces many more variables, and many more nonzero coefficients, which may slow down the linear programming code. The disadvantages of the second approach are that it introduces more rows and that it does not allow precise representation of two features of the situation. These two features are the fact that the time for loading and unloading a ship depend on whether or not it is carrying more than the minimum loading of freight and ammunition, and the fact that port handling capacities are measures in kilotons per day. The first of these features can be included by defining separate x_p, w_T and w_Q variables, and separate extra constraints, for vehicle-carrying and freight-carrying voyages. The second feature can only be handled correctly by a very large expansion of the model.

We have adopted a compromise between these two approaches. This increases the complexity of the formulation but gives less than half the disadvantages of each approach. Our compromise is to use modes but not to distinguish at this stage between Load types 1, 2 and 3. These loads are sorted out using w_T and w_Q variables.

Our more general formulation problem concerns the representation of continuous time in a model working with discrete time periods. This type of model inevitably creates some approximations in the capacities of fixed facilities such as ports and airports as well as in the capacities of moving facilities such as ships.

Port capacities are measured in kilotons per day. But the port capacity during a time period is not really measured by the number of kilotons per day multiplied by the number of days in the time period, because a port of destination cannot start operating until the first ship arrives, and neither type of port can contribute to the operation unless the material loaded or unloaded can reach a concentration area before the end of the last time period. There is no simple exact solution to this problem. But we have introduced an approximate solution by calculating, for each fixed facility, the number of hours after the start of the first time period when the facility could first be used and the number of hours before the end of the last time period when the facility could last be used. The sum of these two times is substracted from the total length of the first time period to determine the effective duration of this time period for the facility. This correction is only a first approximation, but it can easily be made in a matrix generator without any additional input data, and it is therefore worth making. We call the solution that calculates these times CPM, since the work is identical to the calculation of early and late event times in Critical Path Method calculations.

The calculation of ship capacities is rather more difficult. To a first approximation we can simply divide the duration of a time period by the duration of a round trip along a particular route to determine the number of voyages the ship can make along this route in the time period. But this calculation is unsatisfactory if the resulting number is very small. We therefore proceed as follows.

We first calculate the duration of a one-way trip, including loading and unloading. If this exceeds the duration of the time period, then this type of ship cannot usefully undertake this type of voyage at all. Otherwise the number of voyages is taken to be one plus the number of complete round trips that can be undertaken within the time period in addition to this first one-way trip. So the number of voyages is always an integer, although it may be only 1. But we then allow for the fact that the ship may be at sea returning from one of these voyages at the end of the first time period. It cannot then be assigned to second time period duties until it has returned home, and in these circumstances we insert a coefficient into the constraint for ship availability in Time Period 2, representing the number of hours lost before the ship returns to its port of embarkation.

Similar considerations determine the calculation of ship carrying capacity for stock piled material from non UK ports of embarkation. We assume that

a) only military ships will be used for this work.

b) all ships start in the United Kingdom, and their first voyage will be to a port of destination.

c) that any military ship arriving at a port of destination that can be supplied from non-UK ports of embarkation will, after it has been unloaded, start to move material from those non-UK ports of embarkation, provided that at least one round trip can be made.

In this way we compute the total number of ship days available for operations from non UK ports of embarkation to each port of destination. There is then a constraint saying that the number of ship-days required for all voyages from these non UK ports must not exceed the number of days available.

If a ship is so employed in Time Period 1, it may take some time to sail back to the UK at the beginning of Time Period 2. This time is therefore deducted from the number of ship days available in Time Period 2. But we have introduced the variable e_{Dqk} representing the number of ship-days for voyages from non-UK ports of embarkation to Port q given up to allow the ship to return early to the UK. This has a negative coefficient in the ship capacity row for Time Period 2 equal to the ratio of the number of days in Time Period 2 saved by allowing the ship to return immediately after the first onloading at Port q divided by the number of days given up in Time Period 1 by this action.

All these calculations are easily carried out without any further input data. They are therefore worth while as a fair second approximation to the true transport capacities under the conditions specified by the model.

6. COMPUTER PROGRAMMING ASPECTS

The matrix generator and output analyser are written in FORTRAN rather than using a more specialized matrix generator facility such as Scicon's Matrix Generator MGG. The reasons for this are as follows:

a) MGG is only available for the Univac 1108 computer, and it is desirable to be able to run this model on a range of computers.

b) At the time this project was started, MGG generated matrices only in Share standard input format, using 6 character row and column names. Since some variables have 6 subscripts it is desirable to use 8 character names, i.e. MPS format. This format is now acceptable to all major mathematical programming systems.

c) With the considerable amount of input data required for the model, it seemed very important to allow these to be presented, checked and printed out in the way most convenient to the user.

It was therefore desirable to hand code the input section of the program in FORTRAN in any case. This in itself is entirely compatible with the use of MGG to generate the matrix. But the model inevitably requires a substantial number of specially written subroutines to calculate the coefficients. We therefore do not believe that any higher level programming language could have helped us significantly, although MGG is a very useful tool with simpler models, and indeed with any model for people with less experience at writing matrix generators. MGG also has the considerable advantage of being self-documenting, provided it is rerun whenever the model formulation changes.

The form of the model was decided in July and August 1971; but no one was working full time on the project in this period, although one of us (GCB) was extremely busy with computer running on the old model. Serious programming started in September 1971, and the first two months were devoted to the input section. This simply reads the data, checks them for consistency, prints them out in a form suitable for the user to check them for accuracy, and stores them for later use. During this time, DOAE collected data for a test problem using our formats; and as noted earlier this process led to a moderate number of changes to the formats and extensions to the model.

We are sure that it was right to concentrate on this side of the work first, even to the extent of taking great care over the layout on the printed page of both the normal output and the error messages. When it was finished we had something that could have been criticized as an impressive facade with nothing behind it. But such criticism overlooks the fact that - unless there are serious mathematical difficulties over the formulation - data capture is always the most time-consuming part of the development of a matrix generator.

The development of the matrix generator section of the program took about another month. It required a lot of FORTRAN, and in particular a large number of subroutines since we use one subroutine for each different type of variable. But as usual much of the coding is repetitive, and overlays can easily be used to avoid taking an excessive amount of space in the core of the computer.

The output analyser took a further two months, which was rather longer than we had expected. This could have been speeded up if necessary, but the problem was more difficult than we had foreseen for reasons that may be of some general interest.

The requirements were apparently quite straightforward: to specify the military facilities used (aircraft, ship and stockpiles), the civil facilities used under each option, and, for each option, the number of kilotons sent in each time period to each region from the UK by each type of military aircraft, by civil aircraft, by military ship and by civil ship. The number of Kilotons supplied from stockpiles is also required, and the totals are compared with the requirements. In the Central Region, the amounts sent to each concentration area are classified as arriving either by rail, by road, by air direct from the UK or from stockpiles in the area.

The difficulties in implementing this were not fundamental. But the program is long because there are many different types of variable, and different types of constraint, that have to be analysed. Furthermore the fact that the matrix generator automatically suppresses unnecessary rows and columns adds slight difficulties. But the most interesting difficulaties arose from the fact that the mathematical formulation allows oversupply in some circumstances. For example if it is necessary to use a certain type of ship to supply tanks to some port, then there may be additional space in the ship that cannot be used for tanks and which cannot usefully be filled with anything else. It seems undesirable to define modes of loading ships that leave them partly empty, so in these circumstances the program implicitly sends unnecessary material to the port of destination. The problem of devising and programming suitable general rules for editing out these unnecessary shipments in the output analyser proved rather more troublesome than we had expected.

REFERENCES

C.N. Beard and C.T. McIndoe (1970) "The determination of the least cost mix of transport aircraft, ships and stockpiled material to provide British forces with adequate strategic mobility in the future". pp 392-405 of Applications of Mathematical Programming Techniques (Ed. E.M.L. Beale) English Universities Press London.

Max Shaw (1970) "Review of Computational Experience in solving large mixed integer programming problems", pp 406-412 of Applications of Mathematical Programming Techniques (Ed. E.M.L. Beale) English Universities Press London.

E.M.L. Beale(1968) Mathematical Programming in Practice Pitmans London.

TABLE 1. Order of Presentation for a Mathematical Programming Model

1.	Subscripts	Use lower-case letters.
2.	Sets	(Which may be needed to give a precise definition of ranges of summation, or conditions under which a constraint, a variable or a constant is defined). Use Capital letters.
3.	Constants	Use Capital letters.
4.	Variables	Use lower-case letters not used as subscripts.
5.	Constraints	

If there are too few different letters to give suitably mnemonic symbols, then use Capital letters as Literal Subscripts to create additional composite letters.

Literal subscripts can be used to make the names of constants correspond to their FORTRAN names.

Lower case subscripts should be reserved for quantities that take numerical values.

TABLE 2. Subscripts Used in the Reinforcement and Redeployment Model

a	Aircraft Type	
b	Road Bottleneck	
c	Concentration Area	
d	Demand Type (i.e. Load Type)	1 = Tanks 2 = APCs 3 = Heavy Lorries 4 = Medium Lorries 5 = Light Lorries 6 = Freight 7 = Ammunition 8 = Helicopters 9 = Personnel
f	Airfield of Embarkation	
g	Airfield of Destination	
h	Type of Railway Train	
i	Railway Route	
j	Railway	
k	Option	
m	Mode of loading ship	
n	Non UK Port of Embarkation	
p	UK Port of Embarkation	
q	Port of Destination	
r	Road Route	
s	Stockpile	
t	Time Period (1 or 2 only)	
v	Ship Type	

TABLE 3. Variables in the Reinforcement and Redeployment Model

Variable	Description
C_{IVFRT}	No. of millions of ton miles of peacetime military air freighting carried in civil aircraft during costing period.
C_{IVTRP}	No. of millions of passenger miles of peacetime military air trooping carried in civil aircraft during costing period.
$^{e}{}_{Aa}$	No. of existing aircraft retained, in addition to the minimum number to be operated.
$^{e}{}_{Ccdk}$	No. of kilotons arriving in Concentration Area in Period 1 above minimum requirements.
$^{e}{}_{Dqk}$	No. of ship-days for voyages from non-UK ports of embarkation given up to allow military ship to return early to U.K.
$^{e}{}_{Gdgk}$	No. of kilotons left at Airfield of Detination at end of Period 1 for later use.
$^{e}{}_{Qdkq}$	No. of kilotons left at Port of Destination at end of Period 1 for later use.
$^{e}{}_{Sd0}$	UK stockpile capacity retained (KT) (d = 1, 2 or 6).
$^{e}{}_{Sds}$	Overseas stockpile capacity retained (KT) (d = 1, 2 or 6).
$^{e}{}_{V}$	No of existing military ships retained.
$^{m}{}_{Fa}$	No. of aircraft assigned to air freightingin peactime.
$^{m}{}_{Ta}$	No. of aircraft assigned to air trooping in peactime.
$^{s}{}_{Ccdkst}$	No. of kilotons of Load Type dassigned to concentration area c. Only defined when stockpile is in concentration area c but has onward rail or road links.
$^{S}{}_{Gcdgkt}$	No. of kilotons of Load Type d, originating from a stockpile either at the airfield of destination g or from a Non UK stockpile supplying the airfield, assigned to Concentration Area c. Only defined when airfield is in concentration area c but has onward rail or road links.

$^{S}Hdhikt$	No. of train journeys carrying stockpiled material of Load type d.
$^{S}Rdkrt$	No. of kilotons of stockpiled material of Load type d sent by road.
$^{w}Adgk0t$	No. of kilotons flown from UK (d = 6 or 7 only).
$^{w}Adgkst$	No. of kilotons flown from overseas stockpile site s (d = 6 or 7 only).
$^{w}Nd1knq$	No. of kilotons of Load Type d that could have been shipped in vehicle-carrying voyage omitted to provide space for Load Type (d + 1). (d = 1, 2, 3 or 4 only).
$^{w}Nd6knq$	No. of kilotons of Load Type d that could have been shipped in freight-carrying voyages omitted to provide space for Load Type (d + l) (d = l, 2, 3, 4 or 5 only).
$^{w}Qdkqtv$	No. of kilotons of Load Type d carried from UK in ships of Type v (d = l or 2 only).
$^{w}Tdkqtv$	No. of kilotons of Load Type d that could have been shipped from UK omitted to provide space for Load Type (d + l) (d = l or 2 only).
$^{x}Aadfgkt$	No. of flights by aircraft assigned to Load Type d (d = l, 2,3,4,5,6,8 or 9).
$^{x}Gcdgkt$	No. of kilotons of Load Type d assigned to concentration area c. Only defined when stockpile is in concentration area c but has onward rail or road links.
$^{x}Hdhikt$	No. of train journeys starting from ports or airfields of destination carrying Load Type d.
$^{x}N1knq$	No. of Vehicle-carrying voyages.
$^{x}N6knq$	No. of freight-carrying voyages.
$^{x}Pkmpqtv$	No. of voyages made.

x_{Rdkrt} No. of kilotons of Load Type d originating in the UK, sent from a port or airfield of destination on Road Route r.

x_{S1knq} No. of kilotons of Load Type d sent in Vehicle-carrying voyages (originating from stockpile).

x_{Sd6knq} No. of kilotons of Load Type d sent in Freight-carrying voyages (originating from stockpile).

y_{Aa} No. of new aircraft purchased, in addition to any excess of minimum no. over exisiting number.

y_{Sd0} Additional UK stockpile capacity purchased (KT) (d = 1, 2 or 6).

y_{Sds} Additional stockpile capacity purchased at Site s (KT) (d = 1, 2 or 6).

y_V No. of new military ships purchased.

z_{Ddst} No. of kilotons of double-purchased material of Type d held in Stockpile Site s for use in Period t.

z_{Sd0} No. of kilotons of single-purchased material of Type d held in UK stockpiles for use in Period 2.

z_{Sds} No. of Kilotons of single-purchased material of Type d, in addition to the minimum permitted stockpile, held in Stockpile Site s for use in Period 2. (It is assumed that no single-purchased material can be used in Period 1, other than the minimum stockpile $D_{SMINdst}$ at each site.)

TABLE 4. Types of Constraint in the Reinforcement and Redeployment Model

1. Maximum permitted foreign exchange cost.

2. Peacetime trooping requirement.

3. Peacetime freighting requirement.

4. Allocation of military aircraft to peacetime trooping or freighting.

5. Material balance constraints defining the total amount of material for Time Period 2 that must be stockpiled somewhere.

6. Material balance constraints for stockpile sites not at airfields of destination.

7. Material balance constraints for stockpiled material at airfields of destination.

8. Material balance constraints at ports of destination.

9. Material balance constraints at airfields of destination.

10. Demand constraints at concentration areas for total deliveries.

11. Demand constraints at concentration areas for mandatory air loads.

12. Stockpile capacity constraints.

13. Port capacity constraints.

14. Ship capacity constraints.

15. Capacity constraints defining numbers of voyages needed to carry loads on routes starting at Non UK Ports of Embarkation.

16. Airfield capacity constraints.

17. Aircraft capacity constraints.

18. Railway station capacity constraints.

19. Train capacity constraints.

20. Road Bottleneck Capacity constraints.

21. Road Vehicle Capacities.

TABLE 5. Types of Data Card in the Reinforcement and Redeployment Model

1. Title Card

2. GENINF card giving numbers of possible values of subscripts.

3. 1 card for each option, giving its code letter and name and the number of days in each time period, and also indicating whether civil ships or aircraft may be used in Period 1.

4. Up to 11 cards for each Aircraft Type.

5. Cards for each Airfield of Embarkation. The header gives data on the airfield itself. It is followed by 1 card for each air route starting from this airfield.

6. 1 card for each Airfield of Destination.

7. 4 or 5 cards for each Ship Type.

8. Cards for each Port of Embarkation. As for airfields of embarkation.

9. 1 card for each Port of Destination.

10. 2 cards for each type of train.

11. 1 card for each railway route.

12. 1 card for each railway station.

13. 7 cards on road transport, one for each load type that can go by road, i.e. 1,2,3,4,5,6, and 9. It is assumed that the requirements for Load Type 7 are the same as for Load Type 6.

14. 1 card for each road route.

15. 1 card for each road bottleneck.

16. Cards for each concentration area. The header gives the area name and the maximum permitted distance of any airfield of destination delivering mandatory air loads. It is followed by 9 cards for each option defining the requirements for each load type.

17. 7 cards giving costs for double-purchased stockpiled material, and total requirements for single-purchased material for Time Period 2. These 7 cards are for Load Types 1 to 7 respectively.

18. 3 cards giving capacities and costs for UK stockpile sites. These are for Load Types 1,2 and 6 respectively. It is assumed that Load Types 3, 4 and 5 are added to Load Type 2 to determine total storage capacity needed for vehicles other than tanks. Similarly Load Type 7 is added to Load Type 6 to determine total storage capacity needed for freight.

19. 6 cards for each overseas stockpile site. The first gives code letter and name of site, and which port, airfield or concentration area it is in. The next 3 are as for UK stockpile sites. The last 2 give minimum permitted stockpile sizes, represented by the constants $D_{SMINdst}$.

20. 1 card giving requirements and cost data for peacetime operations.

21. 1 card for each "region" defining the concentration areas in the region. These data are used only in the output analyser.

Mathematical Programming in Theory and Practice,
P.L. Hammer and G. Zoutendijk, (Eds.)

CHOOSING INVESTMENT PORTFOLIOS WHEN THE RETURNS HAVE STABLE DISTRIBUTIONS*

by

W. T. Ziemba
University of British Columbia

ABSTRACT

This paper presents an efficient method for computing approximately optimal portfolios when the returns have symmetric stable distributions and there are many alternative investments. The procedure is valid, in particular, for independent investments and for multivariate investments of the classes introduced by Press and Samuelson. The algorithm is based on a two stage decomposition of the problem and is analogous to the procedure developed by the author that is available for normally distributed investments utilizing Lintner's reformulation of Tobin's separation theorem.

---

* Presented by invitation at the NATO Advanced Study Institute on Mathematical Programming in Theory and Practice, Figueira da Foz, Portugal, June 12-23, 1972. This research was partially supported by the National Research Council of Canada grant NRC-A7-7147, the Samuel Bronfman Foundation, The Graduate School of Business, Stanford University, and Atomic Energy Commission, Grant AT 04-3-326-PA #18.

Without implicating them, I would like to thank S.L. Brumelle, W.E. Diewert, J.L. Evans, J. Ohlson, C.E. Sarndal, C. Swoveland, and R. Vickson for some useful information and helpful discussions related to this work.

A tradeoff analysis between mean (μ) and dispersion (d) is valid since an investment choice maximizes expected utility if and only if it lies on a μ-d efficient curve. When a risk free asset is available one may find the efficient curve, which is a ray in μ-d space by solving a fractional program. The fractional program always has a pseudo-concave objective function and hence may be solved by standard non-linear programming algorithms. Its solution, which is generally unique, provides optimal proportions for the risky assets that are independent of the unspecified concave utility function. One must then choose optimal proportions between the risk free asset and a risky composite asset utilizing a given utility function. The composite asset is stable and consists of a sum of the random investments weighted by the optimal proportions. This problem is a stochastic program having one random variable and one decision variable. Symmetric stable distributions have known closed form densities only when α, the characteristic exponent, is 1/2 (the arc sine), 1 (the cauchy) or 2 (the normal). Hence, there is no apparent algorithm that will solve the stochastic program for general $1 < \alpha < 2$. However, one may obtain a reasonably accurate approximate solution to this program utilizing tables recently compiled by Fama and Roll. Standard nonlinear programming algorithms may be adapted to approximately solve the portfolio problem when the risk free asset assumption is not made. Such a direct approach is also available when the risk free asset assumption is made, however, the computations in the two stage approach would generally be much less formidable.

1. Introduction

Stable distributions have increasing interest for the empirical explanation of asset price changes and other economic phenomenon. An important reason for this is that all limiting sums (that exist) of independent identically distributed random variables are stable. Thus it is reasonable to suspect that empirical variables which are sums of random variables conform to stable laws. Such an observation has led to a substantial body of literature concerned with the estimation of the distributions of stock price changes[1] and other economic variables[2] utilizing stable distributions. Much of this literature focuses on the fact that the empirical distributions have more "outliers" and hence "fatter" tails than one would expect to be generated by a normal distribution. This leads to the conclusion that variance does not exist and that a normal distribution will not adequately fit the data.[3] The normal distribution has an important place in the theory of portfolio selection because the Markowitz theory[20] is then consistent with expected utility maximization. In this case it is well known that a portfolio maximizes expected utility if and only if it is mean-variance efficient. Since the normal is only one member of the stable family it is of interest to generalize the Markowitz theory to apply for more general classes of stable distributions. Samuelson [29] and Fama [5] have shown how this may be accomplished utilizing a mean-dispersion (μ-d) analysis when the random variables are independent or when they follow a Sharpe-Markowitz diagonal model (see [30]), respectively. Utilizing the fact that linear combinations of multivariate stable distributions are univariate stable it is shown here that the μ-d analysis is valid as long as the mean vector exists.

The calculation of the μ-d curve is generally quite difficult because one must solve a parametric concave program. However, when a risk free asset exists the efficient surface is a ray in μ-d space and a generalization of Tobin's separation theorem [31] obtains. One may then calculate the optimal proportions of the risky assets by solving a fractional program. The character of the fractional program depends, of course, upon the assumptions made about the joint distribution of the stable random variables. However, in fairly general circumstances the fractional program has a pseudo-concave objective function and hence may be solved via a standard nonlinear programming algorithm. Typically the optimal solution is unique. This calculation comprises stage one of a two stage procedure that will efficiently solve the portfolio problem when there are many random investments. In the second stage one introduces the investor's utility function and an optimal ratio between a stable composite asset and the risk free asset must be chosen. The composite asset is a sum of the random investments weighted by the optimal proportions found in stage one. The problem to solve is a stochastic program having a single random variable and a single decision variable. Such problems are generally easy to solve if the density of the random variable is known (as it is in the normal distribution case). Unfortunately the density of the stable composite asset is known only in a few special cases. However, Fama and Roll [7], using series approximations due to Bergstrom, have tabulated, at discrete points, the density and cumulative distributions of a standardized symmetric stable distribution. These tables may be used to obtain a very good nonlinear programming approximation to the stochastic program. The solution of the nonlinear program generally provides a good approximation to the optimal solution of the stochastic program and hence of the portfolio problem.

Section 2 discusses the case when the random returns are independent. Some sufficient conditions for the expected utility and expected marginal utility to be bounded are given. The fractional program in this case is shown to have a strictly pseudo-concave objective and it has a unique solution. The nonlinear programming approximation of the stochastic program is also described. Section 3 considers a class of multivariate stable distributions introduced by Press [22, 23]. This class generalizes the independent case and decomposes the dependence of the random variables into several independent subsets. Thus partial and full dependence may be handled in a convenient fashion. The optimization procedure described in section 2 may be utilized for this class and a class of multivariate stable random variables introduced by Samuelson [29] as well. Section 4 shows how one may find a portfolio that approximately maximizes expected utility when the risk free asset assumption is not made, by utilizing standard nonlinear programming algorithms. The calculations in this direct approach are generally much more formidable than in the two stage approach hence it is generally preferable to use the latter approach if it is valid.

2. The Independent Case

We consider an investor having one dollar[4] to invest in assets $i=0,1,\ldots,n$. Assets $1,2,\ldots,n$ are random and they exhibit constant returns to scale so that if x_i is invested in i then $\xi_i x_i$ is returned at the end of the investment period. The ξ_i are assumed to have independent stable distribution functions $F_i(\xi_i;\bar{\xi}_i,S_i,\beta,\alpha)$ $i=1,\ldots,n$. A distribution function $F(y)$ is said to be stable if and only if for all positive numbers a_1 and a_2 and all real numbers b_1 and b_2 there exists a positive number a and a real number b such that

$$F\left(\frac{y-b_1}{a_1}\right) * F\left(\frac{y-b_2}{a_2}\right) = F\left(\frac{y-b}{a}\right) \tag{1}$$

where the * indicates the convolution operation. Equation (1) formalizes the statement that the stable family is precisely that class of distributions that is closed under addition of independent and identically distributed random variables. The F_i are unimodal, absolutely continuous and have continuous densities f_i. The parameter $-1 \leq \beta \leq 1$ is related to the skewness of the distribution. When $\beta > 0 (<0)$ the distribution is skewed to the right (left). It is convenient for our purposes to assume that the distribution F_i is symmetric, in which case $\beta = 0$[5]. The parameter α is termed the characteristic exponent and absolute moments of order $<\alpha$ exist, where $0<\alpha\leq 2$. When $\alpha = 2$ and $\beta = 0$, F is the normal distribution and all moments exist. It will be convenient to assume that $1 < \alpha \leq 2$ so that absolute first moments always exist and that the value of α is the same for each F_i. The parameter $\bar{\xi}_i$ corresponds to the central tendency of the distribution which is the mean if $\alpha > 1$. The parameter S_i, assumed to be positive, refers to the dispersion of the distribution. It will be convenient to differentiate between S_i and $S_i^{1/\alpha}$. We will follow a suggestion of E.M.L. Beale and call this latter quantity the α-dispersion. When F_i is normal S_i is one-half the variance,,in other cases it is approximately equal to the semi-interquartile range. In general $S_i^{1/\alpha}$ is proportional to the mean absolute deviation $E|\xi_i-\bar{\xi}_i|$.

The f_i are known in closed form in only very special cases hence the most convenient way to study the stable family utilizes the characteristic function

$$\psi_y(t) \equiv E\, e^{ity} = \int_{\infty}^{\infty} e^{ity} f(y)\, dy,$$

where $i = \sqrt{-1}$.

The log characteristic function for a symmetric stable distribution $F_i(\xi_i;\bar{\xi}_i, S_i, 0,\alpha)$ is

$$\ln \psi_{\xi_i}(t_i) = i\bar{\xi}_i t_i - S_i|t_i|^{\alpha}$$

Asset 0 is riskless and returns $\bar{\xi}_0$ with certainty and exhibits constant returns to scale so that if x_0 is invested the return is $\bar{\xi}_0 x_0$. $x_0 < 0$, corresponds to borrowing at the risk free rate. It is assumed that the investor may borrow or lend any amount at the constant rate of $\bar{\xi}_0$. This asset may be considered to have the degenerate stable distribution $F_0(\xi_0;\ \bar{\xi}_0,\ 0,\ 0,\alpha)$.

It is supposed that the investor wishes to choose the x_i to maximize the expectation of a utility function u of wealth $w = \xi'x$, where $\xi' \equiv (\xi_0,\xi_1,\ldots,\xi_n)$, $x' \equiv (x_0,x_1,\ldots,x_n)$ and primes denote transposition. Suppose that u is non-decreasing, continuously differentiable and concave. (The continuously differentiable assumption is not crucial for the theory although it is useful for the algorithmic development).

The investors problem is

$$\text{(1)}\quad \text{maximize } Z(x) \equiv E_{\xi}\, u(\xi'x),$$

$$\text{s.t.}\qquad e'x = 1,\; x_i \geq 0\,,\; i = 1, \ldots, n,\; x_0 \text{ unconstrained},$$

where E_{ξ} represents expectation with respect to ξ, and $e \equiv (1,\ldots,1)'$.

The portfolio $\xi'x$ is known (see e.g. [5]) to have the stable distribution $F(\xi'x) = F(\xi'x;\, \bar{\xi}'x,\, \sum_{i=0}^{n} s_i|x_i|^{\alpha}\, ,\, 0,\, \alpha)$. It will be convenient to begin with some conditions on u that will guarantee that expected utility and expected marginal utility are bounded.

Theorem 1: Suppose w has the stable distribution

$F(w;\ \bar{w},\ S_w,\ 0,\alpha)$, where $|\bar{w}| < \infty$, $0 < S_w < \infty$ and $1 < \alpha \leq 2$.

(a) If $|u(w)| \leq L_1|w|^{\beta_1}$ for some $\beta_1 < \alpha$ and some $0 < L_1 < \infty$

then $|E\ u(w)| < \infty$.

(b) If $|u'(w)| \leq L_2|w|^{\beta_2}$ for some $\beta_2 < \alpha$ and some $0 < L_2 < \infty$

then $|E\ u'(w)| < \infty$.

(c) If $|u'(w)| \leq L_3|w|^{\beta_3}$ for some $\beta_3 < \alpha - 1$ and some

$0 < L_3 < \infty$ and[6] $w \equiv \xi'x \sim F(\xi'x;\ \bar{\xi}'x,\ \sum_{i=0}^{n} S_i x_i^{\alpha},\ 0,\ \alpha)$

then $$\left|\frac{\partial E_w u(\xi'x)}{\partial x_i}\right| = \left|E_w\left(\frac{\partial u(\xi'x)}{\partial x_i}\right)\right| < \infty\ ,\ i=0,1,\ldots,n,$$

and the computation of the partials involves only a univariate integration utilizing a normalized variable $\tilde{w} \sim F(\tilde{w};\ 0,1,0,\alpha)$.

(d) The results in (a)-(c) remain valid if the hypotheses are modified to read, $\lim_{|w|\to\infty} \frac{|u(w)|}{|w|^{\beta_1}} = v_1$ for some $0 \leq v_1 < \infty$

and some $\beta_1 < \alpha$, $\lim_{|w|\to\infty} \frac{|u'(w)|}{|w|^{\beta_2}} = v_2$ for some $0 \leq v_2 < \infty$

and some $\beta_2 < \alpha$ and $\lim_{|w|\to\infty} \frac{|u'(w)|}{|w|^{\beta_3}} = v_3$ for some $0 \leq v_3 < \infty$

and some $\beta_3 < \alpha - 1$, respectively.

Proof: see the Appendix.

Remark: Most common utility functions have either unbounded expected utility or they are undefined over portions of the range of w, which is R. However one may modify many of these utility functions by adding appropriate linear segments so that the utility functions are concave, non-decreasing, defined over the entire range of w and have bounded expected utility. For the logarithmic, power and exponential forms such modified utility functions are:

$$\text{a) } u(w) = \begin{cases} \log \tau w & \text{if } w \geq w_0, \\ \log \tau w_0 - \tau) + w(\frac{\tau}{w_0} & \text{if } w < w_0 \end{cases} \qquad w_0 > 0,\ \tau > 0$$

$$\text{b) } u(w) = \begin{cases} w^{\delta} & \text{if } w \geq w_0 \\ (1-\delta)w_0^{\delta} + (\delta w_0^{\delta-1})\, w & \text{if } w < w_0 \end{cases} \qquad w_0 > 0,\ 0<\delta<1$$

$$\text{c) } u(w) = \begin{cases} -e^{-\tau w} & w \geq w_0 \\ -(1+\tau w_0)e^{-\tau w_0} + (\tau e^{-\tau w_0})\, w & \text{if } w < w_0 \end{cases} \qquad 0>w_0>-\infty_0\ \tau>0$$

Note that all polynomial utility functions of order two or more have unbounded expected utility unless the distribution of w is truncated from above and below.

It is our purpose to solve (1) using the two step procedure: i) find an efficient surface independent of the utility function u; and ii)given a particular u find a maximizing point on this surface. Such a procedure for the case when the random returns are normally distributed was suggested by Tobin's separation theorem [31] and implemented by Ziemba et.al. [35] utilizing Lintner's [15] reformulation of the separation theorem. The analysis here is analogous and utilizes a mean-dispersion efficient surface.

The following theorem indicates that a point cannot solve (1) unless it lies on a mean-dispersion efficiency curve.

Theorem 2: Let G and H be two distinct distributions with finite means μ_1 and μ_2 and finite positive dispersions D_1^{α}, and D_2^{α}, $(1 < \alpha \le 2)$, respectively, such that $G(x) = H(y)$ whenever $\frac{x-\mu_1}{D_1} = \frac{y-\mu_2}{D_2}$. Let $\mu_1 \ge \mu_2$ and $G(z) > H(z)$ for some z. Then $E\, u(w)dG(w) \ge E\, u(w)dH(w)$ for all concave non-decreasing u if and only if $D_1 \le D_2$.

Proof: Sufficiency: Case (i): $\mu_1 = \mu_2 = o$.

Let $k \equiv D_2/D_1 \ge 1$.

Then $$\int_{-\infty}^{\infty} u(w)\ dH(w) - \int_{-\infty}^{\infty} u(w)\ dG(w) = \int_{-\infty}^{\infty} [u(kw)-u(w)]dG(w)$$

(since G is symmetric)

$$= \int_0^{\infty} \underbrace{\{(u(kw)-u(w)) - (\mu(-w)-u(-kw))\}}_{\le 0 \text{ by concavity}} dG(w) \le 0$$

Case (ii): $\mu_1 \ge \mu_2$.

Let $\widetilde{G}$ and $\widetilde{H}$ be the cumulative distribution functions for x and y, respectively, when their means are translated to zero. Let $\epsilon \equiv \mu_1 - \mu_2 \ge o$.

$$\int u(w)dH(w) = \int u(w+\mu_2)d\widetilde{H}(w) \overset{(\text{by } i)}{\le} \int u(w+\mu_2)d\widetilde{G}(w)$$

(since $\epsilon \ge o$ and u is non-decreasing)

$$\le \int u(w+\mu_2+\epsilon)d\widetilde{G}(w) = \int u(w)dG(w).$$

Necessity: Let $\mu_1 = \mu_2 = o$, $D_1 > D_2$ and $u(w) = \begin{cases} w^{\frac{1}{2}} \\ w \end{cases}$ if $w \gtreqless 0$.

Then $$\underbrace{\int_0^{\infty} [(kw)^{\frac{1}{2}} - w^{\frac{1}{2}}]\ dG(w)}_{>0} + \underbrace{\int_{-\infty}^{0} [kw - w]\ dG(w)}_{=0} > 0, \text{ where } k \equiv \left(\frac{D2}{D1}\right)^{\frac{1}{\alpha}} < 1.$$

For any other $\mu_1 \ge \mu_2$ having $D_1 < D_2$ and a given α a u can be found such that $E\, u(w)dG(w) < E\, u(w)dH(w)$, since G and H cross only once, see Theorem 3 in [12].

The proof is a generalization of the proof given by Hillier[13a] for the mean-variance case ($\alpha = 2$). The monotonicity assumption is not crucial, however, dropping the monotonicity assumption does not seem to add any apparent generality as free disposal of wealth always seems possible. A proof of Theorem 2 without the monotonicity assumption may be obtained by letting D_1^{α} replace σ_1^2 in Hanoch and Levy's proof for the mean variance case [12].

The theorem indicates that it is sufficient to limit consideration to only those points that lie on the mean-α-dispersion (μ-d) efficient surface. Let the risky assets $(x_1,\ldots,x_n) \equiv \hat{x}$, and let $\hat{\bar{\xi}} \equiv (\bar{\xi}_1,\ldots,\bar{\xi}_n)$. Then the (μ-d) surface corresponding to assets $(\xi_1,\ldots,\xi_n)$ may be obtained by solving[7]

$$
\begin{aligned}
(2) \qquad \phi(\beta) \equiv \; & \max \; \hat{\bar{\xi}}' \hat{x}, \\
\text{s.t.} \; & e'\hat{x} = 1, \quad \hat{x} \geqq 0, \\
& \{\sum_{i=1}^{n} S_i x_i^{\alpha}\}^{1/\alpha} \leq \beta,
\end{aligned}
$$

for all $\beta > 0$.

Now (2) is a parametric concave program which would generally be difficult to solve, for all $\beta > 0$. It is easier to bypass the calculation of the μ-d curve if the risk free asset is not available, see section 4, below. However, when the risk free asset does exist, as we are assuming, it is convenient to consider points that are linear combinations of $x_0 = 1$ (total investment in the risk free asset) and points that lie on $\phi(\beta)$. These combinations correspond to straight lines in μ-d space as well because the mean is linear and the α-dispersion measure is positively homogeneous (i.e., $f(\lambda\hat{x}) = \lambda f(\hat{x})$ for all $\lambda \geq 0$) in the x_i. Clearly the best points will lie on the line L that is a support to the concave function $\phi(\beta)$ (proof below); see Figure 1. Such heuristic arguments indicate that the μ-d efficient surface is now L, and one gets the analogue of the Tobin separation theorem for normal distributions which has the important implication that $x_i^*/\sum_{j=1}^{n} x_j^*$, $i \neq 0$, is independent of u and of initial wealth. The result may be stated as the

Separation theorem: Let the efficiency problem be

$$\text{(2')} \qquad \max \ \bar{\xi}' x,$$
$$\text{s.t. } f(\hat{x}) \leq \beta,$$
$$e'x = w, \ \hat{x} \geq o, \ x_o \text{ unconstrained.}$$

In (2') f is the α-dispersion measure, w is the initial wealth and explicit consideration of the risk free asset (i=o) is allowed. Suppose that asset i=o has no dispersion and that borrowing or lending any amount is possible at the fixed rate $\bar{\xi}_o$. Assume that f is convex and homogeneous of degree one and that u is concave and non-decreasing.

a) Total Separation: If $x_o^* \neq 0$ the relative proportions invested in the risky assets, namely $x_i^*/\sum_{j=1}^{n} x_j^*$, $i \neq o$, are independent of u and initial wealth w.

b) Partial Separation: If $x_o^* = 0$ all investment is in the risky assets and $\sum_{i=1}^{n} x_j^* = w$ and the x_j^* are independent of u.

Proof: a) Suppose $x_o^* \neq 0$. Since x^* solves (2') it must satisfy the Kuhn-Tucker conditions:

$$\text{i.} \qquad f(\hat{x}) \leq \beta, \quad \hat{x} \geq o, \quad e'x = w,$$

$$\text{ii. a)} \qquad \bar{\xi}_i - \lambda \frac{\partial f}{\partial x_i} - \mu \quad \leq o, \quad i=1,\ldots,n,$$

$$\text{b)} \qquad \bar{\xi}_o - \mu \quad = o, \quad \lambda \geq 0,$$

$$\text{iii.} \qquad x_i \left(\bar{\xi}_i - \lambda \frac{\partial f}{\partial x_i} - \mu\right) = o, \quad i=1,\ldots,n.$$

It will suffice to show that for all $\sigma > o$ there exists a $\gamma \neq o$ such that $x^{**} \equiv (\gamma x_o^*, \sigma x_1^*, \ldots, \sigma x_n^*)$ also solves the Kuhn-Tucker conditions (which are necessary and sufficient) for all $\beta^{**} \geq \beta\sigma$. Condition (i) is satisfied because:

$$f(\hat{x}^*) \leq \beta \Rightarrow f(\sigma\hat{x}^*) = \sigma f(\hat{x}^*) \leq \sigma\beta \leq \beta^{**}$$

$$\hat{x}^* \geq o \Rightarrow \sigma\hat{x}^* \geq o$$

$$x_o^* + \sum_{i=1}^n x_i^* = 1 \Rightarrow \gamma x_o^* + \sigma\sum_{i=1}^n x_i^* = 1 \quad \text{for} \quad \gamma \equiv \frac{1 - \sigma\sum_{i=1}^n x_i^*}{x_o^*} \neq o .$$

Now $\partial f / \partial x_i$ is homogeneous of degree zero hence

$$\bar{\xi}_i - \lambda \frac{\partial f(\sigma\hat{x}^*)}{\partial x_i} \leq \bar{\xi}_o \quad \text{iff} \quad \bar{\xi}_i - \lambda \frac{\partial f(\hat{x}^*)}{\partial x_i} \leq \bar{\xi}_o$$

hence (ii).

If $x_i^* = o$ (iii) is trivially satisfied.

If $x_i^* > o$ (iii) is equivalent to (2) which is satisfied for $\sigma\hat{x}^*$ iff $\hat{x}^*$ satisfies (2) .

By Theorem 2 an optimal solution to (2') must solve (1) hence $\sum_{i=1}^n x_i^* = w$ and the x_i^* are independent of the u .

The proof and statement of the separation theorem given here is similar in spirit to that given by Breen [2]. Breen considered an efficiency problem in which one minimizes α-dispersion given that expected return is a stipulated level as well as some alternative assumptions regarding the risk free asset.

For the analysis indicated in Figure 1 to be valid it is necessary that the α-dispersion measure $f(\hat{x}) \equiv \{\sum_{i=1}^{n} S_i x_i^{\alpha}\}^{1/\alpha}$ be convex and that ϕ be a concave function of β. The function f is actually strictly convex as we now establish using

Theorem 3: (Minkowski's Inequality): Suppose $M_r(y) \equiv \{\frac{1}{n} \sum_{i=1}^{n} y_i^r\}^{1/\alpha}$, $n \geqq r > 1$

$n < \infty$ and that a and b are not proportional (i.e., constants q_1 and q_2 not both zero do not exist such that $q_1 a = q_2 b$) then

$$M_r(a) + M_r(b) > M_r(a+b)$$

Proof: See e.g. [13, p.30].

Let $P \equiv \{\hat{x} \mid \hat{x} \geqq 0,\ e'\hat{x} = 1\}$

Lemma 1: $f(\hat{x}) \equiv \{\sum_{i=1}^{n} S_i x_i^{\alpha}\}^{1/\alpha}$ is a strictly convex function of $\hat{x}$ on P if $2 \geq \alpha > 1$ and $S_i > 0$, $i=1,\dots,n$.

Proof: Let $a_i = \lambda S_i x_i^1$ and $b_i = (1-\lambda)S_i x_i^2$, $i=1,\dots,n$ where $\hat{x}^1 \neq \hat{x}^2$, and $0 < \lambda < 1$. Then by Minkowski's Inequality

$$\{\frac{1}{n}\sum_{i=1}^{n} a_i^{\alpha}\}^{1/\alpha} + \{\frac{1}{n}\sum_{i=1}^{n} b_i^{\alpha}\}^{1/\alpha} > \{\frac{1}{n}\sum_{i=1}^{n} (a_i+b_i)^{\alpha}\}^{1/\alpha}$$

$$\Rightarrow \{\sum_{i=1}^{n} [\lambda S_i x_i^1]^{\alpha}\}^{1/\alpha} + \{\sum_{i=1}^{n} [(1-\lambda)S_i x_i^2]^{\alpha}\}^{1/\alpha} > \{\sum_{i=1}^{n} [\lambda S_i x_i^1 + (1-\lambda)S_i x_i^2]^{\alpha}\}^{1/\alpha}$$

Hence $\lambda f(\hat{x}^1) + (1-\lambda)f(\hat{x}^2) > f\{\lambda\hat{x}^1 + (1-\lambda)\hat{x}^2\}$.

Thus f is strictly convex unless a and b are proportional. But this requires that there exist constants q_1 and q_2 not both zero such that $q_1 a = q_2 b$ or that $q_1 \lambda S_i x_i^1 = q_2(1-\lambda) S_i x_i^2 \Rightarrow q_1\lambda x_i^1 = q_2(1-\lambda)x_i^2$ (since $S_i > 0$)

or that x_i^1 and x_i^2 are proportional since $q_1' \equiv q_1\lambda$ and $q_2' \equiv q_2(1-\lambda)$ are both not zero. However, the condition that $\sum_{i=1}^{n} x_i^1$ and $\sum_{i=1}^{n} x_i^2 = 1$ means that x^1 and x^2 cannot be proportional unless $x_i^1 = x_i^2$ for all i which is a contradiction.

Remark: f is convex but not strictly convex on $M \equiv \{\hat{x} \mid \hat{x} \geqq 0\}$ because f is linear on every ray that contains the origin.

Lemma 2: ϕ is a concave function of $\beta > \beta_L$, where $\beta_L > 0$ is defined below, if $2 \geqq \alpha > 1$ and $S_i \geqq 0$, $i=1,\ldots,n$.

Proof: Let $K_\beta \equiv \{\hat{x} | e'\hat{x} = 1, \quad \hat{x} \geqq 0, \ f(\hat{x}) \leq \beta\}$.

Let $\beta_L > 0$ be the smallest β for which $K_\beta \neq \phi$. Clearly $\beta_L \leq \delta$ where $\delta^\alpha \equiv \min_i S_i$, and $K_\beta \neq \phi$ if and only if $\beta \geqq \beta_L$.

Choose $\beta_1 \geqq \beta_L$ and $\beta_2 \geqq \beta_L$. Since (2) has a linear objective function and a non-empty compact convex feasible region there is an optimal solution for all $\beta \geq \beta_L$. Let optimal solutions when β equals β_1 and β_2 be $\hat{x}_1$ and $\hat{x}_2$, respectively.

Consider $\quad \beta_\lambda = (1-\lambda)\beta_1 + \lambda\beta_2, \qquad 0 \leqq \lambda \leqq 1,$

$\qquad\qquad \hat{x}_\lambda = (1-\lambda)\beta_1 + \lambda\beta_2.$

Now $\hat{x}_\lambda$ is feasible when $\beta = \beta_\lambda$ because the constraint set of (2) is convex. By the concavity of $h(\hat{x}) \equiv \hat{\bar{\xi}}'\hat{x}$, $h(\hat{x}_\lambda) \geqq (1-\lambda)\ h(\hat{x}_1) + \lambda h(\hat{x}_2)$ (i)

But $h(\hat{x}_i) = \phi(\beta_i)$ (ii) since $\hat{x}_i$ is optimal, $\quad i=1,2,$

and $\phi(\beta_\lambda) \geqq h(\hat{x}_\lambda)$ (iii) since $\phi(\beta_\lambda)$ denotes the maximum when $\beta = \beta_\lambda$. Combining (i)-(iii) gives

$$\phi(\beta_\lambda) \geqq (1-\lambda)\phi(\beta_1) + \lambda\phi(\beta_2).$$

One may find the slope of L and the point M by maximizing

$$(3) \qquad g(\hat{x}) \equiv \frac{\sum_{i=1}^{n} \bar{\xi}_i x_i - \bar{\xi}_0}{\{\sum_{i=1}^{n} S_i x_i^\alpha\}^{1/\alpha}}$$

$$\text{s.t. } \hat{x} \geqq 0, \ e'\hat{x} = 1 .$$

By letting $\bar{\bar{\xi}}_i = \bar{\xi}_i - \bar{\xi}_0$, $i=1,\ldots,n$, and utilizing $e'\hat{x} = 1$,

$$g(\hat{x}) = \{\sum_{i=1}^{n} \bar{\bar{\xi}}_i' x_i\} \Big/ \{\sum_{i=1}^{n} s_i x_i^{\alpha}\}^{1/\alpha}.$$

We will assume that $\bar{\bar{\xi}}'\hat{x} > 0$ for all "interesting" feasible $\hat{x}$. This is a very minor assumption since the fact that the $x \geqq 0$ and $e'\hat{x} = 1$ implies that there always exists an $\hat{x}$ such that $\bar{\bar{\xi}}'\hat{x} > 0$ unless all $\bar{\bar{\xi}}_i \leqq 0$ in which case it is optimal to invest entirely in the risk free asset. It will now be shown that (3) has a unique solution.

A differentiable function $\theta(x)$: $\Lambda \to R$ is said to be strictly pseudo-concave on $\Lambda \subset R^n$ if for all $x, \bar{x} \in \Lambda$, $x \neq \bar{x}$

$$(x-\bar{x})' \nabla\theta(\bar{x}) \leqq 0 \implies \theta(x) < \theta(\bar{x})$$

(∇ denotes the gradient operator so $\nabla\theta = (\partial\theta/\partial x_1, \ldots, \partial\theta/\partial x_n)$). Geometrically functions are strictly pseudo-concave if they strictly decrease in all directions that have a downward or horizontal pointing directional derivative. A normal distribution is such a function in R.

Theorem 4: Suppose the differentiable functions Ψ and ψ are defined on Λ, a convex subset of R^n, and that $\Psi > 0$ is concave and $\psi > 0$ is strictly convex. Then $\theta = \Psi/\psi$ is strictly pseudo-concave on Λ.

Proof: $\nabla\theta = \{\frac{\psi\nabla\Psi - \Psi\nabla\psi}{\psi^2}\}$. Let $\bar{x} \in \Lambda$.

Thus $\nabla\theta(\bar{x})'(x-\bar{x}) = \{\frac{\psi(\bar{x})\nabla\Psi(\bar{x}) - \Psi(\bar{x})\nabla\psi(\bar{x})}{[\psi(\bar{x})]^2}\}'(x-\bar{x}) \leqq 0 \quad x \in \Lambda, x \neq \bar{x}$

$\Longrightarrow \{\psi(\bar{x})\nabla\Psi(\bar{x}) - \Psi(\bar{x})\nabla\Psi(\bar{x})\}'(x-\bar{x}) \leq 0$ since $\psi(\bar{x}) \neq 0$

$\psi(\bar{x})[\Psi(x)-\Psi(\bar{x})] - [\Psi(\bar{x})\nabla\psi(\bar{x})]'(x-\bar{x}) \leq$ since Ψ is concave and $\psi(\bar{x}) > 0$

$\psi(\bar{x})[\Psi(x) - \Psi(\bar{x})] + \Psi(\bar{x})[\psi(\bar{x}) - \psi(x)] <$ since ψ is strictly convex and $\Psi(\bar{x}) > 0$

$\psi(\bar{x})\Psi(x) - \Psi(\bar{x})\psi(x) =$, which implies that

$\theta(x) - \theta(\bar{x}) < 0$ since $\psi(\bar{x}) > 0$ and $\psi(x) > 0$

Lemma 3: Suppose $f: \Lambda \to R$, where Λ is a convex subset of R^n, is differentiable at x and there is a direction d such that $\nabla f(x)'d > 0$. Then a $\sigma > 0$ exists such that for all τ, $\sigma \geq \tau > 0$ $f(x+\tau d) > f(x)$.

Proof: [32, p.24].

Theorem 5: Suppose $\theta(x): \Lambda \to R$, where Λ is a closed convex subset of R^n, is strictly pseudo-concave. Then the maximum of θ if it exists is attained at most one point $x \in \Lambda$.

Proof: a) Case 1: $\exists\ x \in \Lambda$ such that $\nabla\theta(x)=0$.

Then by the strict pseudo-concavity of θ,

i.e. $\nabla\theta(x)'(y-x) \leq 0 \Rightarrow \theta(y) < \theta(x)$ for all $y \in \Lambda$, x is the unique maximizer.

b) Case 2: $\nexists\ x \in \Lambda$ such that $\nabla\theta(x) = 0$

Suppose x maximizes θ over Λ.

Then $\forall y \neq x, \nabla\theta(x)'(y-x) \leq 0 \Rightarrow \theta(y) < \theta(x)$ and x is clearly the unique maxima unless $\exists y \in \Lambda$ such that $\nabla\theta(x)'(y-x) > 0$. But by lemma 3 that would imply that there exists a $\tau > 0$ such that $\theta(x+\tau(y-x)) > \theta(x)$, which contradicts the assumed optimality of x. (Note that the point $[x+\tau(y-x)] = [\tau y+(1-\tau)x] \in \Lambda$, by the convexity of Λ).

Theorem 6: (3) has a unique solution

Proof: The function g, may be seen to be strictly pseudo-concave on P by letting $\Psi(\hat{x}) = \bar{\xi}'\hat{x}$ which is $>$ 0 and concave and $\psi(\hat{x}) = \{\sum_{i=1}^{n} S_i x_i^{\alpha}\}^{1/\alpha}$ which is >0 and strictly convex. Hence by Theorem 5 it has at most one maximizer. But P is compact and g is continous hence g is maximized at a unique point $x^* \varepsilon P$.

Suppose that the fractional program (3) has been solved to obtain $\hat{x}^*$. The problem then is to determine the optimal ratios of risky to non-risky assets. The risky asset is

$$R = \hat{\xi}'\hat{x}^* \sim F(\hat{\xi}'\hat{x}^*;\ \bar{\xi}'\hat{x}^*,\ \sum_{i=1}^{n} S_i(\hat{x}_i^*)^{\alpha},\ 0,\alpha) \equiv F(R;\ \bar{R}\ ,\ S_R,\ 0,\alpha),$$

and the best combination of R and the risk free asset may be found by solving

$$(4) \qquad \max_{-\infty<\lambda\leq 1} \Psi(\lambda) \equiv E_R\, u[\lambda\bar{\xi}_0 + (1-\lambda)R].$$

Problem (4) is a stochastic program with one random variable (R) and one decision variable (λ). Since u is concave in w it follows that ψ is concave in λ. Under the assumptions of Theorem 1 $\psi(\lambda)$ and $d\psi(\lambda)/d\lambda$ will be bounded.

Now

$$(5) \qquad \Psi(\lambda) = \int_{-\infty}^{\infty} u[\lambda\bar{\xi}_0 + (1-\lambda)R]\, f(R)dR$$

$$= \int_{-\infty}^{\infty} u[\lambda\bar{\xi}_0 + (1-\lambda)\{\bar{r} + (S_R)^{1/\alpha}\tilde{R}\}]\, f(\tilde{R})\, d\tilde{R},$$

where the standardized stable variate,

$$\tilde{R} \equiv \frac{R - \bar{R}}{(S_R)^{1/\alpha}} \sim F(\tilde{R};\, 0,\, 1,\, 0, \alpha).$$

The continuous derivative of Ψ is

$$(6)\quad \frac{d\Psi(\lambda)}{d\lambda} = \int_{\infty}^{\infty} \{\bar{\xi}_0 - \bar{R} - (S_R)^{1/\alpha}\hat{R}\}\; \frac{du[\lambda\bar{\xi}_0 + (1-\lambda)\{\bar{R} + (S_R)^{1/\alpha}\tilde{R}\}]}{dw}\; f(\tilde{R})d\tilde{R}\,.$$

Problems having the general form of (4) are generally easy to solve by combining a univariate numerical integration scheme with a search procedure that uses function or derivative evaluations, see [35]. The difficulty here is that the density $f(\tilde{R})$ is not known in closed form except for very special cases such as the normal ($\alpha = 2$) and cauchy ($\alpha=1$) distributions. However, Fama and Roll [7] have utilized series expansions of $f(\tilde{R})$, due to Bergstrom, to tabulate approximate values of $f(\tilde{R})$ and $F(\tilde{R})$. The tables consider $\alpha=1.0, 1.1, \ldots, 2.0$ and $\tilde{R} = \pm\, 0,\ 05, \ldots, 1.0,\ 1.1, \ldots,\ 2.0,\ 2.2, \ldots, 4.0,\ 4.4, \ldots, 6.0,\ 7.0,\ 8.0,\ 10.0,\ 15.0$ and 20.0 (a grid of 50 points). One may then utilize the tables to get a good approximation to (5) of the form

$$(7)\quad \Psi(\lambda) \tilde{=} \sum_{j=1}^{m} P_j\, u[\lambda\bar{\xi}_0 + (1-\lambda)\{\bar{R} + (S_R)^{1/\alpha}\tilde{R}_j\}],$$

where $P_j \equiv \Pr\{\tilde{R} = \tilde{R}_j\} > 0$, $\sum_{j=1}^{m} P_j = 1$ and $m = 100$ (recall that $\tilde{R}$ is symmetric).

The approximation to (6) is

$$(8)\quad \frac{d\Psi(\lambda)}{d\lambda} \simeq \sum_{j=1}^{m} P_j\{\bar{\xi}_0-\bar{R}-(S_R)^{1/\alpha}\tilde{R}_j\}\ \frac{du[\lambda\bar{\xi}_0+(1-\lambda)\{\bar{R}+(S_R)^{1/\alpha}\tilde{R}_j\}]}{dw}.$$

One may then obtain an approximate solution λ^a to (5) via a golden section search using (7) or a bisecting search using (8) (see [32] for details on these search methods). The approximate optimal portfolio is then[8]

$$x_0^a = \lambda^a \text{ and } x_i^a = (1-\lambda^a)\,\hat{x}_i,\ i = 1,\ldots, n.$$

3. The Dependent Case

A set of random variables $r \equiv (r_1,\ldots,r_n)$ is said to be multivariate stable if every linear combination of r is univariate stable. The characteristic function of r is

$$\psi_r(t) \equiv \psi_r(t_1,\ldots,t_n) \equiv E^{it'r} = \int_{-\infty}^{\infty} \cdots \int_{-\infty}^{\infty} e^{it'r} g(r)dr_1\cdots dr_n,$$

where g is the joint density of $(r_1,\ldots,r_n)$. If $g(r)$ is symmetric then the log characteristic function (see Ferguson [10]) is

$$\ell n\, \psi_r(t) = i\,\delta(t) - \gamma(t),$$

where the dispersion measure $\gamma(\lambda t_1,\ldots,\lambda t_n) = |\lambda|^{\alpha}\, \gamma(t_1,\ldots,t_n)$, i.e. positive homogeneous of degree α, $\gamma > 0$ and the central tendency measure δ is linear homogeneous. The characteristic exponent α is assumed to satisfy $0<\alpha \leq 2$. With specific choices of δ and γ one may generate many classes of (symmetric) multivariate stable laws.

Press [22] has developed a statistical theory along with associated estimation procedures [23] for particularly convenient choices of δ and γ. Let $\bar{r} \equiv (\bar{r}_1,\ldots,\bar{r}_n)$ be the mean of r, which will exist if $\alpha > 1$. Press sets $\delta(t) = \bar{r}'t$ and $\gamma(t) = \frac{1}{2}\sum_{j=1}^{m}(t'\Omega_j t)^{\alpha/2}$ where $n \geqq m \geqq 1$ and each Ω_j is a positive semi-definite matrix of order nxn. The log characteristic function is then

$$\text{(9)} \qquad \ell n \psi_r(t) = i\,\bar{r}'t - \frac{1}{2}\sum_{j=1}^{m}(t'\,\Omega_j t)^{\alpha/2}.$$

The m in (9) indicates the number of independent partitions of the random variables $(r_1,\ldots,r_n)$. Hence if m = 1 all the random variables are dependent (as long as Ω_1 is positive definite). Then if $\alpha = 2$, the case of normal distributions,

$$\gamma(t) = \frac{1}{2} t' \Omega_1 t$$

which is precisely half the variance of the linear sum t'r. If m = n the random variables are independent. In this case with $\alpha = 2$ one may interpret $\gamma(t)$ as half the variance by setting the jjth coefficient of Ω_j equal to $\omega_j > 0$, (and all other coefficients equal to zero). Then $\gamma(t) = \frac{1}{2} \sum_{j=1}^{m} (t' \Omega_j t) = \frac{1}{2} \sum_{j=1}^{n} \omega_j t_j^2$.

Thus the class of distributions defined by (9) allows for the decomposition of the $(r_1,\ldots,r_n)$ into independent parts in a way consistent with and motivated by the way such a decomposition might be utilized for joint normally distributed random variables.

If $y \equiv (y_1,\ldots,y_n)$ then the linear combination r'y has the symmetric univariate stable distribution[9] $F(r'y;\ \bar{r}'y,\ S_{r'y},\ 0,\ \alpha)$, where

$$S_{r'y} \equiv \frac{1}{2} \sum_{j=1}^{m} (y' \Omega_j y)^{\alpha/2}.$$ The standardized variate

$$r_0 \equiv \frac{r'y - \bar{r}'y}{\{\frac{1}{2} \sum_{j=1}^{m} (y' \Omega_j y)^{\alpha/2}\}^{1/\alpha}} \sim F(r_0;\ 0,\ 1,\ 0, \alpha).$$

We will now suppose that the random returns in our portfolio problem $\hat{\xi} \equiv (\xi_1, \ldots, \xi_n)$ have the log characteristic function

$$(10) \quad \ell n \; \psi_\xi(t) = i \; \bar{\hat{\xi}}'t - \frac{1}{2} \sum_{j=1}^{m} (t' \; \Omega_j t)^{\alpha/2},$$

where $1 < \alpha \leq 2$, the Ω_j are positive semi-definite matrices of order nxn and $\bar{\hat{\xi}}$ denotes the mean vector of $\hat{\xi}$. We wish to show that a two stage optimization procedure analogous to that in section 2 is valid when the random variables have the characteristic function (10) and the risk free asset exists. Now Theorem 2 is valid in this case so that a mean-dispersion analysis may be used. The μ-d efficient surface may be found by solving

$$(11) \quad \phi_1(\beta) \equiv \max \; \bar{\hat{\xi}}'\hat{x}$$
$$\text{s.t.} \quad e'\hat{x} = 1, \; \hat{x} \geqq 0,$$
$$\left\{\frac{1}{2} \sum_{j=1}^{m} (\hat{x}' \Omega_j \hat{x})^{\alpha/2}\right\}^{1/\alpha} \leqq \beta,$$

for all $\beta > 0$. Now it is clear from lemma 2 that $\phi_1(\beta)$ will be a concave function of $\beta > 0$ if $\Psi(\hat{x}) \equiv \{\frac{1}{2} \sum_{j=1}^{m} (\hat{x}' \; \Omega_j \hat{x})^{\alpha/2}\}^{1/\alpha}$ is a convex function of $\hat{x}$.

The pertinent facts relating to the convexity of Ψ are summarized in

Theorem 7: a) (Independent Case). Suppose $m = n$ and the jjth element of Ω_j is $\omega_j > 0$ (and all other elements are zero) then Ψ is strictly convex on P.

b) (Totally Dependent Case). Suppose $m = 1$ and Ω_1 is positive definite then f is strictly convex on any convex subset of R^n.

c) (General Case). Suppose $n \geqq m \geqq 1$ and each nxn matrix is positive semi-definite and for at least one j, $\hat{x}'\Omega_j\hat{x} > 0$ if $x \neq 0$. Then f is convex on M.

Proof: a) Let $S_j > 0$ be defined by $\omega_j = \frac{1}{\sqrt{2}} S_j^{\alpha/2}$ then

$$\Psi(\hat{x}) = \{\frac{1}{2}\sum_{j=1}^{m}(\hat{x}'\Omega_j\hat{x})^{\alpha/2}\}^{1/\alpha} = \sum_{j=1}^{m} S_j x_j^{\alpha},$$ which was shown to be strictly convex on P in Lemma 1.

b) $$\Psi(\hat{x}) = \{\sum_{j=1}^{m}(\hat{x}'\Omega_j\hat{x})^{\alpha/2}\}^{1/\alpha} = \{\frac{1}{2}(\hat{x}'\Omega_1\hat{x})^{\alpha/2}\}^{1/\alpha}$$

$$= (\frac{1}{2})^{1/\alpha}(\hat{x}'\Omega_1\hat{x})^{1/2} \equiv (\frac{1}{2})^{1/\alpha}\phi_2(\hat{x}).$$

Hence Ψ will be strictly convex if ϕ_2 is strictly convex. Choose $\hat{x}^1, \hat{x}^2 \neq 0$, $\hat{x}^1 \neq \hat{x}^2$ and $0 < \lambda < 1$. Let $\hat{x}^{\lambda} = \lambda\hat{x}^1 + (1-\lambda)\hat{x}^2$ and suppose $z \equiv a'\hat{x}$ has variance $\hat{x}'\Omega_1\hat{x}$. Then

$$v(z^{\lambda}) = \lambda^2 v(z^1) + (1-\lambda)^2 v(z^2) + 2\lambda(1-\lambda)\mathrm{cov}(z^1, z^2),$$
$$< \lambda^2 v(z^1) + (1-\lambda)^2 v(z^2) + 2\lambda(1-\lambda)\{v(z^1)v(z^2)\}^{1/2},$$

since the fact that Ω_1 is positive definite implies that the correlation coefficient between z^1 and z^2 is <1. Taking the square root of both sides of this expression gives

$$\{v(z^{\lambda})\}^{1/2} < \lambda\{v(z^1)\}^{1/2} + (1-\lambda)\{v(z^2)\}^{1/2}.$$

c) The proof requires some preliminary definitions that lead to a theorem whose application will yield the result.

Definitions: A set $C \subset R^n$ is a convex cone if for any points c_1 and c_2 in C and any non-negative numbers Δ_1 and Δ_2 the point

$\Delta_1 c_1 + \Delta_2 c_2 \in C$.

A function ψ defined on a convex cone C is essentially positive if for all $c \in C$ such that $c \neq 0$, $\psi(c) > 0$.

Theorem 8: If a non-negative function ψ is defined on a convex cone c and is positively homogenous and essentially positive then it is convex if and only if it is quasi-convex.

Proof: see Newman [21]

Now Ψ is essentially positive and positively homogeneous on the convex cone M. Now Ψ is quasi-convex if and only if the set $\{\hat{x} \mid \{\frac{1}{2} \sum_{j=1}^{m} (\hat{x}'\Omega_j\hat{x})^{\alpha/2}\}^{1/\alpha} \leqq q\}$ is convex for all $q \geqq 0$. Raising both sides of the inequality to the αth power gives

$\{\hat{x} \mid \sum_{j=1}^{m} (\hat{x}'\Omega_j\hat{x})^{\alpha/2} \leqq 2^{\alpha}q^{\alpha}\}$ and this set is convex for all $q' \equiv 2^{\alpha}q^{\alpha} \geq 0$ since Press [22] has shown that $\sum_{j=1}^{m} (\hat{x}'\Omega_j\hat{x})^{\alpha/2}$ is convex if $2 \geq \alpha \geq 1$.

The fractional program to be solved in stage one is then

$$\text{(12)} \qquad \max\ g_1(\hat{x}) \equiv \frac{\bar{\bar{\xi}}'\hat{x}}{\Psi(\hat{x})}$$

$$\text{s.t.} \quad \hat{x} \geqq 0, \quad e'\hat{x} = 1,$$

where each $\bar{\bar{\xi}}_i = \bar{\xi}_i - \bar{\xi}_0$.

A differentiable function $\theta(x)\colon \Lambda \to R$ is said to be a pseudo-concave [32] on $\Lambda \subset R^n$ if for all x, $\bar{x} \in \Lambda$, $(x-\bar{x})'\nabla\theta(\bar{x}) \leq 0 \Rightarrow \theta(x) \leq \theta(\bar{x})$. Geometrically functions are pseudo-concave if they are non-increasing in all directions that have downward or horizontal pointing directional derivatives. Local maxima of pseudo-concave functions constrained over convex sets satisfying a constraint qualification are global maxima. Now g_1 is the ratio of an (assumed) non-negative differentiable concave function to a strictly positive convex function and hence is pseudo-concave

(the proof is analogous to that of Theorem 4). An optimal solution to (12) always exists, because the constraint region is compact, and may be found using a standard algorithm [32]. For cases (a) and (b) of Theorem 7 the solution will be unique since g_1 will be strictly pseudo-concave. It has not been possible to prove that $\Psi(\hat{x})$ is generally strictly-convex (although one suspects that Ψ will usually have this property) hence (12) may have multiple solutions.

Once (12) has been solved to obtain $\hat{x}^*$, the problem is then to obtain the optimal ratios of risky to non-risky assets. The risky asset is $R = \hat{\xi}'\hat{x}^* \sim F(\hat{\xi}'\hat{x}^*; \bar{\hat{\xi}}'\hat{x}^*, \{\Psi(\hat{x}^*)\}^\alpha, 0, \alpha) \equiv F(R; \bar{R}, S_R, 0, \alpha)$. The analysis is then precisely analogous to that in section 2, see equations (4)-(8) and the accompanying discussion.

The essential features that one needs to utilize the two stage decomposition are the risk free asset assumption, and the assumption that the random returns are multivariate symmetric stable variates having the same characteristic exponent α $(1 < \alpha \leq 2)$ and that the dispersion measure is convex and positively homogeneous.
The assumption that each x_i be non-negative is not crucial.

An additional class of multivariate stable distributions that has these features was suggested by Samuelson [29]. Suppose each $\xi_i = \sum_{j=1}^{k} a_{ij} Y_j$, where each Y_j has the independent stable distribution $F(Y_j;\ \bar{Y}_j,\ S_{Y_j},\ 0,\ \alpha)$ and the a_{ij} are known constants. Then the portfolio $\hat{x}'\hat{\xi}$ has the univariate stable distribution $F(\hat{x}'\hat{\xi};\ \hat{x}'\bar{\hat{\xi}},\ S_{\hat{x}'\hat{\xi}}, 0, \alpha)$,

where $$S_{\hat{x}'\hat{\xi}} \equiv \sum_{j=1}^{k} \left| \sum_{i=1}^{n} a_{ij} x_i \right|^{\alpha} (S_{Y_j})^{\alpha}.$$

It must be shown that $$\Psi_1(\hat{x}) \equiv \left\{ \sum_{j=1}^{k} \left| \sum_{i=1}^{n} a_{ij} x_i \right|^{\alpha} (S_{Y_j})^{\alpha} \right\}^{1/\alpha}$$

is convex. Let $c_j \equiv (S_{Y_j})^{\alpha} > 0$ and $b_j' \equiv (a_{j1}, \ldots, a_{jn})$,

then $\Psi_1(x) = \{ \sum_{j=1}^{k} c_j\ |\hat{x}'b_j|^{\alpha} \}.^{1/\alpha}$

Suppose that Ψ_1 is defined on R^n and that for some j, $|\hat{x}'b_j| \neq 0$ if $\hat{x} \neq 0$. Hence Ψ_1 is essentially positive and positively homogeneous on the convex cone R^n. Thus by Theorem 8 Ψ_1 is convex if and only if it is quasi-convex. Now Ψ_1 is quasi-convex if and only if

$$N \equiv \{\hat{x} \mid \{\sum_{j=1}^{k} c_j |\hat{x}' b_j|^{\alpha}\}^{1/\alpha} \leqq q\} \text{ is convex for all } q \geq 0,$$

or equivalently if $\{\hat{x} \mid \sum_{j=1}^{k} c_j |\hat{x}\, b_j| \leq q'\}$ is convex for all $q' \equiv q^{\alpha} \geq 0$.
Since each $|\hat{x}\, b_j|^{\alpha}$ is convex and the $c_j > 0$ the sum $\sum_{j=1}^{k} c_j |\hat{x}\, b_j|^{\alpha}$ is convex, hence the set N is convex.

4. Dropping the Risk Free Asset Assumption

Let us reconsider problem

$$\text{(1)} \quad \text{maximize} \quad Z(x) \equiv E_{\xi}\, u(\xi' x),$$
$$\text{s.t.} \quad e'x = 1, \; x \geqq 0.$$

Suppose that $\xi = (\xi_0, \ldots, \xi_n)$ has a symmetric multivariate stable distribution G with characteristic exponent α $(1 < \alpha \leq 2)$, having the log characteristic function

$$\ell n\, \psi_{\xi}(t) = i\, \bar{\xi}'\, t - \gamma(t),$$

where $\gamma(t)$ is positive homogeneous of degree α. Suppose that a risk free asset does not exist then one may (approximately) solve (1) as follows.[10]

For any feasible x, say x^k, $\xi' x$ has the univariate symmetric stable distribution $F(\xi' x^k;\, \bar{\xi}' x^k,\, \gamma(x^k),\, 0, \alpha)$.

$$\text{Now } Z(x^k) = \int \cdots \int u(\xi' x^k)\, dG(\xi)$$

$$\text{(13)} \quad = \int u[\bar{\xi}' x^k + \{\gamma(x^k)\}^{1/\alpha} \tilde{R}\,]\; f(\tilde{R}')\, d\tilde{R}',$$

where the standardized variate

$$\tilde{R} \equiv \frac{\xi' x^k - \bar{\xi}' x^k}{[\gamma(x^k)]^{1/\alpha}} \sim F(\tilde{R}\colon 0, 1, 0, \alpha)$$

The continuous partial derivatives of Z are

$$(14)\quad \frac{\partial Z(x^k)}{\partial x_i} = \int \{\bar{\xi}_i + \frac{1}{\alpha}[\gamma(x^k)]^{\frac{1-\alpha}{\alpha}} \frac{\partial\gamma(x^k)}{\partial x_i}\tilde{R}\}\frac{du[\bar{\xi}'x^k + \{\gamma(x^k)\}^{1/\alpha}\tilde{R}]}{dw} f(\tilde{R})d\tilde{R},$$

$i = 0,\ldots,n.$

Now using the Fama-Roll tables, as in section 2, one may obtain a good approximation to (13) and (14) via

$$(15)\quad Z(x^k) \cong \sum_{j=1}^{m} P_j\, u\,[\bar{\xi}'x^k + \{\gamma(x^k)\}^{1/\alpha}\tilde{R}_j], \quad \text{and}$$

$$(16)\quad \frac{\partial Z(x^k)}{\partial x_i} \cong \sum_{j=1}^{m} P_j\{\bar{\xi}_i + \frac{1}{\alpha}[\gamma(x^k)]^{\frac{1-\alpha}{\alpha}} \frac{\partial\gamma(x^k)}{\partial x_i}\tilde{R}_j\}.$$

$$\frac{du[\bar{\xi}'x^k + \{\gamma(x^k)\}^{1/\alpha}\tilde{R}_j\}}{dw}, \quad i=0,\ldots,n, \text{ respectively,}$$

where $P_j \equiv \Pr\{\tilde{R} = \tilde{R}_j\} > 0$, $\sum_{j=1}^{m} P_j = 1$ and $m = 100$.

One may then apply any standard nonlinear programming algorithm that uses function values and/or partial derivates to approximately solve (1) utilizing (15) and/or (16). If one uses an algorithm that utilizes only function values, such as the generalized programming algorithm, see [33], then for each evaluation of (15) one merely performs m function evaluations of the form u(.) and adds them up with weights P_j. The evaluation of the portfolio is more complicated since for each i (i=1,...,n) one must perform m function evaluations of the form $\{\cdot\}du(.)/dw$ and add these up with the weights P_j.

One would suspect that it would be economically feasible to solve such approximate problems when there are say 40-60 investments the grid m is say 20-50 points and u, γ and k are reasonably convenient. It is possible, of course, to apply this direct solution approach even when the risk free asset exists. However, the two stage decomposition approach appears simpler because one must solve a fractional program in n variables plus a nonlinear program in one variable that has m terms. In the direct approach one must solve one nonlinear program in n variables having m terms that may fail to be concave or pseudo-concave (because u is concave and γ is convex). Some numerical results will appear in [34].

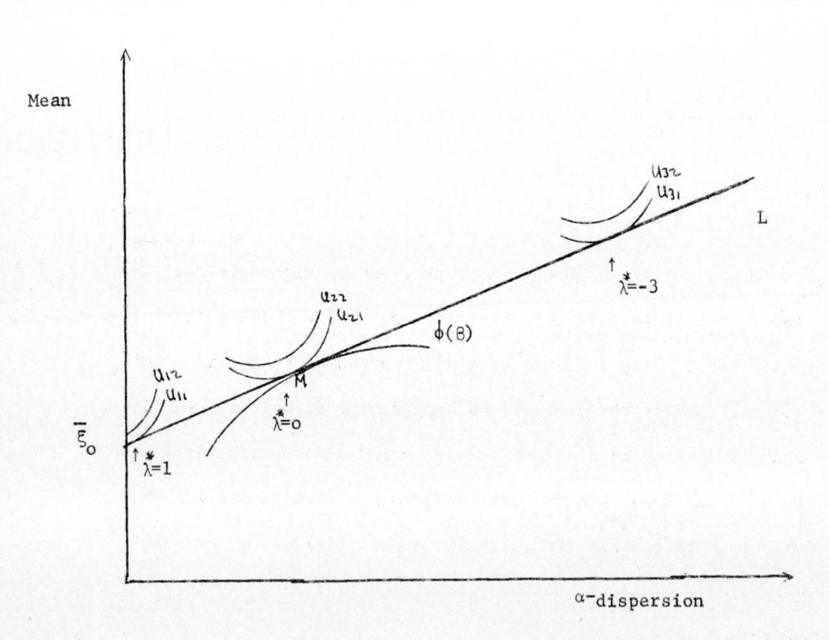

Figure 1: Geometry of the Generalized Tobin Separation Theorem

APPENDIX

PROOF OF THEOREM 1:

$$\text{a)}\ \left|\int u(w)dF(w)\right| \overset{(i)}{\leq} \int |u(w)|dF(w) \overset{(ii)}{\leq} \int L_1|w|^{\beta_1}dF \overset{(iii)}{<} \infty$$

(iii) is a well known property of stable distributions, see e.g. Feller [9]. (ii) is by assumption, while (i) follows because the middle term is finite, see Loève [16].

$$\text{b)}\ \left|\int \frac{du(w)}{dw}\,dF(w)\right| \overset{(i)}{\leq} \int \left|\frac{du(w)}{dw}\right| dF(w) \overset{(ii)}{\leq} \int L_2|w|^{\beta_2} \overset{(iii)}{<} \infty$$

(i)-(iii) follow for the same reasons as in (a) since (b) is a special case of (a).

c) Now $w \equiv \xi'x \sim F(\xi'x;\ \bar{\xi}'x,\ \sum_{i=0}^{n} S_i x_i^{\alpha},\ 0, \alpha)$ if and only if

$$\tilde{w} \equiv \frac{w-\bar{\xi}'x}{\{\sum_{i=1}^{n} S_i x_i^{\alpha}\}^{1/\alpha}} \sim F(\tilde{w};\ 0,1,0,\alpha).$$ Notice that the F corresponding to the standardized variable w does not depend upon x. Now

$$E_w u(w) = \int u(w)\ dF(w) = \int u(\bar{\xi}'x + s(x)\tilde{w}\ )\ dF(\tilde{w};\ 0,1,0,\alpha)$$

where $s(x) \equiv \{\sum_{i=0}^{n} S_i x_i^{\alpha}\}^{1/\alpha}$

Now if it is legitimate to differentiate under the integral sign

$$\frac{\partial E\ u(w)}{\partial x_i} = \int \frac{du[\bar{\xi}'x + s(x)\tilde{w}]}{dw} (\bar{\xi}_i + s_i'(x)\tilde{w}\ )\ dF\ (\tilde{w};\ 0,1,0,\alpha)$$

$$= \bar{\xi}_i \int \frac{du[\bar{\xi}'x + s(x)\tilde{w}]}{dw}\ dF(\tilde{w};\ 0,1,0,\alpha)$$

$$+\ s_i'(x) \int \tilde{w}\ \frac{du[\bar{\xi}'x + s(x)\tilde{w}]}{dw}\ dF\ (\tilde{w};\ 0,1,0,\alpha),$$

where $s_i'(x) \equiv \{\sum_{i=0}^{n} S_i x_i^{\alpha}\}^{\frac{1-\alpha}{\alpha}} x_i^{\alpha-1}$.

Now under the assumptions $\bar{\xi}_i$ and s_i' are finite hence $\frac{\partial Eu(w)}{\partial x_i}$ is finite if

the latter two integrals are finite, which there are using an argument as in (a) and the assumption on u'. Note that $|u'(w)w| \leq |w|^{\beta}$ if and only if $|u'(w)| \leq |w|^{\beta-1}$ and that $|u'(w)| \leq |w|^{\beta-1}$ implies that $|u'(w)| \leq |w|^{\beta}$

for $\beta > 1$. Also one may differentiate under the integral sign because u is continuously differentiable and the absolute expected marginal utility is bounded, see Loève [16].

d) The proofs remain valid as long as the appropriate absolute integrals remain finite. We show here for case (a) that $\int |u(w)| \, dF(w) < \infty$, similar arguments may be used to establish (b) and (c).

Suppose $v_1 > 0$ then without loss of generality we may take $v_1 = 1$, (and $L_1 = 1$). Now $\lim_{|w|\to\infty} \frac{|u(w)|}{|w|^{\beta}} = 1$ if and only if for all ε such

that $0 < \varepsilon < 1$, $\exists N_1 > 0$

such that $\forall |w| \geq N_1$, $\left| \frac{|u(w)|}{|w|^{\beta_1}} - 1 \right| < \varepsilon$ or $\left| |u(w)| - |w|^{\beta_1} \right| < \varepsilon |w|^{\beta_1}$

hence $|u(w)| < (1+\varepsilon)|w|^{\beta_1}$ (i).

By the Cauchy convergence criterion, see Loève [16],

$$\int_{-\infty}^{\infty} |w|^{\beta_1} \, dF(w) < \infty \text{ if and only if } \exists\, N_2 \text{ such that}$$

$$\forall n \geq N_2, \quad \int_{n}^{\infty} |w|^{\beta_1} \, dF(w) < \frac{\varepsilon}{2} \quad \text{and} \quad \forall n \leq -N_2, \quad \int_{-\infty}^{n} |w|^{\beta_1} \, dF(w) < \frac{\varepsilon}{2} .$$

Let $N \geq \max(N_1, N_2)$.

$$\text{Then} \quad \int_{N}^{\infty} |u(w)| \, dF(w) \overset{\text{(by i)}}{<} (1+\varepsilon) \int_{N}^{\infty} |w|^{\beta_1} \, dF(w) \overset{\text{(since } \varepsilon<1)}{<} 2 \int_{N}^{\infty} |w|^{\beta_1} \, dF(w) < \varepsilon,$$

$$\text{and} \quad \int_{-\infty}^{-N} |u(w)| \, dF(w) < (1+\varepsilon) \int_{-\infty}^{-N} |w|^{\beta_1} \, dF(w) < 2 \int_{-\infty}^{-N} |w|^{\beta_1} \, dF(w) < \varepsilon .$$

$$\text{Thus} \quad \int_{-\infty}^{\infty} |u(w)| \, dF(w) < \infty$$

When $v = 0$ the same proof applies by letting ε and 1 replace $(1+\varepsilon)$ and (2), respectively.

FOOTNOTES

[1]See, for example, Blume [1], Fama [4] and Roll [27] and Fama and Roll [7,8], A survey of this and other related research appears in Fama [6].

[2]See, for example, Mandelbrot [17, 18]. The estimates in these papers and those mentioned in footnote one were generally performed on the logarithms of the changes in the economic variables. The theory in this paper is applicable only to data regarding absolute changes. However, if the changes are small say of the order of fifteen percent or less, then these variables and their logarithms are approximately equal and the results here are approximately valid for such data.

[3]There are other distributions besides the stable class that can be used to explain such data, see Press [24, 25, 26] and Mandelbrot and Taylor [19]. However, the distributions in these papers are not particularly convenient for the analysis of portfolio problems because they are not closed on addition.

[4]Without loss of generality, we may normalize the investment returns so that initial wealth is one dollar.

[5]The analysis in this section is valid though for $\beta \neq 0$ as long as β is the same for each F_i.

[6]For convenience we will suppose, in (c) that $x_i \geq 0$ for all i. If any x_i such as x_0 is unconstrained then the modification that

$\frac{\partial}{\partial x_i} S_i|x_i|^\alpha = \text{Sign}(x_i)\,\alpha S_i|x_i|^{\alpha-1}$ may be used. Note that this partial derivative is continuous at $x_i = 0$, since $\alpha > 1$.

[7]Samuelson [29] analyzes a problem similar to (2) in which one minimizes dispersion subject to the mean equalizing a given parameter. He notes that his problem is a convex program and he analyzes it using the Kuhn-Tucker conditions. His paper also contains some illustrative graphical results for some special cases. See also Fama [5] for a similar analysis, presented in the context of the Sharpe-Markowitz diagonal model, that is particularly concerned with diversification questions.

[8]Since the ξ_i are independent it is known, see Samuelson [28], that $x_i^* > 0$ for each i whose $\bar{\xi}_i \geq \min_{1 \leq j \leq n} \bar{\xi}_j$.

[9]Press [23], also proves that Ar + b is multivariate symmetric stable of form(9) if A is any mxn constant matrix and b is any n vector of constants.

[10]It may be noted that the following procedure is also applicable when the risk free asset does exist but the lending and borrowing rates differ. Let $x_0 \geq 0$ and $x_1 \geq 0$ be the levels of borrowing and lending activities at constant rates $\bar{\xi}_0$ and $\bar{\xi}_1$ respectively. Then in (1) $\xi'x$ becomes $-\bar{\xi}_0 x_0 + \bar{\xi}_1 x_1 + \xi_2 x_2 + \cdots + \xi_n x_n$ and $e'x$ becomes $-x_0 + x_1 + \cdots + x_n$.

REFERENCES

[1] M. Blume, "Portfolio Theory: A Step Toward its Practical Application," Journal of Business, XLIII (1970), 152-173.

[2] W. Breen, "Homogeneous Risk Measures and the Construction of Composite Assets," Journal of Financial and Quantitative Analysis, III (1968), 405-413.

[3] J. Chipman, "On the Ordering of Portfolios in Terms of Mean and Variance", Review of Economic Studies, forthcoming.

[4] E.F. Fama, "The Behavior of Stock Market Prices," Journal of Business, XXXVIII (1965), 34-105.

[5] ________, "Portfolio Analysis in a Stable Paretian Market," Management Science, XI (1965), 404-419.

[6] ________, "Efficient Capital Markets: A Review of Theory and Empirical Work", Journal of Finance, XXV (1970), 383-417.

[7] ________, and R. Roll, "Some Properties of Symmetric Stable Distributions," Journal of the American Statistical Association, LXIII (1968), 817-836.

[8] ________, and ______, "Parameter Estimates for Symmetric Stable Distributions," Journal of the American Statistical Assoc., LXVI, (1971), 331-338.

[9] W. Feller, An Introduction to Probability Theory and Its Applications Vol. II, New York: John Wiley and Sons, Inc., 1966.

[10] T. Ferguson, "On the Existence of Linear Regression in Linear Structural Relations", University of California Publications in Statistics, II (1955), 143-166.

[11] B.V. Gnedenko and A.N. Kolmogorov, Limit Distributions for Sums of Independent Random Variables, Reading, Mass.: Addison-Wesley Publishing Co., Inc., 1954.

[12] F. Hanoch and H. Levy, "The Efficiency Analysis of Choices Involving Risk", Review of Economic Studies, XXXVI (1969), 335-346.

[13] G.H. Hardy, J.E. Littlewood and G. Poyla Inequalities, Cambridge, England, Cambridge University Press, 2nd edition, 1964.

[13a] F.S. Hillier, The Evaluation of Risky Interrelated Investments, Amsterdam: North-Holland Publishing Co., 1969.

[14] P. Levy, Theorie de l'Addition des Variables Aleataires, Paris: Gauthier Villurs, 2nd edition, 1954.

[15] J. Lintner, "The Valuation of Risk Assets and the Selection of Risky Investments in Stock Portfolios and Capital Budgets," Review of Economics and Statistics, XLVII (1965), 13-37.

[16] M. Loève, Probability Theory, Princeton, N.J.: D. Van Nostrand Co. Inc., 3rd edition, 1963.

[17] B. Mandelbrot, "The Variation of Certain Speculative Prices", Journal of Business, XXXVI (1963), 394-419.

[18] ____________, "New Methods of Statistical Economics," Journal of Political Economy, LXI (1963), 421-440.

[19] ____________, and H.M. Taylor, "On the Distribution of Stock Price Differences", Operations Research , XV (1967), 1057-62.

[20] H.M. Markowitz, Portfolio Selection: Efficient Diversification of Investments, New York: John Wiley & Sons, Inc., 1959.

[21] P. Newman, "Some Properties of Concave Functions," Journal of Economic Theory, I (1969), 291-314.

[22] S.J. Press, "Multivariate Stable Distributions", Journal of Multivariate Analysis, II (1972), 444-463.

[23] __________, "Estimation in Univariate and Multivariate Stable Distributions", Journal of the American Statistical Association, LXVII (1972), 842-846.

[24] __________, "A Compound Events Model for Security Prices", Journal of Business, XL (1967), 317-335.

[25] __________, "A Modified Compound Poisson Process with Normal Compounding", J. Amer. Stat. Assoc. LX (1968), 607-613.

[26] __________, "A Compound Poisson Process for Multiple Security Analysis," in Random Counts in Scientific Work, G. Patil (ed.) Pennsylvania State Univ. Press, 1970.

[27] R. Roll, The Behavior of Interest Rates, New York: Basic Books, Inc., 1970.

[28] P.A. Samuelson,"General Proof that Diversification Pays," Journal of Financial and Quantitative Analysis, II (1967), 1-13.

[29] ____________, "Efficient Portfolio Selection for Pareto-Levy Investments", Journal of Financial and Quantitative Analysis, II (1967), 107-122.

[30] W.F. Sharpe, "A Simplified Model for Portfolio Analysis", Management Science, IX (1963), 277-293.

[31] J. Tobin, "Liquidity Preferences as Behavior Towards Risk", Review of Economic Studies, XXV (1958), 65-86.

[32] W.I. Zangwill, Nonlinear Programming: A Unified Approach, Englewood Cliffs, N.J.: Prentice-Hall, Inc., 1969.

[33] W.T. Ziemba, "Solving Nonlinear Programming Problems with Stochastic Objective Functions", Journal of Financial and Quantitative Analysis , VIII (1972), 1809-1827.

[34] __________, "Choosing Investment Portfolios when the Returns have Stable' Distributions, Part II: Computational Results", paper in preparation.

[35] __________, F.J. Brooks-Hill and C. Parkan, "Calculating Investment Portfolios when the Returns are Normally Distributed," Management Science, forthcoming.

302546810T